Theory and Applications
of Natural Language Processing

Series editors
Julia Hirschberg
Eduard Hovy
Mark Johnson

Aims and Scope

The field of Natural Language Processing (NLP) has expanded explosively over the past decade: growing bodies of available data, novel fields of applications, emerging areas and new connections to neighboring fields have all led to increasing output and to diversification of research.

"Theory and Applications of Natural Language Processing" is a series of volumes dedicated to selected topics in NLP and Language Technology. It focuses on the most recent advances in all areas of the computational modeling and processing of speech and text across languages and domains. Due to the rapid pace of development, the diversity of approaches and application scenarios are scattered in an ever-growing mass of conference proceedings, making entry into the field difficult for both students and potential users. Volumes in the series facilitate this first step and can be used as a teaching aid, advanced-level information resource or a point of reference.

The series encourages the submission of research monographs, contributed volumes and surveys, lecture notes and textbooks covering research frontiers on all relevant topics, offering a platform for the rapid publication of cutting-edge research as well as for comprehensive monographs that cover the full range of research on specific problem areas.

The topics include applications of NLP techniques to gain insights into the use and functioning of language, as well as the use of language technology in applications that enable communication, knowledge management and discovery such as natural language generation, information retrieval, question-answering, machine translation, localization and related fields.

The books are available in printed and electronic (e-book) form:

* Downloadable on your PC, e-reader or iPad
* Enhanced by Electronic Supplementary Material, such as algorithms, demonstrations, software, images and videos
* Available online within an extensive network of academic and corporate R&D libraries worldwide
* Never out of print thanks to innovative print-on-demand services
* Competitively priced print editions for eBook customers thanks to MyCopy service http://www.springer.com/librarians/e-content/mycopy

More information about this series at http://www.springer.com/series/8899

Kemal Oflazer • Murat Saraçlar
Editors

Turkish Natural Language Processing

 Springer

Editors
Kemal Oflazer
Carnegie Mellon University Qatar
Doha-Education City, Qatar

Murat Saraçlar
Electrical and Electronic Engineering
Boğaziçi University
Istanbul-Bebek, Turkey

ISSN 2192-032X   ISSN 2192-0338   (electronic)
Theory and Applications of Natural Language Processing
ISBN 978-3-030-07949-9   ISBN 978-3-319-90165-7   (eBook)
https://doi.org/10.1007/978-3-319-90165-7

Preface

Turkish has proved to be a very interesting language for natural language processing techniques and applications. There has been a significant amount of work on Turkish since the early 1990s on introducing and/or adapting fundamental techniques, compiling resources, and developing applications.

The idea for this book came after one of us gave an invited talk at the LREC Conference held in Istanbul, Turkey, in 2012. Since then, the authors and we have worked hard to bring this effort to fruition. This book brings together most of the work done on Turkish in the last 25 years or so. After a bird's-eye overview of relevant aspects of Turkish, it covers work on morphological processing and disambiguation, statistical language modeling, speech processing, named-entity recognition, dependency, and deep parsing. It then continues with statistical machine translation from English to Turkish and from Turkic languages to Turkish and sentiment analysis for Turkish, a topic that has recently been quite popular with the advent of social media. Finally, the book covers the most important natural language processing resources that have been developed for Turkish including the Turkish WordNet, the Turkish Treebank, Turkish National Corpus, and Turkish Discourse Bank.

We hope that this book helps other researchers in advancing the state of the art for Turkish and possibly other Turkic languages that share nontrivial similarities with Turkish.

Doha, Qatar
Istanbul, Turkey
July, 2017

Kemal Oflazer
Murat Saraçlar

Acknowledgements

The work presented in the following chapters has been supported by grants from various institutions: Early work was funded by a grant by NATO Science for Stability Program. The Turkish Scientific and Technological Research Council (TÜİTAK) has funded much of the more recent work in the last 15–20 years through many grants. Turkish Academy of Sciences (TUBA) provided support through its Outstanding Young Scientist Award program (TUBA-GEBIP). Boğaziçi University Research Fund (BU-BAP) also funded the language modeling and speech recognition research through grants. Additional funding was provided by the EU Framework 5 program.

A number of people have been very supportive of our work throughout these years: Lauri Karttunen has encouraged us and provided us with the Xerox Finite State Toolkit since the beginning, in order for us to build the very basic resources we needed. Other colleagues (then) at XRCE/PARC, notably Ken Beesley and Ron Kaplan, have supported our work by helping with the intricacies of the toolkit and by including our team in the ParGram Project. We thank them. The FSM Library and AMTools software packages developed by AT&T Labs-Research also have been essential in the early work on Turkish large vocabulary speech recognition. These were superseded by the OpenFST and Kaldi toolkits. We thank the authors of these toolkits and their organizations.

During these years, many graduate students and/or research assistants contributed to the work described in these chapters with their theses. Without their contributions, most of the work reported in the following chapters would not have been possible. We won't attempt to list them, lest we forget some. We thank them also.

Finally, we owe a lot to our families who supported us through these years. We cannot thank them enough.

Contents

About the Authors

Eşref Adalı received his B.Sc. and Ph.D. from Istanbul Technical University, Turkey, in 1971 and 1976, respectively. Currently, he is the Dean of the Faculty of Computer Engineering and Informatics, which he founded in 2010 after founding the Department of Computer Engineering in 1998. He was a Visiting Research Fellow at Case Western University, Cleveland, OH, USA, in 1977–1978; a Visiting Professor at Carleton University, Ottawa, Canada, in 1979; and a Visiting Professor at Akron University, Akron, OH, USA, in 1985. Adalı worked at the TÜBİTAK Marmara Research Center as the Founding Director of Informatics Group in 1990–1991. Although his more recent work has been on Turkish natural language processing, he has worked on microprocessors, system analysis, and design in the past and published two books on microcomputers and real-time systems.

Mustafa Aksan is Professor of Linguistics at Mersin University, Turkey, where he is also the Head of the Turkish Center for Corpus Studies. His main research interests are in corpus linguistics, lexical semantics, pragmatics, and morphology. He is the Principal Researcher in the construction of the *Turkish National Corpus* and also a coauthor of *A Frequency Dictionary of Turkish* (Routledge 2016). Recently, Aksan started a project on affix ordering in Turkish.

Yeşim Aksan is Professor of Linguistics at Mersin University, Turkey. Her main research interests are in corpus linguistics, lexical semantics, cognitive semantics, and pragmatics. She is the Project Director of the *Turkish National Corpus* and a coauthor of *A Frequency Dictionary of Turkish* (Routledge 2016). Previously, Aksan conducted corpus-based and corpus-driven studies on various aspects of Turkish.

Ebru Arısoy received her B.Sc., M.Sc., and Ph.D. degrees from the Electrical and Electronics Engineering Department, Boğaziçi University, Istanbul, Turkey, in 2002, 2004, and 2009, respectively. From 2010 to 2013, she worked as a postdoctoral researcher at IBM T. J. Watson Research Center, Yorktown Heights, NY, USA. She then moved to IBM Turkey, where she had a key role in developing language modeling approaches for voice search and mobile dictation applications. Since 2014, she has worked as an Assistant Professor at the Electrical and Electronics

Engineering Department of the MEF University, Istanbul, Turkey. Her research interests include automatic speech recognition, statistical language modeling, as well as speech and language processing for educational technologies.

Orhan Bilgin is a translator, editor, and a dictionary publisher based in Istanbul, Turkey. He holds a B.A. in Economics and an M.A. in Cognitive Science from the Boğaziçi University, Istanbul, Turkey. His master's thesis was on frequency effects in the processing of morphologically complex nouns in Turkish. Between 2001 and 2004, he worked as a lexicographer on the Turkish team at the Sabancı University, Istanbul, Turkey, as part of the Balkanet project: an EU-funded project that designed and developed medium-sized word nets for six Balkan languages, including Turkish. Bilgin is the founder of zargan.com, an online English-Turkish dictionary that has been active since 2001, as well as a partner of Banguoglu Ltd., a translation company specializing in the translation of legal documents.

Cem Bozşahin works on the learning and projection of structure, including argument structure and constituent structure, at the intersection of computer science, linguistics, and philosophy. He holds a Ph.D. in Computer Science from the Arizona State University, Tempe, AZ, USA. He worked at the Ohio University (1990–1992) before permanently joining the Middle East Technical University, Ankara, Turkey, in 1992. He held visiting research assignments at the University of Edinburgh, Scotland, UK (2002–2017), at the Boğaziçi University, Istanbul, Turkey (2011), and at the University of Lisbon, Portugal (2015–2016).

Ruket Çakıcı is a lecturer at the Department of Computer Engineering of the Middle East Technical University, Ankara, Turkey. She received her Ph.D. from the School of Informatics of the University of Edinburgh, Scotland, UK, after receiving her M.Sc. and B.Sc. degrees in Computer Engineering from the Middle East Technical University. Her research interests are in empirical and data-driven methods for natural language processing, in particular (multilingual) morphological, syntactic, and semantic parsing; combinatory categorical grammar; automatic description generation for images and videos; and natural language generation.

Özlem Çetinoğlu is a postdoctoral researcher at the IMS, University of Stuttgart, Germany. She received her Ph.D. in Computer Science from the Sabancı University, Istanbul, Turkey, in 2009, where she developed a large-scale Turkish LFG grammar as part of her thesis in the context of the ParGram project. From 2009 to 2011, she worked at the CNGL, Dublin City University, Ireland, on automatically labeling English with deep syntactic information and on parsing noncanonical data. Çetinoğlu's research interests include deep grammars, statistical dependency parsing, morphologically rich languages, and code switching.

Işın Demirşahin holds a Ph.D. in Cognitive Science from the Middle East Technical University, Ankara, Turkey. Her research focuses on discourse structures in general and Turkish discourse and information structure in particular. Demirşahin currently works for Google UK Ltd. as a computational linguist and focuses on internationalizing end-to-end dialogue systems.

İlknur Durgar-El Kahlout is currently Chief Researcher at the TÜBİTAK-BİLGEM Research Center in Gebze, Turkey. She received her B.Sc. degree in Computer Engineering from the Başkent University, Ankara, Turkey, in 1997 and her M.Sc. and Ph.D. degrees in Computer Science and Engineering from the Sabancı University, Istanbul, Turkey, in 2003 and 2009 respectively. She worked as a postdoctoral fellow in the Quero project at LIMSI-CNRS, Orsay, France. Durgar-El Kahlout's current research interests are in natural language processing and statistical machine translation, especially of morphologically complex languages.

Gülşen Eryiğit obtained her M.Sc. and Ph.D. degrees from the Computer Engineering Department of the Istanbul Technical University, Turkey, in 2002 and 2007, respectively, where she is currently a faculty member. In addition, she is a founding member and coordinator of the Natural Language Processing Group and a member of the Learning from Big Data Group. In 2006, she worked as a Visiting Student Researcher at Växjö University, Sweden, in the first-ranked team in CoNLL shared tasks on multilingual dependency parsing. In 2007, Eryiğit won the Siemens Excellence Award for her Ph.D. thesis. Her current research focuses on natural language processing of Turkish on which she has authored and coauthored publications in prestigious journals and conferences. She represented Turkey in the CLARIN Project (EU 7th Framework Program, Common Language Resources and Technology Infrastructure) and in PARSEME (EU Cost Action, Parsing and Multiword Expressions). Recently, Eryiğit acted as the Principal Investigator of two research projects "Parsing Web 2.0 Sentences" (funded by EU Cost Action and TÜBİTAK) and "Turkish Mobile Personal Assistant" (funded by the Turkish Ministry of Science, Industry and Technology). Moreover, she is also the NLP coordinator of the interdisciplinary research project "A Signing Avatar System for Turkish-to-Turkish Sign Language Machine Translation" (funded by TÜBİTAK). Eryiğit also used her research to consult several domestic and international IT companies on Turkish natural language processing.

Gizem Gezici received her B.Sc. and M.Sc. degrees in Computer Science and Engineering from the Sabancı University, Istanbul, Turkey, in 2011 and 2013. She is currently a fourth-year Ph.D. student at the Sabancı University and is involved in a project in the emerging research area of *bias in search*. Gezici's research interests include sentiment analysis, machine learning, and data mining.

Dilek Zeynep Hakkani-Tür is a Research Scientist at Google. Prior to joining Google, she was a researcher at Microsoft Research (2010–2016), the International Computer Science Institute (2006–2010), and AT&T Labs-Research (2001–2005). She received her B.Sc. degree in Computer Engineering from the Middle East Technical University, Ankara, Turkey, in 1994, and her M.Sc. and Ph.D. degrees in Computer Engineering from the Bilkent University, Ankara, Turkey, in 1996 and 2000, respectively. Her research interests include natural language and speech processing, spoken dialogue systems, and machine learning for language processing. She coauthored more than 200 papers in natural language and speech processing and is the recipient of three Best Paper Awards for her work on active learning for dialogue systems from IEEE Signal Processing Society, ISCA, and EURASIP. She was an associate editor of *IEEE Transactions on Audio, Speech, and Language Processing* (2005–2008), a member of the IEEE Speech and Language Technical Committee (2009–2014), and an area editor for speech and language processing for Elsevier's *Digital Signal Processing Journal* and *IEEE Signal Processing Letters* (2011–2013). Since 2014, she has been a Fellow of IEEE and ISCA and currently serves on the ISCA Advisory Council (2015–2019). In addition, Hakkani-Tür was granted over 50 patents for her work.

Joakim Nivre is Professor of Computational Linguistics at Uppsala University, Sweden. He holds a Ph.D. in General Linguistics from the University of Gothenburg, Sweden, and a Ph.D. in Computer Science from Växjö University, Sweden. His research focuses on data-driven methods for natural language processing, in particular for syntactic and semantic analysis. Nivre is one of the main developers of the transition-based approach to syntactic dependency parsing, described in his book *Inductive Dependency Parsing* (Springer 2006) and implemented in the widely used MaltParser system. In addition to his current position of President of the Association for Computational Linguistics, he is one of the founders of the "Universal Dependencies" project, which aims to develop cross-linguistically consistent treebank annotation for many languages and currently involves over 50 languages and over 200 researchers around the world. As of July 2017, Nivre was cited more than 11,000 times and produced over 200 scientific publications.

Kemal Oflazer received his Ph.D. in Computer Science from Carnegie Mellon University in Pittsburgh, PA, USA, and his M.Sc. in Computer Science and B.Sc. in Electrical and Electronics Engineering from the Middle East Technical University, Ankara, Turkey. He is currently a faculty member at Carnegie Mellon University in Doha, Qatar, where he is also the Associate Dean for Research. He held visiting positions at the Computing Research Laboratory of the New Mexico State University, Las Cruces, USA, and at the Language Technologies Institute, Carnegie Mellon University. Prior to joining CMU-Qatar, he worked in the faculties of the Sabancı University in Istanbul, Turkey (2000–2008), and the Bilkent University in Ankara, Turkey (1989–2000). He has worked extensively on developing natural language processing techniques and resources for Turkish. Oflazer's current research interests are in statistical machine translation into morphologically complex languages, the use of NLP for language learning, and machine learning for computational morphology. In addition, he was a member of the editorial boards of *Computational Linguistics, Journal of Artificial Intelligence Research, Machine Translation*, and *Research on Language and Computation* and was a book review editor for *Natural Language Engineering*. Apart from having been a member of the Nomination and Advisory Boards for EACL, he served as the Program Co-chair for ACL 2005, an area chair for COLING 2000, EACL 2003, ACL 2004, ACL 2012, EMNLP 2013, and the Organization Committee Co-chair for EMNLP 2014. Currently, he is an editorial board member of both *Language Resources and Evaluation* and *Natural Language Engineering* journals and is a member of the advisory board for "SpringerBriefs in Natural Language Processing."

Murat Saraçlar received his B.Sc. degree in 1994 from the Electrical and Electronics Engineering Department of the Bilkent University, Ankara, Turkey; his M.S.E. degree in 1997; and Ph.D. degree in 2001 from the Electrical and Computer Engineering Department of the Johns Hopkins University, Baltimore, MD, USA. From 2000 to 2005, he was with the multimedia services department of the AT&T Labs Research. In 2005, he joined the Electrical and Electronic Engineering Department of the Boğaziçi University, Istanbul, Turkey, where he is currently Full Professor. He was a Visiting Research Scientist at Google Inc., New York, NY, USA (2011–2012) and an Academic Visitor at IBM T. J. Watson Research Center (2012–2013). Saraçlar was awarded the AT&T Labs Research Excellence Award in 2002, the Turkish Academy of Sciences Young Scientist (TUBA-GEBIP) Award in 2009, and the IBM Faculty Award in 2010. He published more than 100 articles in journals and conference proceedings. Furthermore, he served as an associate editor for *IEEE Signal Processing Letters* (2009–2012) and *IEEE Transactions on Audio, Speech, and Language Processing* (2012–2016). Having been editorial board member of *Language Resources and Evaluation* from 2012 to 2016, Saraçlar is currently an editorial board member of *Computer Speech and Language* as well as a member of the IEEE Signal Processing Society Speech and Language Technical Committee (2007–2009, 2015–2018).

Mark Steedman is Professor of Cognitive Science at the School of Informatics of the University of Edinburgh, Scotland, UK. Previously, he taught at the Department of Computer and Information Science of the University of Pennsylvania, Philadelphia, PA, USA, which he joined as Associate Professor in 1988. His Ph.D. on artificial intelligence is from the University of Edinburgh. He is a Fellow of the American Association for Artificial Intelligence, the British Academy, the Royal Society of Edinburgh, the Association for Computational Linguistics, the Cognitive Science Society and a member of the European Academy. Steedman's research interests cover issues in computational linguistics, artificial intelligence, computer science and cognitive science, including syntax and semantics of natural language, wide-coverage parsing and open-domain question answering, comprehension of natural language discourse by humans and by machine, grammar-based language modeling, natural language generation, and the semantics of intonation in spoken discourse. Much of his current NLP research addresses probabilistic parsing and robust semantics for question answering using the CCG grammar formalism, including the acquisition of language from paired sentences and meanings by child and machine. Some of his research also concerns the analysis of music by humans and machines. Steedman occasionally works with colleagues in computer animation where these theories are used to guide the graphical animation of speaking virtual or simulated autonomous human agents.

Umut Sulubacak is a research and teaching assistant at the Department of Computer Engineering of the Istanbul Technical University, Turkey. As part of his B.Sc. and M.Sc. studies, his research focused on the morphological and syntactic analysis of Turkish, using both rule-based and data-driven methods and optimizing morpho-syntactic annotation processes for Turkish dependency treebanks. He was involved in the construction of the Turkish treebank as part of the "Universal Dependencies" project and has remained an active contributor to the project ever since. In addition to his teaching responsibilities, he currently pursues his Ph.D. degree at the same institution with research in treebank linguistics and machine learning for Turkish language processing.

A. Cüneyd Tantuğ is currently Associate Professor at the Faculty of Computer and Informatics Engineering of the Istanbul Technical University, Turkey, where he completed his Ph.D. and has been a faculty member since 2009. His research areas include natural language processing, machine translation, and machine learning.

Gökhan Tür is a computer scientist focusing on human/machine conversational language understanding systems. He was awarded his Ph.D. in Computer Science from the Bilkent University, Ankara, Turkey, in 2000. Between 1997 and 1999, he was a Visiting Scholar at the Language Technologies Institute, Carnegie Mellon University, Pittsburgh, PA, USA; then at the Johns Hopkins University, Baltimore, MD, USA; and at the Speech Lab of SRI, Menlo Park, CA, USA. He was at AT&T Research (2001–2006), working on pioneering conversational systems like "How May I Help You?" Later at SRI, he worked for the DARPA GALE and CALO projects (2006–2010). He was a founding member of the Microsoft Cortana team, focusing on deep learning methods (2010–2016), and was the Conversational Understanding Architect at the Apple Siri team (2014–2015). Tür is currently with the Deep Dialogue team at Google Research. Apart from frequent presentations at conferences, he coauthored more than 150 papers in journals and books. He is also a coeditor of the book *Spoken Language Understanding: Systems for Extracting Semantic Information from Speech* (Wiley 2011) and of a special issue on Spoken Language Understanding of the journal *Speech Communication*. He is also the recipient of the Best Paper Award of *Speech Communication* by ISCA for 2004–2006 and by EURASIP for 2005–2006. Tür was the organizer of the HLT-NAACL 2007 Workshop on Spoken Dialog Technologies, the HLT-NAACL 2004, and the AAAI 2005 Workshops on Spoken Language Understanding. He also served as the area chair for spoken language processing for IEEE ICASSP conferences from 2007 to 2009 and IEEE ASRU Workshop in 2005, as the spoken dialog area chair for the HLT-NAACL conference in 2007, and as an organizer of the SLT Workshop in 2010. Having been a member of the IEEE Speech and Language Technical Committee (SLTC) (2006–2008) and the IEEE SPS Industrial Relations Committee (2013–2014) as well as an associate editor for the *IEEE Transactions on Audio, Speech, and Language Processing* (2010–2014) and *Multimedia Processing* (2014–2016) journals, Tür is currently a senior member of IEEE, ACL, and ISCA.

Berrin Yanıkoğlu is Professor of Computer Science at the Sabancı University, Istanbul, Turkey. She received a double major in Computer Science and Mathematics from the Boğaziçi University, Istanbul, Turkey, in 1988 and her Ph.D. in Computer Science from the Dartmouth College, Hanover, NH, USA, in 1993. Her research interests lie in the areas of machine learning, with applications to image and language understanding, currently focusing on multimodal deception detection, sentiment analysis, online handwriting recognition, and object recognition from photographs. Yanıkoğlu received an IBM Research Division award in 1996 and first place positions in several international signature verification competitions in collaboration with her students and colleagues.

Reyyan Yeniterzi received her B.Sc. and M.Sc. degrees in Computer Science and Engineering from the Sabancı University, Istanbul, Turkey, in 2007 and 2009, and her M.Sc. and Ph.D. degrees from the Language Technologies Institute of Carnegie Mellon University, Pittsburgh, PA, USA, in 2012 and 2015. Since 2015, she has worked as an Assistant Professor at the Computer Science Department of the Özyeğin University, Istanbul, Turkey. Previous to her various visiting research positions at the International Computer Science Institute (ICSI), Berkeley, CA, USA; at Vanderbilt University, TN, USA; and at Carnegie Mellon University in Doha, Qatar, Yeniterzi gained practical experience as an intern at Google. Her main research interests include natural language processing, text mining, social media analysis, information retrieval, search engines, and statistical machine translation.

Deniz Yuret is Associate Professor of Computer Engineering at the Koç University in Istanbul, Turkey, where he has worked at the Artificial Intelligence Laboratory since 2002. Previously, he was at the MIT AI Lab, Cambridge, MA, USA (1988–1999), and later cofounded Inquira, Inc., a company commercializing question answering technology (2000-2002). Yuret worked on supervised and unsupervised approaches to syntax, morphology, lexical semantics, and language acquisition. His most recent work is on grounded language acquisition and understanding.

Deniz Zeyrek is Professor of Cognitive Science and Director of the Informatics Institute at the Middle East University in Ankara, Turkey. She holds a Ph.D. in Linguistics from the Hacettepe University, Ankara, Turkey. Her broad research interests are in Turkish discourse and pragmatics, development of annotation schemes for recording discourse, and pragmatic phenomena on corpora. She contributed to the development of the METU Turkish Corpus, a corpus of modern written Turkish, and is the principal developer of Turkish Discourse Bank, a discourse corpus annotated in the PDTB style. Zeyrek's research mainly focuses on discourse relations and means of expressing discourse relations in Turkish and expands to discourse structure as revealed by discourse relations.

Chapter 1
Turkish and Its Challenges for Language and Speech Processing

Kemal Oflazer and Murat Saraçlar

Abstract We present a short survey and exposition of some of the important aspects of Turkish that have proved to be interesting and challenging for natural language and speech processing. Most of the challenges stem from the complex morphology of Turkish and how morphology interacts with syntax. Finally we provide a short overview of the major tools and resources developed for Turkish over the last two decades. (Parts of this chapter were previously published as Oflazer (Lang Resour Eval 48(4):639–653, 2014).)

1.1 Introduction

Turkish is a language in the Turkic family of Altaic languages which also includes Mongolic, Tungusic, Korean, and Japonic families. Modern Turkish is spoken mainly by about 60M people in Turkey, Middle East, and in Western European countries. Turkic languages comprising about 40 languages some of which are extinct are spoken as a native language by 165–200M people in a much wider geography, shown in Fig. 1.1. Table 1.1 shows the distribution of Turkic speakers to prominent members of the Turkic family.

Turkish and other languages in the Turkic family have certain features that pose interesting challenges for language processing. Turkish is usually used as a textbook example while discussing concepts such as agglutinating morphology or vowel harmony in morphophonology, or free constituent order in syntax. But there are many other issues that need to be addressed for robust handling language processing tasks.

K. Oflazer (✉)
Carnegie Mellon University Qatar, Doha-Education City, Qatar
e-mail: ko@cs.cmu.edu

M. Saraçlar
Boğaziçi University, Istanbul, Turkey
e-mail: murat.saraclar@boun.edu.tr

© Springer International Publishing AG, part of Springer Nature 2018
K. Oflazer, M. Saraçlar (eds.), *Turkish Natural Language Processing*,
Theory and Applications of Natural Language Processing,
https://doi.org/10.1007/978-3-319-90165-7_1

Fig. 1.1 The geography of Turkic languages (Source: Wikipedia), https://en.wikipedia.org/wiki/Turkic_languages, accessed 26 April 2018

Table 1.1 Distribution of speakers of Turkic languages (Data source: Wikipedia, https://en.wikipedia.org/wiki/Turkic_languages, accessed 26 April 2018)

Language	Percentage (%)
Turkish	30.3
Azerbaijani	11.7
Uzbek	10.2
Kazakh	4.3
Uyghur	3.6
Tatar	2.2
Turkmen	1.3
Kyrgyz	1.0
Other	35.4

Despite being the native language of over 60M speakers in a wide geography, Turkish has been a relative late-comer into natural language processing and development of tools and resources for Turkish natural language processing has only been attempted in the last two decades. Yet Turkish presents unique problems for almost all tasks in language processing ranging from tag-set design to statistical language modeling, syntactic modeling, and statistical machine translation, among many others. On the other hand, solutions to problems observed for Turkish when appropriately abstracted turn out to be applicable to a much wider set of languages. Over the years many tools and resources have been developed but many more challenges remain: For example, there are no natural sources of parallel texts where one side is Turkish (akin to say Europarl parallel corpora), so researchers working on statistical machine translation can only experiment with rather limited data which will not increase to the levels used for pairs such as English-Chinese or English-Arabic any time soon. Other more mundane issues such as drifting away from a one-to-one correspondence between orthography and pronunciation due to the recent wholesale import of words from other languages such as English with their native orthography *and* pronunciation, cause rather nasty problems even for the basic stages of lexical processing such as morphology. For example, one usually sees words like *serverlar* (servers) where, as written, the vowels violate the harmony constraints, but as pronounced, they don't, because of a bizarre assumption by the writers of such words that the readers will know the *English* pronunciation of the root words for the vowel harmony to go through!

Nevertheless, despite these difficulties the last several years have seen a significant increase of researchers and research groups who have dedicated efforts into building resources and addressing problems and the future should be quite bright moving forward.

In this introductory chapter we present a bird's eye view of relevant aspects of Turkish important from a language and speech processing perspective. Readers interested in Turkish grammar from more of a linguistics perspective may refer to, e.g., Göksel and Kerslake (2005).

1.2 Turkish Morphology

Morphologically Turkish is an agglutinative language with morphemes attaching to a root word like "beads-on-a-string." There are no prefixes and no productive compounding (e.g., as found in German) and most lexicalized compounds have non-compositional semantics (e.g., *acemborusu*, literally *Persian pipe*, actually is the name of a flower.)

Words are formed by very productive affixations of multiple suffixes to root words from a lexicon of about 30K root words excluding proper names. The noun roots do not have any classes nor are there any markings of grammatical gender in morphology and syntax. The content word root lexicons have been heavily influenced by Arabic, Persian, Greek, Armenian, French, Italian, German

Fig. 1.2 Two examples of
the cascaded operation of
vowel harmony (Oflazer
2014) (Reprinted with
permission)

ev+ler+de+ydi
(they were in the houses)

oku+yabil+iyor+du
((s)he was able to read)

and recently English, owing to the many factors such as geographical, cultural, commercial, and temporal proximity. Literally overnight, the alphabet used for writing the language was switched from the Arabic alphabet to a Latin alphabet in 1928, and this was followed by a systematic replacement of words of Arabic, Persian, and sometimes western origins, with native Turkish ones, but many such words still survive.

When used in context in a sentence, Turkish words can take many inflectional and derivational suffixes. It is quite common to construct words which correspond to almost a sentence in English:

yap+abil+ecek+se+k → if we will be able to do (it)

Almost all morphemes have systematic allomorphs that vary in respective vowels and sometimes in boundary consonants. For example, in

paket+ten (from the package) vs. *araba+dan* (from the car)

we see an example of a consonant assimilating at the morpheme boundaries and vowels in morphemes "harmonizing" with the previous vowel. Vowel harmony in fact operates from left-to-right in a cascaded fashion as shown in Fig. 1.2. Oflazer (1994) presents details of Turkish morphophonology as implemented in a two-level morphology setting (Koskenniemi 1983). Many relevant aspects of Turkish morphology will be covered in Chap. 2.

Multiple derivations in a given word are not an uncommon occurrence. Arısoy (2009) cites the word *ruhsatlandırılamamasındaki* as a word with nine morphemes, observed in a large corpus she worked with. The word roughly means *related to (something) not being able to acquire certification*, and is used as a modifier of some noun in context. Internal to the word, there are five derivations as shown in Fig. 1.3, where we start with a root word *ruhsat* (certification) and after five derivations end up as a modifier.

But in general things are saner: The average number of bound and unbound morphemes in a word in running text is about three but this is heavily skewed. Also, on the average, each word has about two different morphological interpretations due to root having multiple parts-of-speech, homography of some suffixes, and multiple segmentations of a given word into morphemes.

Fig. 1.3 Derivations in a complex Turkish word (Oflazer 2014) (Reprinted with permission)

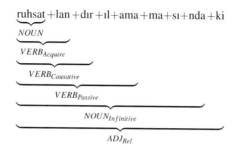

Table 1.2 Morpheme count and morphological ambiguity in the most frequent 20 Turkish words (Oflazer 2014) (Reprinted with permission)

	Word	Morphemes	Ambiguity		Word	Morphemes	Ambiguity
1	bir	1	4	11	kadar	1	2
2	bu	1	2	12	ama	1	3
3	da	1	1	13	gibi	1	1
4	için	1	4	14	ol+an	2	1
5	de	1	2	15	var	1	2
6	çok	1	1	16	ne	1	2
7	ile	1	2	17	sonra	1	2
8	en	1	2	18	ise	1	2
9	daha	1	1	19	o	1	2
10	ol+arak	2	1	20	ilk	1	1

Table 1.2 shows the 20 most frequent words in a large Turkish corpus, along with the number of morphemes in the word and morphological ambiguity for each. We can estimate from these numbers that, since the more frequent words have just one morpheme, many of the lower frequency words have more than three or more morphemes. Also, most of the high-frequency words have relatively high morphological ambiguity, which, for words with one morpheme, corresponds to having different root parts-of-speech. Hence an average of two morphological interpretations mentioned above means that morphological ambiguity for words with many morphemes (owing usually to, for example, segmentation ambiguity) is actually less.

Another aspect of Turkish morphology is the heavy use for derivational morphemes in word formation as exemplified in Fig. 1.3. Table 1.3 shows the number of possible word forms (including inflected variants) that can be generated from only *one* noun or a verb root using zero, one, two, and three derivational morphemes, with the zero case counting only the basic inflectional variants. The total column shows the cumulative number of word forms with up to the number of derivations

Table 1.3 Number of words that can be derived using 0, 1, 2, or 3 derivational morphemes (Oflazer 2014) (Reprinted with permission)

Root	# derivations	# words	Total
masa	0	112	112
(Noun, (*table*))	1	4663	4775
	2	49,640	54,415
	3	493,975	548,390
oku	0	702	702
(Verb, (*read*))	1	11,366	12,068
	2	112,877	124,945
	3	1,336,266	1,461,211

on the same row.[1] It is certain that many of these derived words are never used but nevertheless, the generative capacity of the morphological processes can generate these. The fact that a given verb root can give rise to about 1.5M different word forms is rather amazing.[2] To tame this generative capacity, the derivational processes need to be semantically constrained which is extremely hard to do in a morphological analyzer.

Sak et al. (2011) present statistics from a large corpus of Turkish text of close to 500M Turkish words collected from mainly news text. They find about 4.1M unique words in this corpus, with the most frequent 50K/300K word forms covering 89%/97% of the words, respectively, and 3.4M word form appearing less than 10 times and 2M words appearing only once. The most crucial finding is that while increasing the corpus size from 490M to 491M by adding a text of 1M words, they report encountering 5539 new word forms not found in the first 490M words!

Figure 1.4 from Sak et al. (2011) shows the number of distinct stems and the number of distinct morpheme combinations that have been observed in this corpus. One can see that at around 360M words in the corpus, the number of distinct morpheme combination observed reaches around 46K *and* exceeds the number of distinct stems observed. This leads to an essentially infinite lexicon size and brings numerous challenges in many tasks.[3]

---

[1]These numbers were counted by using the *xfst*, the Xerox finite state tool (Beesley and Karttunen 2003), by filtering through composition by restricting output by the respective root words and with the number of symbols marking a derivational morpheme, and then counting the number of possible words.

[2]See Wickwire (1987) for an interesting take on this.

[3]It turns out that there are a couple of suffixes that can at least theoretically be used iteratively. The causative morpheme is one such morpheme, but in practice up to three could be used and even then it is hard to track who is doing what to whom.

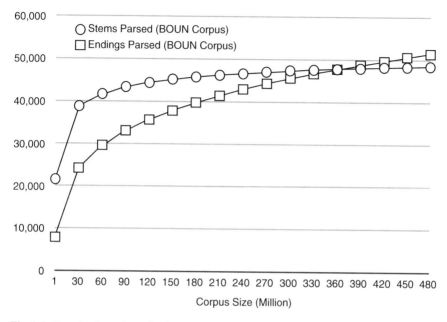

Fig. 1.4 Growth of number of unique stems and endings with corpus size (Sak et al. 2011) (Reprinted with permission)

1.3 Constituent Order and Morphology-Syntax Interface

The unmarked constituent order in Turkish is *Subject–Object–Verb* with adjuncts going in more or less freely anywhere. However all six constituent orders are possible with minimal constraints.[4] As is usual with other free constituent order languages, the freeness comes with the availability of case marking on the nominal arguments of the verbs.

The following are examples of constituent order variations along with the contextual assumptions when they are used. In all cases, the main event being mentioned is *Ekin saw çağla*, with the variations encoding the discourse context and assumptions.

- Ekin Çağla'yı gördü. (*Ekin saw Çağla.*)
- Çağla'yı Ekin gördü. (*It was Ekin who saw Çağla.*)
- Gördü Ekin Çağla'yı. (*Ekin saw Çağla (but was not really supposed to see her.)*)
- Gördü Çağla'yı Ekin. (*Ekin saw Çağla (and I was expecting that)*)

---

[4]One constraint usually mentioned is that indefinite (and nominative marked) direct objects move with the verb, but there are valid violations of that observed in speech (Sarah Kennelly, personal communication).

- Ekin gördü Çağla'yı. (*It was Ekin who saw Çağla (but someone else could also have seen her.)*)
- Çağla'yı gördü Ekin. (*Ekin saw Çağla (but he could have seen someone else.)*)

Handling these variations in the usual CFG-based formalisms is possible (though not necessarily trivial or clean). Çetinoğlu's large scale LFG grammar for Turkish (Çetinoğlu 2009), developed in the context of the Pargram Project (Butt et al. 2002), handled these variations in a principled way but did not have a good way to encode the additional information provided by the constituent order variations.

A more interesting impact of complex morphology especially derivational morphology is on modeling syntactic relationships between the words. Before elaborating on this, let's describe an abstraction that has helped us to model these relationships.

The morphological analysis of a word can be represented as a sequence of tags corresponding to the morphemes. In our morphological analyzer output, the tag ^DB denotes derivation boundaries. We call the set of morphological features encoded between two derivations (or before the first of after the last, if any) as an *inflectional group* (IG). We represent the morphological information in Turkish in the following general form:

$$\text{root}+\text{IG}_1 + {}^{\wedge}\text{DB}+\text{IG}_2 + {}^{\wedge}\text{DB}+\cdots + {}^{\wedge}\text{DB}+\text{IG}_n.$$

where each IG_i denotes the relevant sequence of inflectional features including the part-of-speech for the root (in IG_1) and for any of the derived forms.[5] A given word may have multiple such representations depending on any morphological ambiguity brought about by alternative segmentations of the word, and by ambiguous interpretations of morphemes.

For instance, the morphological analysis of the derived modifier uzaklaştı-rılacak ("(the one) that will be sent away," literally, "(the one) that will be made to be far,") would be:[6]

```
uzak+Adj

         ^DB+Verb+Become
         ^DB+Verb+Caus
         ^DB+Verb+Pass+Pos
         ^DB+Adj+FutPart+Pnon
```

---

[5]Although we have written out the root word explicitly here, whenever convenient we will assume that the root word is part of the first inflectional group.

[6]*uzak* is far/distant; the morphological features other than the obvious part-of-speech features are: +Become: become verb, +Caus: causative verb, +Pass: passive verb, +Pos: Positive Polarity, +FutPart: Derived future participle, +Pnon: no possessive agreement.

spor   arabanızdaydı

sports car-your-in DB it-was   *spor*   *arabanızda* DB *ydı*

Fig. 1.5 Relation between inflectional groups

The five IGs in this word are:

1. uzak+Adj
2. +Verb+Become
3. +Verb+Caus
4. +Verb+Pass+Pos
5. +Adj+FutPart+Pnon

The first IG indicates that the root is a simple adjective. The second IG indicates a derivation into a verb whose semantics is "to become" the preceding adjective (equivalent to "to move away" in English). The third IG indicates that a causative verb (equivalent to "to send away" in English) is derived from the previous verb. The fourth IG indicates the derivation of a passive verb with positive polarity, from the previous verb. Finally, the last IG represents a derivation into future participle which will function as a modifier of a nominal in the sentence.

We can make two observations about IGs: (1) the syntactic relations are NOT between words, but rather between IGs of different words, and (2) the role of a given word in the sentence is determined by its last IG! To further motivate this, we present the example in Fig. 1.5. The second word in the phrase *spor arabanızdaydı* ("it was in your sports car") has a second/final IG which happens to have the part-of-speech of a verb. However there is also the adjective-noun construction *spor araba-* (sports car), where the word *spor* acts as a modifier of *araba*. So the modification relation is between (the last IG of) *spor* and the first IG of the next word (which has the part-of-speech noun) and not with the whole word whose final part-of-speech is a verb. In fact, different IGs of a word can be involved in multiple relations with different IGs of multiple words as depicted in a more comprehensively annotated sentence in Fig. 1.6.[7] In Fig. 1.6, the solid lines denote the words and the broken lines denote the IGs in the words. Note that in each case, a relation from a dependent emanates from the last IG of a word, but may land on any IG as the head. The morphological features encoded in the IGs are listed vertically under each IG with different IGs' features separated by vertical dashed lines. For instance, if we zoom into the three words in the middle of the sentence (shown in Fig. 1.7), we can note the following: The word *akıllısı* is composed of three IGs; it starts as noun *akıl* ("intelligence"), and with the derivational suffix *+li*, becomes an adjective ("with intelligence/intelligent") and then through a zero derivation becomes again a noun ("one who is intelligent"). The word *öğrencilerin* (of the students) and this final IG of *akıllısı* have the necessary morphological markings and agreement features to

---

[7]Here we show surface dependency relations, but going from the dependent to the head.

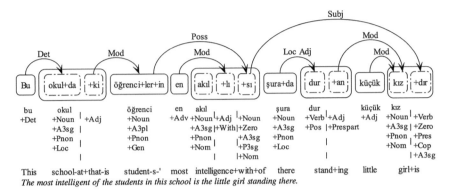

Fig. 1.6 Relations between IGs in a sentence (Oflazer 2014) (Reprinted with permission)

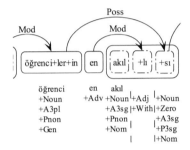

Fig. 1.7 Multiple syntactic relations for a word in Fig. 1.6 (Oflazer 2014) (Reprinted with permission)

form a possessor/possessee noun compound, and this is indicated by the relation by *Poss*. The more interesting example is the adverbial intensifier *en* ("most") modifying the intermediate IG with the part-of-speech adjective—it cannot have any other relationship, adverbials modify adjectives and not nouns. Thus we get a noun phrase meaning "the most intelligent of the students."

We have used IGs as a convenient abstraction in both statistical and rule-based contexts: Hakkani-Tür et al. (2002) modeled morphological disambiguation in terms of IGs. Çetinoğlu (2009) used IGs as basic units when modeling LFG syntax. Eryiğit et al. (2008) used IGs in the context of dependency parsing. The Turkish Treebank (Oflazer et al. 2003) has been encoded in terms of relations between IGs.

1.4 Applications

In this section we review some natural language and speech applications for Turkish, highlighting the challenges presented by Turkish in the context of these applications together with proposed solutions. While the applications span a wide spectrum, the challenges and solutions mostly follow a common theme. The complex morphology in combination with free word order and morphology–syntax interface summarized in the previous sections underlie the challenges. The solutions make use of morphological and morphosyntactic analysis to alleviate the challenges.

Spelling Checking and Correction Methods that rely on a finite list of words or a list of root words with some fixed number of affixes cannot capture lexicon of Turkish. We have developed efficient spelling correction algorithms for languages like Turkish based on error tolerant finite state recognition, operating on a finite state recognizer model of morphology that can encode an infinite number of words (Oflazer 1996).

Tagset Design It is not possible to fully represent the morphosyntactic information encoded in morphology with a finite set of tags. The data in Fig. 1.4 already hints at this. There are of course a small number of root part-of-speech categories but with multiple inflectional and derivational affixes affixed, the word may end up having many morphological features including multiple parts-of-speech, all of which may have syntactic implications. See Hakkani-Tür et al. (2002) for statistics on the number of different possible tags.

Syntactic Modeling As we saw in the previous section, derivational morphemes have interesting implications in syntactic modeling using either constituency based formalisms or dependency based formalisms. These will be discussed in more detail in Chaps. 7 and 9.

Statistical Language Modeling A large vocabulary size almost always leads to a data sparseness problem in word-based language modeling. This is especially important when the text corpora used for language model estimation are not extremely large. One approach to limit the vocabulary size and hence combat data sparseness is to use sub-lexical units instead of words in language modeling. Traditional n-gram language models predict the next unit given the history consisting of $n - 1$ units. There is a trade-off between the length and the predictive power of the units used for traditional n-gram language models. On the one hand, the shorter the units, the more common they are. So data sparseness is less of an issue for shorter units. On the other hand, for shorter units, the history needs to include many more units for the same level of predictive power. This is easy to see when one compares letter-based language models with word-based language models.

Arısoy (2009) and Sak (2011) have investigated using sub-lexical units in language modeling for Turkish. Both morphological analyzers and unsupervised statistical word segmentation techniques yield sub-lexical units that improve the coverage and performance of statistical language models.

Although morphological analysis provides meaningful units useful for language modeling, it also has some issues. First, building a wide coverage morphological analyzer is costly and requires expert knowledge. Second, the coverage of the morphological analyzer is limited by the root lexicon and this is especially important for proper nouns. Finally, when using morphological analysis to obtain the sub-lexical language modeling units, an important issue is morphological disambiguation. For statistical language modeling, consistency in disambiguation can be as important as accuracy.

On the other hand, unsupervised word segmentation techniques typically require only a word list to come up with the sub-lexical language modeling units. However, these units do not necessarily correspond to actual morphemes and may not be as meaningful and informative as those obtained by morphological analysis. Further unsupervised statistical processing such as clustering can provide a way of improving the predictive power of these units.

In addition to the traditional language models that predict the next unit given the units in the history, feature based language models allow easy integration of other information sources. For Turkish, Arısoy (2009) incorporated morphological and syntactic features in language modeling both for lexical and sub-lexical units.

Details of these approaches will be covered in Chap. 4.

Pronunciation Modeling Applications that aim a conversion between text and speech require a way of determining how words are pronounced. For limited vocabulary applications, a hand-crafted pronunciation lexicon that simply lists the pronunciations of the words in the vocabulary is adequate. However, for Turkish, the large vocabulary size implies that a list of pronunciations for use in speech applications is rather inadequate.

Oflazer and Inkelas (2006) describe a computational pronunciation lexicon capable of determining the position of the primary stress. Their implementation uses a series of finite state transducers including those for two level morphological analysis, grapheme-to-phoneme mapping, syllabification, and stress computation. They also report that for a corpus of about 11.6 million tokens, while the average distinct morphological parses per token is 1.84, the average distinct pronunciations per token is 1.11 when taking stress into account and only 1.02 ignoring stress. The implications of this analysis for speech applications will be discussed below.

Automatic Speech Recognition (ASR) In addition to the challenges related to statistical language modeling, ASR (or STT) systems also have to deal with issues related to pronunciation modeling. In particular, the mainstream ASR systems make use of phone-based acoustic models that require a pronunciation lexicon to map words into phone sequences. While information about the position of stress can improve the acoustic models, the use of stress is not vital and common for ASR.

As mentioned above, while a pronunciation lexicon can be built by hand for medium vocabulary sizes, large vocabulary continuous speech recognition (LVCSR) requires an automatic process for building the pronunciation lexicon such as the one implemented by Oflazer and Inkelas (2006). Although the process of mapping the graphemic representation to the phonetic representation is not overly complicated, it does require morphological analysis. Their observation that over 98% of the tokens have a single pronunciation when the position of the primary stress is ignored, and that the remaining tokens have only two alternative pronunciations (differing mostly in vowel length and consonant palatality), suggests that pronunciation disambiguation is not really necessary for ASR.

An alternative approach uses grapheme-based acoustic models and lets the context-dependent graphemic acoustic models implicitly take care of pronunciation modeling. While graphemic acoustic modeling might seem somewhat simplistic, it works quite well in practice for languages where the orthography is not far from the pronunciations.

Using a sub-lexical language model further complicates pronunciation modeling. When morphological analysis is used to obtain the sub-lexical units, it is not possible to determine the pronunciation of a sub-lexical item without looking at its context, the vowels of most suffixes are determined by vowel harmony and adding a suffix may change the pronunciation of the root. Therefore, the pronunciation lexicon will have to include multiple pronunciations complicating the system and allowing for incorrect pronunciations. This issue is even more dramatic when unsupervised word segmentation is used to obtain the sub-lexical units. Some units may not even be pronounced. As graphemic acoustic models do not require a phonetic representation, no further complications arise from using sub-lexical units for language modeling.

Acoustic confusability is another issue that needs to be considered when using sub-lexical units. Longer units are less confusable than shorter units simply because their acoustic neighborhood is less populated. Acoustic confusability and the trade-off for language modeling discussed above suggest that short units are not preferable for ASR.

For the Turkish broadcast news transcription task (Saraçlar 2012), using context-dependent grapheme-based acoustic models, and a language model based on a vocabulary of 76K sub-lexical units with an average unit length of 7–8 letters gives a very good coverage and the lowest word error rate percentage (Arısoy et al. 2009).

Speech Retrieval Speech retrieval systems combine ASR with information retrieval (IR). The IR component typically forms an index from the output of the ASR system and searches this index given the user query. While obtaining a simple text output from the ASR system makes it possible to directly leverage text retrieval techniques, using alternative speech recognition hypotheses in the form of a lattice has been shown to significantly improve retrieval performance (Chelba et al. 2008).

For Turkish, Arısoy et al. (2009) investigated spoken term detection (or keyword search) for Turkish broadcast news. Parlak and Saraçlar (2012) further extended this work and also built a spoken document retrieval system for the same task.

Since queries tend to include rare words, the frequency of queries containing words that are outside the vocabulary of the ASR system can be quite high, especially for Turkish. In order to deal with these queries it is common to make use of sub-lexical units even when the ASR system produces word-based outputs. Of course, the same sub-lexical units used for ASR can also be used for indexing and search. Arısoy et al. (2009) have shown that the best performance for Turkish broadcast news retrieval is obtained by combining the output of systems based on word and sub-word units.

Another common technique utilized especially for spoken document retrieval is stemming. While it is possible to determine the stem using full morphological analysis, stemming is actually an easier task. For both text and speech document retrieval using the first five characters of a word was shown to perform well (Can et al. 2008; Parlak and Saraçlar 2012).

Speech Synthesis or Text-to-Speech (TTS) Text-to-Speech systems require a text analysis step in order to obtain a phonetic representation enriched with stress and prosodic markers for a given input text. Determining the pronunciation of a word sequence together with the required stress and prosodic information is more involved than building a pronunciation lexicon for ASR.

Oflazer and Inkelas (2006) report that, when taking the primary stress into account, about 90% of the tokens have a single pronunciation, about 9% have two distinct pronunciations and the rest have three or more pronunciations. Therefore, pronunciation disambiguation is a required component for the text analysis component of a TTS system. Külekçi (2006) analyzed the pronunciation ambiguities in Turkish and suggested that morphological disambiguation (MD), word sense disambiguation (WSD), and named entity recognition (NER) can be used for pronunciation disambiguation.

Statistical Machine Translation Just as with statistical language modeling, a large vocabulary implies sparseness in statistical machine translation, which is compounded by the fact that no really large parallel corpora involving Turkish exist to offset this. Thus approaches exploiting morphology in various ways have been proposed with good improvements over word-based baseline.

At this point, it should be clear that morphology is bound to create problems for three components of a statistical machine translation systems for Turkish. Let's look at a rather contorted but not that unreasonable example of a hypothetical process of how an English phrase becomes a Turkish word in the ideal case. Figure 1.8 shows how different parts of the English phrase (mostly function words) are scrambled around and then translated into morphemes which when concatenated gives us a single word *sağlamlaştırabileceksek*. One can immediately see that the process of alignment—the starting point for training SMT systems—is bound to

Fig. 1.8 How English becomes Turkish in translation (Oflazer 2014) (Reprinted with permission)

| if | we | will | be | able | to | make | ... | become | strong |

| if | we | will | be | able | to | make | ... | become | strong |

| ... | strong | become | to | make | be | able | will | if | we |

| ... | sağlam | +laş | +tır | +abil | +ecek | +se | +k |

... **sağlamlaştırabileceksek**

have problems, especially if the unit of translation is a word on both sides. In this example, a (discontinuous) sequence of nine words ideally has to align with one Turkish word if word-to-word alignment is done. This clearly stresses the alignment process. This is especially due to the data sparseness problem since without any consideration of the internal structure of the words, the commonalities between variants of the same word will not be exploited. Another problem is that, if one gets a single morpheme wrong, the Turkish word cannot be constructed properly and this hurts in evaluation (even if the rest of the word is OK.)

An obvious thing to do is to make Turkish more like English and segment Turkish words in their overt morphemes (and maybe even do some segmentation on the English side separating suffixes like the plural or verb suffixes). For example, one would use the following representations on the two sides of a parallel sentence pair:

E: *I would not be able to read* . . .   T: . . . *oku +yama +yacak +tı +m*

This approach was experimented with by Oflazer (2008), Durgar-El Kahlout (2009), Durgar-El Kahlout and Oflazer (2010) in a phrase-based SMT setting, using the Moses tool (Koehn et al. 2007), with some additional techniques. Significant improvements over a word-based baseline was obtained but clearly other fundamental problems came to surface:

- When Turkish words are segmented, the number of tokens per (Turkish) sentence increases by almost a factor of 3. This increase causes problems with alignment tools.
- The decoder of the SMT tool is now tasked with getting both the word *and* the morpheme order right. Yet these both are fundamentally different processes. Since the decoder has no notion of morphology, the morphology frequently gets mangled in the resulting word structures.

An alternative approach is to make English side more like Turkish based on the observation that many kinds of specific (possibly discontinuous) phrases in English actually correspond solely to morphology on the Turkish side. Yeniterzi and Oflazer (2010) experimented with a so-called *syntax-to-morphology* scheme in which English sentences were syntactically processed to identify various phrases or parts of phrases and transform those so that the resulting structures "look like Turkish." So if the original English sentence had a phrase like

. . . in their economic relations . . .

a parser would identify the preposition *in* and possessive pronoun *their* are related to the word *relations* and convert this to a representation like

. . . economic relation+s+their+in . . .

essentially treating these function words as if they were morphemes on *relation*. Morphological preprocessing on the Turkish side also gives

. . . ekonomik ilişki+ler+i+nde . . .

Using factored phrase-based SMT, Yeniterzi and Oflazer (2010) concatenated "morphemes" on both sides into a single morphology factor. Alignment was then performed *only based on the root words*—morphology on the Turkish side and the function words tacked on to the words as pseudo-morphology on the English-side, were assumed to align if the root words aligned. These transformations reduced the number of tokens on the English side by about 30% hence alignment could be run on much shorter sentences (of just the remaining root words on both sides). Again significant improvements over a baseline were obtained. This method is probably a bit more promising but the recall for finding English side patterns to map to Turkish morphological structures was not very high, as the patterns and the corresponding rules for extracting them were hand-crafted.

1.5   State-of-the-Art Tools and Resources for Turkish

Many tools and resources for Turkish language processing have been developed over the last two decades starting essentially from scratch. In this section, we present brief overview of these, with pointers to literature and web (whenever applicable) for researchers interested in following up.

1. *Morphological Analysis*: Oflazer (1994) describes a wide-coverage morphological analyzer for Turkish based on the two-level formalism. This has been implemented using the Xerox finite state tools. This analyzer has also been used to build a more general system that produces both the morphological analyses and possible pronunciations of a Turkish word (phonemes encoded in SAMPA, syllable boundaries and primary stress location) (Oflazer and Inkelas 2006).
2. *Morphological Disambiguation*: In addition to early work by Oflazer and Kuruöz (1994), Oflazer and Tür (1996), Hakkani-Tür et al. (2002), the disambiguators by Sak et al. (2007) and Yuret and Türe (2006) are the more recent efforts for this problem, and in practice they have performed satisfactorily in various applications they have been used.
3. *Statistical Dependency Parsers*: There have been several dependency parsers for Turkish that have been trained on the Turkish Treebank (Oflazer et al. 2003). Eryiğit and Oflazer (2006) describe a statistical dependency parser that estimates IG-to-IG linking probabilities and then among possible dependency structures for a sentence, finds the most probable. Eryiğit et al. (2008) use the MaltParser framework (Nivre et al. 2007) to develop a deterministic dependency parser for Turkish.[8]
4. *An LFG-based parser*: A large-scale wide-coverage deep grammar for Turkish using the lexical functional grammar formalism was built by Çetinoğlu (2009), in

---

[8]The pre-trained MaltParser model and configuration files for Turkish can be downloaded from web.itu.edu.tr/gulsenc/TurkishDepModel.html (Accessed Sept. 14, 2017).

the context of the ParGram (Parallel Grammars) Project (Butt et al. 2002).[9] The goal was to implement a principled analysis for many linguistic phenomena and produce functional structures that was parallel to a large extent to the functional structures produced by other grammars for the same sentences in respective languages.

5. *The Turkish Treebank*: A treebank of 5635 sentences using a dependency representation with links representing IG-to-IG relations was constructed and made available to the research community (Oflazer et al. 2003).[10] Among many other uses, it has been used in recent CONLL Multilingual Dependency Parsing Competitions (Buchholz and Marsi 2006).

6. *The Turkish Discourse Bank*: A Turkish discourse bank with annotations of discourse connectives in the style of Penn Discourse Treebank was recently developed (Zeyrek et al. 2009).

7. *Turkish WordNet*: A WordNet for Turkish of about 15K synsets (Bilgin et al. 2004) was developed in the context of the Balkanet Project (Stamou et al. 2002) and has been used by tens of researchers for various applications.

8. *Miscellaneous Corpora and Resources*: There have been numerous other efforts for compiling various corpora for Turkish linguistics and language processing work. Notable among these is the Turkish National Corpus (Aksan et al. 2012), recently announced at www.tnc.org.tr/ (Accessed Sept. 14, 2017). Deniz Yuret has compiled at list of resources for Turkish at www.denizyuret.com/2006/11/turkish-resources.html (Accessed Sept. 14, 2017).

Notes

In the following chapters, while we tried, to a large extent, to standardize notation for various linguistic description (e.g., morphological parses) and the definition of certain concepts (e.g., inflectional groups), we nevertheless gave the authors leeway if they wanted to use their own notational conventions and variations of definitions to fit into their narrative.

References

Aksan Y, Aksan M, Koltuksuz A, Sezer T, Mersinli Ü, Demirhan UU, Yılmazer H, Kurtoğlu Ö, Öz S, Yıldız İ (2012) Construction of the Turkish National Corpus (TNC). In: Proceedings of LREC, Istanbul, pp 3223–3227

---

[9]See also ParGram/ParSem. An international collaboration on LFG-based grammar and semantics development: pargram.b.uib.no (Accessed Sept. 14, 2017).

[10]Available at web.itu.edu.tr/gulsenc/treebanks.html (Accessed Sept. 14, 2017).

Arısoy E (2009) Statistical and discriminative language modeling for Turkish large vocabulary continuous speech recognition. PhD thesis, Boğaziçi University, Istanbul

Arısoy E, Can D, Parlak S, Sak H, Saraçlar M (2009) Turkish broadcast news transcription and retrieval. IEEE Trans Audio Speech Lang Process 17(5):874–883

Beesley KR, Karttunen L (2003) Finite state morphology. CSLI Publications, Stanford University, Stanford

Bilgin O, Çetinoğlu Ö, Oflazer K (2004) Building a Wordnet for Turkish. Rom J Inf Sci Technol 7(1–2):163–172

Buchholz S, Marsi E (2006) CoNLL-X shared task on multilingual dependency parsing. In: Proceedings of CONLL, New York, NY, pp 149–164

Butt M, Dyvik H, King TH, Masuichi H, Rohrer C (2002) The parallel grammar project. In: Proceedings of the workshop on grammar engineering and evaluation, Taipei, pp 1–7

Can F, Koçberber S, Balçık E, Kaynak C, Öcalan HC, Vursavaş OM (2008) Information retrieval on Turkish texts. J Am Soc Inf Sci Technol 59(3):407–421

Çetinoğlu Ö (2009) A large scale LFG grammar for Turkish. PhD thesis, Sabancı University, Istanbul

Chelba C, Hazen TJ, Saraçlar M (2008) Retrieval and browsing of spoken content. IEEE Signal Process Mag 25(3):39–49

Durgar-El Kahlout İ (2009) A prototype English-Turkish statistical machine translation system. PhD thesis, Sabancı University, Istanbul

Durgar-El Kahlout İ, Oflazer K (2010) Exploiting morphology and local word reordering in English-to-Turkish phrase-based statistical machine translation. IEEE Trans Audio Speech Lang Process 18(6):1313–1322

Eryiğit G, Oflazer K (2006) Statistical dependency parsing of Turkish. In: Proceedings of EACL, Trento, pp 89–96

Eryiğit G, Nivre J, Oflazer K (2008) Dependency parsing of Turkish. Comput Linguist 34(3):357–389

Göksel A, Kerslake C (2005) Turkish: a comprehensive grammar. Routledge, London

Hakkani-Tür DZ, Oflazer K, Tür G (2002) Statistical morphological disambiguation for agglutinative languages. Comput Hum 36(4):381–410

Koehn P, Hoang H, Birch A, Callison-Burch C, Federico M, Bertoldi N, Cowan B, Shen W, Moran C, Zens R, Dyer C, Bojar O, Constantin A, Herbst E (2007) Moses: open source toolkit for statistical machine translation. In: Proceedings of ACL, Prague, pp 177–180

Koskenniemi K (1983) Two-level morphology: a general computational model for word-form recognition and production. PhD thesis, University of Helsinki, Helsinki

Külekçi MO (2006) Statistical morphological disambiguation with application to disambiguation of pronunciations in Turkish. PhD thesis, Sabancı University, Istanbul

Nivre J, Hall J, Nilsson J, Chanev A, Eryiğit G, Kübler S, Marinov S, Marsi E (2007) MaltParser: a language-independent system for data-driven dependency parsing. Nat Lang Eng 13(2):95–135

Oflazer K (1994) Two-level description of Turkish morphology. Lit Linguist Comput 9(2):137–148

Oflazer K (1996) Error-tolerant finite-state recognition with applications to morphological analysis and spelling correction. Comput Linguist 22(1):73–99

Oflazer K (2008) Statistical machine translation into a morphologically complex language. In: Proceedings of CICLING, Haifa, pp 376–387

Oflazer K (2014) Turkish and its challenges for language processing. Lang Resour Eval 48(4):639–653

Oflazer K, Inkelas S (2006) The architecture and the implementation of a finite state pronunciation lexicon for Turkish. Comput Speech Lang 20:80–106

Oflazer K, Kuruöz İ (1994) Tagging and morphological disambiguation of Turkish text. In: Proceedings of ANLP, Stuttgart, pp 144–149

Oflazer K, Tür G (1996) Combining hand-crafted rules and unsupervised learning in constraint-based morphological disambiguation. In: Proceedings of EMNLP-VLC, Philadelphia, PA

Oflazer K, Say B, Hakkani-Tür DZ, Tür G (2003) Building a Turkish Treebank. In: Treebanks: building and using parsed corpora. Kluwer Academic Publishers, Berlin

Parlak S, Saraçlar M (2012) Performance analysis and improvement of Turkish broadcast news retrieval. IEEE Trans Audio Speech Lang Process 20(3):731–741

Sak H (2011) Integrating morphology into automatic speech recognition: morpholexical and discriminative language models for Turkish. PhD thesis, Boğaziçi University, Istanbul

Sak H, Güngör T, Saraçlar M (2007) Morphological disambiguation of Turkish text with perceptron algorithm. In: Proceedings of CICLING, Mexico City, pp 107–118

Sak H, Güngör T, Saraçlar M (2011) Resources for Turkish morphological processing. Lang Resour Eval 45(2):249–261

Saraçlar M (2012) Turkish broadcast news speech and transcripts (LDC2012S06). Resource available from Linguistic Data Consortium

Stamou S, Oflazer K, Pala K, Christoudoulakis D, Cristea D, Tufis D, Koeva S, Totkov G, Dutoit D, Grigoriadou M (2002) Balkanet: a multilingual semantic network for Balkan languages. In: Proceedings of the first global WordNet conference, Mysore

Wickwire DE (1987) The Sevmek Thesis, a grammatical analysis of the Turkish verb system illustrated by the verb sevmek-to love. Master's thesis, Pacific Western University, San Diego, CA

Yeniterzi R, Oflazer K (2010) Syntax-to-morphology mapping in factored phrase-based statistical machine translation from English to Turkish. In: Proceedings of ACL, Uppsala, pp 454–464

Yuret D, Türe F (2006) Learning morphological disambiguation rules for Turkish. In: Proceedings of NAACL-HLT, New York, NY, pp 328–334

Zeyrek D, Turan ÜD, Bozşahin C, Çakıcı R, Sevdik-Çallı A, Demirşahin I, Aktaş B, Yalçınkaya İ, Ögel H (2009) Annotating subordinators in the Turkish Discourse Bank. In: Proceedings of the linguistic annotation workshop, Singapore, pp 44–47

Chapter 2
Morphological Processing for Turkish

Kemal Oflazer

Abstract This chapter presents an overview of Turkish morphology followed by the architecture of a state-of-the-art wide coverage morphological analyzer for Turkish implemented using the Xerox Finite State Tools. It covers the morphophonological and morphographemic phenomena in Turkish such as vowel harmony, the morphotactics of words, and issues that one encounters when processing real text with myriads of phenomena: numbers, foreign words with Turkish inflections, unknown words, and multi-word constructs. The chapter presents ample illustrations of phenomena and provides many examples for sometimes ambiguous morphological interpretations.

2.1 Introduction

Morphological processing is the first step in natural language processing of morphologically complex languages such as Turkish for downstream tasks such as document classification, parsing, machine translation, etc. In this chapter, we start with an overview of representational issues, and review Turkish morphophonology and morphographemics including phenomena such as vowel harmony, consonant assimilation, and their exceptions. We then look at the root word lexicons and morphotactics, and describe *inflectional groups*, first mentioned in Chap. 1 and are quite important in the interface of morphology with syntax. We then provide numerous examples of morphological analyses highlighting morphological ambiguity resulting from root word part-of-speech ambiguity, ambiguity in segmentation of a word into morphemes, and homography of morphemes.

We then briefly discuss the internal architecture of the finite state transducer that has been built using the two-level morphology approach (Koskenniemi 1983; Beesley and Karttunen 2003), that is the underlying machinery that can be

K. Oflazer (✉)
Carnegie Mellon University Qatar, Doha-Education City, Qatar
e-mail: ko@cs.cmu.edu

© Springer International Publishing AG, part of Springer Nature 2018
K. Oflazer, M. Saraçlar (eds.), *Turkish Natural Language Processing*,
Theory and Applications of Natural Language Processing,
https://doi.org/10.1007/978-3-319-90165-7_2

customized to provide many different analysis representations: for example, as surface morphemes, or as lexical morphemes or as a root word followed by morphological feature symbols. We can also generate more complex representations that encode both the phonological structure of a word (phonemes, syllables, and stress position) and its morphological structure as morphological feature symbols.

Subsequently we discuss issues that arise when processing real texts where one encounters many tokens that present different types of complications. Examples of such phenomena are acronyms, numbers written with digits but then inflected, words of foreign origins but inflected according to Turkish phonological rules and unknown words where the root words are not known but some morphological features can be extracted from any suffixes. We do not cover issues that occur in seriously corrupted sources of text such as tweets where vowels and/or consonants are dropped, capitalization and/or special Turkish characters are haphazardly used or ignored, and character codes that do not occur in Turkish are widely used when text is typed through smartphone keyboards from users in various countries especially across Europe. However our morphological analyzer is very robust in handling many cases that one encounters even in such sources.

Finally we conclude with an overview of multi-word processing covering compound verbs, lexicalized collocations, and non-lexicalized collocations.

2.2 Overview of Turkish Morphology

Morphologically Turkish is an agglutinative language with word forms consisting of morphemes concatenated to a root morpheme or to other morphemes, much like "beads on a string" (Sproat 1992). Except for a very few exceptional cases, the surface realizations of the morphemes are conditioned by various regular morphophonological processes such as vowel harmony, consonant assimilation, and elisions. The morphotactics of word forms can be quite complex when multiple derivations are involved as it is quite possible to construct and productively use words which can correspond to a multiple word sentence or phrase in, say, English. For instance, the derived modifier sağlamlaştırdığımızdaki[1] would be represented as:

```
sağlam+Adj^DB
       +Verb+Become^DB
       +Verb+Caus+Pos^DB
       +Noun+PastPart+A3sg+Pnon+Loc^DB
       +Adj+Rel
```

---

[1]Literally, "(the thing existing) at the time we caused (something) to become strong." Obviously this is not a word that one would use everyday. Turkish words (excluding non-inflecting high-frequency words such as conjunctions, clitics, etc.) found in typical running text average about 10 letters in length. The average number of bound morphemes in such words is about 2.

Starting from an adjectival root *sağlam*, this word form first derives a verbal stem *sağlamlaş*, meaning "to become strong," with the morpheme +*laş*. A second suffix, the causative surface morpheme +*tır* which we treat as a verbal derivation, forms yet another verbal stem meaning "to cause to become strong" or "to make strong." The immediately following participle suffix +*dığ* produces a nominal, which inflects in the normal pattern for nouns (here, for 1st person plural possessor and locative case with suffixes +*ımız* and +*da*). The final suffix, +*ki*, is a relativizer, producing a word that functions as a modifier in a sentence, whose overall semantics was given above modifying a noun somewhere to the right.

The feature form representation above has been generated by a two-level morphological analyzer for Turkish (Oflazer 1994) developed using XRCE finite state tools (Karttunen and Beesley 1992; Karttunen 1993; Karttunen et al. 1996; Beesley and Karttunen 2003). This analyzer first uses a set of morphographemic rules to map from the surface representation to a lexical representation in which the word form is segmented into a series of lexical morphemes. For the word above, this segmented lexical morphographemic representation would be:

```
sağlam+lAş+DHr+DHk+HmHz+DA+ki
```

In this representation, lexical morphemes except the lexical root utilize meta-symbols that stand for a set of graphemes. These graphemes are selected on the surface by a series of morphographemic processes which are originally rooted in the morphophonological processes of the language. We will discuss some of these processes below.

For instance, A stands for back and unrounded vowels *a* and *e* in orthography, H stands for high vowels *ı*, *i*, *u*, and *ü*, and D stands for *d* and *t*, representing alveolar consonants. Thus a lexical morpheme represented as +DHr actually represents 8 possible allomorphs, which appear as one of +*dır*, +*dir*, +*dur*, +*dür*, +*tır*, +*tir*, +*tur*, +*tür* depending on the local morphophonemic/morphographemic context.

Once all segmentations of a word form are produced, they are then mapped to a more symbolic representation where root words are assigned part-of-speech categories from any relevant lexicons, and morphemes are assigned morphosyntactic feature names including default features for covert or zero morphemes, (e.g., if there is no plural morpheme on a noun, then we emit a feature name +A3sg, indicating that word is singular.)

A short listing feature names are provided in Appendix.

2.3 Morphophonology and Morphographemics

Overviews of Turkish phonology can be found in Clements and Sezer (1982), van der Hulst and van de Weijer (1991), and Kornfilt (1997). Turkish has an eight vowel inventory which is symmetrical around the axes of backness, roundness, and height: /i, y, e, 2, a, o, 1, u/ which correspond to *i*, *ü*, *e*, *ö*, *a*, *o*, *ı*, and *u* in

Turkish orthography.[2] Suffix vowels typically harmonize in backness, and (if high) in roundness to the preceding stem vowel (compare, e.g., *ev+ler* /evler/ "houses" to *at+lar* /atlar/ "horses"). But there are several suffixes, e.g., the relativizer *+ki*, whose vowels do not harmonize, as well as others, e.g., progressive suffix *+Hyor*, in which the first vowel harmonizes but the second does not. Many roots are internally harmonic but many others are not; these include loan words (e.g., *kitap* /citap/ "book", from Arabic) as well as some native words (e.g., *anne* /anne/ "mother"). Furthermore, vowel harmony does not apply between the two components of (lexicalized) compounds.

Turkish has 26 consonants: /p, t, tS, k, c, b, d, dZ, g, gj, f, s, S, v, w, z, Z, m, n, N, l, 5, r, j, h, G/. However, orthography uses only 21 letters for consonants: /g/ and its palatal counterpart /gj/ are written as *g*, while /k/ and its palatal counterpart /c/ are written as *k*, /5/ and its palatal counterpart /l/ are written as *l*, /v, w/ are written as *v* and /n/ and its nasal counterpart /N/ are written as *n*. Palatalized segments (/gj, c, l/) contrast with their nonpalatalized counterparts only in the vicinity of back vowels (thus *sol* is pronounced /so5/ when used to mean "left" vs. /sol/ when used to mean the musical note G). In the neighborhood of front vowels, palatality is predictable (*lig* /ligj/ "league"). /G/, written as *ğ*, represents the velar fricative or glide corresponding to the historical voiced velar fricative that was lost in Standard Turkish. When it is syllable-final, some speakers pronounce it as a glide (/w/ or /j/) and others just lengthen the preceding vowel. In morphological processing we treat it as a consonant when it is involved in morphologically induced changes.

Root-final plosives (/b, d, g/) typically devoice when they are syllable-final (thus *kitab+a* /ci-ta-ba/ "to the book," but *kitap* /ci-tap/ "book," *kitap+lar* /ci-tap-lar/ "books".[3] Suffix-initial plosives assimilate in voice to the preceding segment (thus *kitap+ta* /ci-tap-ta/ "in the book" but *araba+da* /a-ra-ba-da/ "in the car".

Velar consonants (/g/ and /k/) reduce to /G/ at most root-suffix boundaries; thus *sokak* /sokak/ "street" *sokak+ta* /so-kak-ta/ "on the street" but *so-ka-ğa* /so-ka-Ga/ "to the street." For more details on the phonology of Turkish words including details of syllable structure and stress patterns, we refer the reader to Oflazer and Inkelas (2006).

We now present relatively informally, a reasonably complete list of phonological phenomena that are triggered when morphemes are affixed to root words or stems. These rules can be implemented in many different ways depending on the

---

[2]For phonological representations we employ the SAMPA representation. The Speech Assessment Methods Phonetic Alphabet (SAMPA) is a computer-readable phonetic script using 7-bit printable ASCII characters, based on the International Phonetic Alphabet (IPA) (see en.wikipedia.org/wiki/Speech_Assessment_Methods_Phonetic_Alphabet (Accessed Sept. 14, 2017) and www.phon.ucl.ac.uk/home/sampa/ (Accessed Sept. 14, 2017)). The Turkish SAMPA encoding convention can be found at www.phon.ucl.ac.uk/home/sampa/turkish.htm (Accessed Sept. 14, 2017).

[3]In this chapter, we use - to denote syllable boundaries and + to denote morpheme boundaries wherever appropriate.

underlying engine for implementing a morphological analyzer.[4] We present our examples through aligned lexical and surface forms in the usual convention of two-level morphology to point out the interactions between phonemes. In the examples below, the first row (L) shows the segmentation of the lexical representation of a word into its constituent lexical morphemes, the second row (S) shows the (aligned) surface form with any morphologically–induced changes highlighted in boldface (where we also use 0 to indicate the empty string resulting from the deletion of a lexical symbol) and the third row indicates the actual orthographical surface form (O) as written in text. At this stage all our representations employ letters in the Turkish alphabet although all these changes are phonologically motivated.[5]

(a) **Vowel Harmony-1**: The lexical vowel A (representing a back and rounded vowel) in a morpheme is realized on the surface as an a if the last vowel on the surface is one of a, ı, o, u, but is realized as an e if the last vowel on the surface is one of e, i, ö, ü. For example:

L	masa+lAr	okul+lAr	ev+lAr	gül+lAr
S	masa**0**lar	okul**0**lar	ev**0**ler	gül**0**ler
O	masalar	okullar	evler	güller

(b) **Vowel Harmony-2**: The lexical vowel H (representing a high-vowel) in a morpheme is realized on the surface as an

- i if the last vowel on the surface is one of e, i,
- ı if the last vowel on the surface is one of a, ı,
- u if the last vowel on the surface is one of o, u,
- ü if the last vowel on the surface is one of ö, ü

For example:

L	masa+yH	okul+yH	ev+yH	sürü+yH	gül+lAr+yH
S	masa**0**yı	okul**00**u	ev**00**i	sürü**0**yü	gül**0**ler**00**i
O	masayı	okulu	evi	sürüyü	gülleri

There are a couple of things to note here. Clearly there are other morphographemic processes going in the second, third, and fifth examples: for example, a lexical y is (concurrently) deleted on the surface (to be discussed below). The fifth example actually shows three processes happening concurrently: the mutually dependent vowel harmony processes take place along with the y in the third morpheme being deleted.

---

[4]For example, Xerox Finite State Tools, available at www.fsmbook.com (Accessed Sept. 14, 2017), FOMA, available at fomafst.github.io/ (Accessed Sept. 14, 2017), HFST available at hfst.sf.net (Accessed Sept. 14, 2017) or OpenFST available at www.openfst.org (Accessed Sept. 14, 2017).

[5]Note that we also explicitly show the morpheme boundary symbol, as in implementation, it serves as an explicit context marker to constrain where changes occur.

While these vowel harmony rules are the dominant ones, they are violated in quite many cases due to vowel quality being modified (usually) as a result of palatalization. For example:

```
L   hilal+lAr                      alkol+yH
S   hilalOler   (not hilalOlar)    alkolOOü   (not alkolOOu)
O   hilaller                       alkolü
```

These cases are for all practical purposes lexicalized, and the internal lexical representations of such cases have to mark them with alternative symbols so as to provide contexts for overriding the default harmony rules.

(c) **Vowel Deletion**: A morpheme initial vowel is deleted on the surface when affixed to a stem ending in a vowel, unless the morpheme is the present progressive morpheme +Hyor in which case the vowel in the stem is deleted. For example,

```
L   masa+Hm      ağla+Hyor
S   masaOOm      ağlOOıyor
O   masam        ağlıyor
```

(d) **Consonant Deletion**: Morpheme-initial s, y, and n are deleted when either of the accusative morpheme +yH or the possessive morpheme +sH or the genitive case morpheme +nHn is attached to a stem ending in a consonant. For example:

```
L   kent+sH      kent+yH      kent+nHn
S   kentOOi      kentOOi      kentOOin
O   kenti        kenti        kentin
```

Note that this can also be seen as insertion of a y, s or an n on the surface when the stem ends in a vowel. As long as one is consistent, this ends up being a representational issue which has no bearing on the computational implementation.

(e) **Consonant Voicing**: A morpheme initial dental consonant (denoted by D representing d or t) will surface as a voiced d, when affixed to a stem ending in a surface vowel or the consonants other than h, ş, ç, k, p, t, f, s. For example:

```
L   kalem+DA     kale+DA
S   kalemOde     kaleOde
O   kalemde      kalede
```

(f) **Consonant Devoicing**: A morpheme-initial D will surface as an unvoiced t, when affixed to a stem ending in the consonants h, ş, ç, k, p, t, f, s. Furthermore stem-final voiced consonants b, c, d with unvoiced counterparts will assimilate by surfacing as their unvoiced counterparts p, ç, t. For example:

```
L   kitab+DA      tad+DHk      saç+DA   kitab
S   kitap0ta      tat0tık      saç+ta   kitap
W   kitapta       tattık       saçta    kitap
```

(g) **Consonant Gemination**: This is a phenomenon that only applies to a set of words imported from Arabic but that set is large enough so that this phenomenon warrants its own mechanism. For this set of words, the root-final consonant is geminated when certain morphemes are affixed. For example:

```
L   tıb0+yH     üs0+sH      şık0+yH     hak0+nHn
S   tıbb00ı     üss00ü      şıkk00ı     hakk00ın
W   tıbbı       üssü        şıkkı       hakkın
```

(h) **Consonant Changes**: A stem-final k will surface as ğ or g depending on the left context, when followed by the accusative case morpheme +yH, the possessive morpheme +sH or the genitive case morpheme +nHn. A stem-final g will surface as ğ under the same conditions. For example:

```
L   tarak+yH     renk+sH      psikolog+yH
S   tarağ00ı     reng00i      psikoloğ00u
W   tarağı       rengi        psikoloğu
```

The phenomena discussed above have many exceptions to them and these are too numerous to cover here in detail. These exceptions are mostly lexicalized and many times some of the rules do not apply when the roots are monosyllabic. For example, even though gök and kök are very similar as far as the affixation boundary is concerned, we have gök+sH → göğü but kök+sH → kökü and not köğü. There are also a set of words, again from Arabic, but ending in vowels where the consonant deletion rule optionally applies and then only in one context but no in another context; e.g., cami+sH could surface as either camisi or as camii, but cami+yH would always surface as camiyi. The orthographic rules for proper nouns also have some bearing on the changes that are reflected to the written forms but they do not impact the pronunciation of those words. The proper noun affix separator ' blocks form changes in the root form. For instance, Işık'+nHn will surface as Işık'ın when written but will be pronounced as /1-S1-G1n/ (note also that when used as a common noun ışık+nHn will surface as ışığın when written and will have the same pronunciation.)

In state-of-the-art finite state formalisms for implementing these rules computationally, one can use either the two-level rule formalism or the cascade-rule formalism, to implement transducers that can map between surface and lexical forms. To implement the exceptions to the rules and many other rare phenomena that we have not covered, one needs to resort to representational mechanisms and tricks to avoid over- and undergeneration. The interested reader can refer to Beesley and Karttunen (2003) for the general formalism-related background and to Oflazer (1994) for details on Turkish two-level implementation.

2.4 Root Lexicons and Morphotactics

In this section we present an overview of the structure of the Turkish words of different root parts-of-speech. Turkish has a rather small set of root words from which very large number of word forms can be generated through productive inflectional and derivational processes. The root parts-of-speech used in Turkish are as follows:

- Nouns
- Verbs
- Numbers
- Postpositions
- Onomatopoeia Words
- Pronouns
- Adjectives
- Adverbs
- Conjunctions
- Determiners
- Interjections
- Question Clitics
- Punctuation

2.4.1 Representational Convention

The morphological analysis of a word can be represented as a sequence of tags corresponding to the overt (or covert) morphemes. In our morphological analyzer output, the tag `^DB` denotes derivation boundaries that we also use to define what we call `inflection groups` (IGs). If we represent the morphological information in Turkish in the following general form:

`root+IG`$_1$` + ^DB+IG`$_2$` + ^DB+`$\cdots$` + ^DB+IG`$_n$`.`

`root` is the basic root word from a root word lexicon and each `IG`$_i$ denotes the relevant sequence of inflectional features including the part-of-speech for the root (in `IG`$_1$) and for any of the derived forms. A given word may have multiple such representations depending on any morphological ambiguity brought about by alternative segmentations of the word, and by ambiguous interpretations of morphemes.

For instance, the morphological analysis of the derived modifier *uzaklaş-tırılacak* (the one) that will be sent away," literally, "(the one) that will be made far") would be:

`uzak+Adj`

> `^DB+Verb+Become`
> `^DB+Verb+Caus`
> `^DB+Verb+Pass+Pos`
> `^DB+Adj+FutPart+Pnon`

The five IGs in this word with root `uzak` are: (1) `+Adj`, (2) `+Verb+Become`, (3) `+Verb+Caus`, (4) `+Verb+Pass+Pos`, (5) `+Adj+FutPart+Pnon`.

The first IG indicates that the root is a simple adjective. The second IG indicates a derivation into a verb whose semantics is "to become" the preceding adjective *uzak* "far," (equivalent to "to move away" in English). The third IG indicates that a causative verb (equivalent to "to send away" in English) is derived from the previous verb. The fourth IG indicates the derivation of a passive verb with positive polarity from the previous verb. Finally, the last IG represents a derivation into future participle which will function as a modifier in the sentence.

2.4.2 Nominal Morphotactics

Nominal stems (lexical and derived nouns) can take up to three morphemes in the order below, that mark

- *Number*: Plural (lack of a number morpheme implies singular—except for mass nouns).
- *Possessive Agreement*: First/second/third person singular/plural (lack of a possessive morpheme implies no possessive agreement).[6]
- *Case*: Accusative, Dative, Ablative, Locative, Genitive, Instrumental, and Equative (lack of a case morpheme implies nominative case).

Thus from a single noun root one can conceivably generate $2 \times 7 \times 8 = 112$ inflected word forms. For instance, the simplest form with the root *ev* "house" is *ev*, which is singular, with no possessive agreement and in nominative case, while one of the more inflected forms would be *evlerimizden* which would be segmented into surface morphemes as ev+ler+imiz+den and would be a plural noun with first person plural possessive agreement and ablative case, meaning *from our houses*. In case we need to mark a noun with plural agreement and third person plural possessive agreement (as would be needed in the Turkish equivalent of toys in *their toys* in English as in the fourth case below), we would need to have a form like *oyuncak+lar+ları*. In such cases the first morpheme is dropped with the final word form being *oyuncakları*. But then in a computational setting such surface forms become four ways ambiguous if one analyzes them into possible constituent (lexical) morphemes:

1. oyuncak+lAr+sH: his toys
2. oyuncak+lAr+yH: toys (accusative)
3. oyuncak+lArH: their toy
4. oyuncak+lArH: their toys

all of which surface as *oyuncakları*.

Nominal inflected forms can undergo many derivations to create words with noun or other parts-of-speech and each of these can further be inflected and derived.

---

[6]There are also very special forms denoting families of relatives, where the number and possessive morphemes will swap positions to mean something slightly different: e.g., *teyze+ler+im* "my aunts" vs. *teyze+m+ler* "the family of my aunt."

2.4.3   Verbal Morphotactics

Verbal forms in Turkish have much more complicated morphotactics. Verbal stems (lexical or derived verbs) will inflect, taking morphemes one after the other in (approximately) the following order, marking:

1. *Polarity*: When this morpheme is present, it negates the verb (akin to *not* in English).
2. *Tense-Aspect-Mood*: There can be one or two such morphemes marking features of the verb such as: Past/Evidential Past/Future Tenses, Progressive Aspect, Conditional/Optative/Imperative Moods. Not all combinations of the two morphemes are allowed when both are present.[7]
3. *Person-Number Agreement*: For agreement with any overt or covert subject in the sentence, finite verbs can take a morpheme marking such agreement. The absence of such a morpheme indicates 3rd singular or plural agreement.
4. *Copula*: This morpheme when present adds *certainty/uncertainty* to the verb semantics depending on the verb context.

With just this much a given verb stem can give rise to about 700 inflected forms.

2.4.4   Derivations

Although the number of word forms quoted above are already impressive, it is the productivity of the derivational morphological processes in Turkish that give rise to a much richer set of word forms. However instead of presenting a full set of details on derivations, we will present a series of examples which we hope will give a feel for this richness, after presenting some rather productive derivations involving verbs.

A verb can have a series of voice markers which have the syntactic effect of changing its argument structure. We treat each such voice as a derivation of a verb from a verb. Thus, for example, a verb can have *reflexive, reciprocal/collective, causative*, and *passive* voice markers.[8] There can be multiple causative markers—two or three are not uncommon, and occasionally, two passive markers. Here is an example of a verbal form that involves four voice markers (with surface morpheme segmentation)

```
yıka+n+dır+t+ıl+ma+mış+sa+m
```

The first morpheme *yıka* is the verbal root meaning "wash/bathe." The next four morphemes mark reciprocal, two causative and passive voice markers. The next four

---

[7]An example below when we discuss derivation will show a full deconstruction of a complex verb to highlight these features.

[8]Obviously the first two are applicable to a smaller set of (usually) transitive verbs.

morphemes are the inflectional morphemes and mark negative polarity, evidential past, conditional mood and 1st person singular agreement respectively. The English equivalent would be (approximately) "if I were not let (by someone) to cause (somebody else) to have me bathe (myself)." Granted, this is a rather contorted example that probably would not be used under any real-world circumstance, it is nevertheless a perfectly valid example that highlights the complexity of verbal derivations. Verbs can also be further derived with modality morphemes to derive compound verbs with a variety of different semantic modifications. Such morphemes modify the semantics of a *verb* in the following ways[9]:

- able to *verb* (e.g., sür+*ebil*+ir, "she can/may drive")
- keep on *verb*ing (sometimes repeatedly) (e.g., yap+*adur*+du+m "I kept doing (it)")
- *verb* quickly/right away (e.g., yap+*ıver*+se+n, "wish you would do it right away")
- have been *verb*ing ever since (e.g., oku+*yagel*+diğ+iniz "that you have been reading since . . .")
- almost *verb*ed but didn't (e.g., düş+*eyaz*+dı+m, "I almost fell")[10]
- entered into/stayed in a *verb*ing state (e.g., uyu+*yakal*+dı+m "(literally) I entered into a sleeping state—I fell asleep")
- got on with *verb*ing (e.g., piş+ir+*ekoy*+du+m, " I got on with cooking (it)")

Some of these derivations are very productive (e.g., the first one above) but most are used rarely and only with a small set of semantically suitable verbal roots.

Verbs can also be derived into forms with other parts-of-speech. One can derive a whole series of temporal or manner adverbs with such derivational morphemes having the following semantics:

- after having *verb*ed (e.g., yap+*ıp* "after doing (it)")
- since having *verb*ed, (e.g., yap+*alı*, "since doing (it)")
- when . . . *verb*(s) (e.g., gel+*ince*, "when . . . come(s)")
- by *verb*ing (e.g., koş+*arak* "by running")
- while . . . *verb*ing (e.g., oku+r+*ken* "while reading . . .")
- as if . . . *verb*ing (e.g., kaç+ar+*casına* "as if . . . running away")
- without having *verb*ed (e.g., bit+ir+*meden* "without having finished")
- without *verb*-ing (e.g., yap+*maksızın*, "without doing")

The final set of derivations from verbs are nominalizations into infinitive or participle forms. After the derivations, the resulting nominalizations inflect essentially like nouns: that is, they can take a plural marker and a possessive marker (which now marks agreement with subjects of the underlying verb), and case marker. Here are some examples:

---

[9]We present the surface morpheme segmentations highlighting the relevant derivational morpheme with italics.

[10]So the next time you are up on a cliff looking down and momentarily lose your balance and then recover, you can describe the experience with the single verb *düşeyazdım*.

- uyu+*mak* "to sleep," uyu+*mak*+tan "from sleeping"
- oku+*ma*+m "(the act of) my reading"
- oku+*yuş*+un "(the process of) your reading"
- oku+*duğ*+u "(the fact) that s/he has read"
- oku+*yacağ*+ı+ndan"from (the fact) that s/he will read"

These forms are typically used for subordinate clauses headed by the verb, that function as a nominal constituent in a sentence.

A similar set of derivations result in forms that head clauses acting as modifiers of nouns usually describing the relation of those nouns to the underlying verb as a argument or an adjunct. These correspond to subject-gapped or non-subject-gapped clauses. For example:

- kitap oku+*yan* adam: "The man reading a book"
- kitap oku+*muş* adam "The man who has read a book"
- adam+ın oku+*duğ*+u kitap "The book the man is reading"
- adam+ın oku+*yacağ*+ı kitap "The book the man will be reading"

We mark these derivations as adjectives as they are used as modifiers of nouns in syntactic contexts but add a minor part-of-speech marker to indicate the nature of the derivation. Additionally in the last two cases, a possessive morpheme marks verbal agreement with the subject of the verb.

Although not as prolific as verbs, nouns and to a much lesser extent adjectives can productively derive stems of same or different parts-of-speech. Instead of giving a comprehensive list of these derivations, we would to list some of the more interesting of such derivations:

- Acquire *noun*: para+*lan*+mak "to acquire money"
- Become *adjective*: zengin+*leş*+iyor+uz "we are becoming rich"
- With *noun*: para+*lı* "with money"
- Without *noun*: para+*sız* "without money"

In addition to these more semantically motivated derivations, nouns and adjectives can be derived (sometimes with zero derivation triggered by a tense/aspect morpheme) into forms that function as nominal/adjectival verbs, adverbs, or clauses in a sentence. For example:

- ev+de+*ydi*+k "we were at home"
- ev+de+*yse*+k "if we are at home"
- mavi+*ydi* "it was blue"
- okul+da+*yken* "while he was at school"

2.4.5 *Examples of Morphological Analyses*

In this section we present several examples of morphological analyses of Turkish words. These examples will also serve to display some of the morphological ambiguity that ambiguous parts-of-speech, segmentation ambiguity or morpheme homography can cause[11]:

1. *bir*

- *bir*   bir+Adverb "suddenly"
- *bir*   bir+Det "a"
- *bir*   bir+Num+Card "one"
- *bir*   bir+Adj "same"

2. *okuma*

- *ok+um+a*   ok+Noun+A3sg+P1sg+Dat "to my arrow"
- *oku+ma*   oku+Verb+Neg+Imp+A2sg "do not read!"
- *oku+ma*   oku+Verb+Pos^DB+Noun+Inf2+A3sg+Pnon+Nom
              "reading"

3. *koyunu*

- *koy+u+nu*   koy+Noun+A3sg+P3sg+Acc "his bay (Accusative)"
- *koy+un+u*   koy+Noun+A3sg+P2sg+Acc "your bay (Accusative)"
- *koyu+n+u*   koyu+Adj^DB+Noun+Zero+A3sg+P2sg+Acc
              "your dark (thing) (Accusative)"
- *koyun+u*   koyun+Noun+A3sg+P3sg+Nom "his sheep"
- *koyun+u*   koyun+Noun+A3sg+Pnon+Acc "sheep (Accusative)"

4. *elmasında*

- *elma+sı+nda*   elma+Noun+A3sg+P3sg+Loc "on his apple"
- *elmas+ı+nda*   elmas+Noun+A3sg+P3sg+Loc "on his diamond"
- *elmas+ın+da*   elmas+Noun+A3sg+P2sg+Loc "on your diamond"

5. *öldürülürken*

- *öl+dür+ül+ür+ken*   öl+Verb^DB+Verb+Caus
                      ^DB+Verb+Pass+Pos+Aor
                      ^DB+Adverb+While
              "while he is being caused to die"

6. *iyileştirilince*

- *iyi+leş+tir+il+ince*   iyi+Adj^DB+Verb+Become
                        ^DB+Verb+Caus
                        ^DB+Verb+Pass+Pos
                        ^DB+Adverb+When
              "when he is made to become well/good"

---

[11]Where meaningful we also give the segmentation of the words form into surface morphemes in italics.

7. *ruhsatlandırılamamasındaki*

- *ruhsat+lan+dır+ıl+ama+ma+sı+nda+ki*
    ```
    ruhsat+Noun+A3sg+Pnon+Nom
     ^DB+Verb+Acquire^DB+Verb+Caus
     ^DB+Verb+Pass^DB+Verb+Able+Neg
     ^DB+Noun+Inf2+A3sg+P3sg+Loc
     ^DB+Adj+Rel
    ```
    "related to (something) not being able to acquire certification"

2.5   The Architecture of the Turkish Morphological Processor

In this section we present a short overview of the implementation of the Turkish morphological processor using the Xerox Finite State Tools (Beesley and Karttunen 2003). These tools take in a description of the morphographemic rules of the language along with the root lexicons and morpheme lexicons and compile them into a (very large) finite state transducer that can map surface forms to multiple analyses. The morphological processor can be customized to produce outputs in different representations, as shown in Fig. 2.1:

- **Morphological Features and Pronunciation**: The output consists of an inter-leaved representation of both the pronunciation and the morphological features of each possible interpretation of the surface word. For the input word *evinde*, one would get

    ```
    - (e - v )ev+Noun+A3sg(i )+P3sg(n - "d e )+Loc
    - (e - v )ev+Noun+A3sg(i n )+P2sg(- "d e )+Loc
    ```

    Here the parts of the representation between (. . .) encode the pronunciation of the word with phonemes in SAMPA, with - denoting syllable boundaries and " indicating the syllable with the primary stress. The following shows a more interesting example where we have three analyses for the surface word *okuma* but only two different pronunciations that only differ in the position of the stressed syllable:

    ```
    - (o - k )ok+Noun+A3sg(u - "m )+P1sg(a )+Dat
    - (o - "k u )oku+Verb(- m a )+Neg+Imp+A2sg
    - (o - k u )oku+Verb+Pos(- "m a )
                        ^DB+Noun+Inf2+A3sg+Pnon+Nom
    ```

- **Surface Morphemes**: The output consists of a set of segmentations of the surface word into surface morphemes. So for the input word *evinde*, one would get

    ```
    - ev+i+nde
    - ev+in+de
    ```

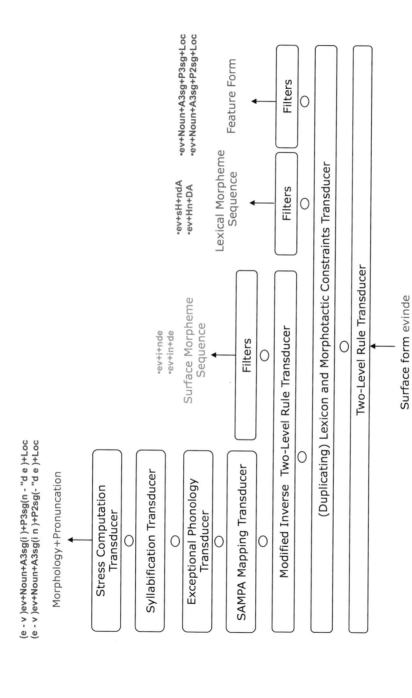

Fig. 2.1 The overall internal transducer architecture of the Turkish Morphological Processor

- **Lexical Morphemes**: The output consists of a set of segmentations of the surface word into lexical morphemes possibly involving meta-symbols denoting miscellaneous subset of phonemes. For the input word *evinde*, one would get

    - `ev+sH+ndA`
    - `ev+Hn+DA`

- **Morphological Features**: The output consists of a set of morphological analyses consisting of the root word followed by a sequence of morphological features encoding the morphological information. So for the input word *evinde*, one would get

    - `ev+Noun+A3sg+P3sg+Loc`
    - `ev+Noun+A3sg+P2sg+Loc`

The root lexicons of the morphological processor comprise about 100,000 root words—about 70,000 being proper names. When fully compiled, the transducer that maps from surface forms to the feature-pronunciation representation ends up having about 10 million states and about 16 million transitions. All transducers are *reversible*, that is, they also be used to generate surface forms from their respective output representations.

Figure 2.1 presents a high-level organizational diagram of the Turkish morphological processor. These transducers are also complemented by a series of helper transducers that help with analysis of misspelled forms, proper nouns violating the use of the apostrophe for separating morphemes, normalizing special characters, etc.

2.6 Processing Real Texts

When one wants to process naturally occurring text from sources ranging from professionally written news text to tweets or blog comments, the basic morphological analyzer proves inadequate for many reasons. In a language like Turkish with almost one to one mapping between phonemes and letters and with morphological processes conditioned by phonological constraints, processing such text requires one to deal with mundane and not so mundane issues brought by tokens encountered in various sources.

2.6.1 Acronyms

An acronym like *PTT* has no vowels as written, but being a noun, it can take suffixes that a normal noun takes. So forms such as *PTT'ye* "to the PTT" or *PTT'den* "from the PTT" are perfectly valid in that the phonological processes are correct *based on the explicit pronunciation of the root PTT*. However forms such as *PTT'ya* or *PTT'ten* are ill-formed as they violate the morphophonological processes. The problem is that the written form is insufficient to condition the morphological processes: there are no vowels as written and we do not know whether the pronunciation of the root ends in a vowel or in a certain type of a consonant. Such cases have to be handled in the root lexicon by carefully adding enough of a set lexical marker symbols in the representation of such roots so that morphophonological constraints can be conditioned: For instance for *PTT*, we would have to indicate that it ends in a vowel and the last vowel in its pronunciation /pe-te-te/ is written as *e*.

2.6.2 Numbers

Numbers when written using numerals can also take suffixes: e.g., *2014'te* "in 2014," *35.si* "the 35th" *1000'den* "from 1000", *2/3'ü* "the two-thirds of" or *2/3'si* "the two-thirds of." Like acronyms, phonological processes proceed based on the *pronunciation* of the number. So for instance the last vowel in the pronunciation of *2014* (iki bin on dört) is /2/ and the pronunciation ends in the unvoiced dental consonant /t/. The vowel harmony and consonant assimilation rules need to access this information which is nowhere in the written form. Thus one needs to really build the equivalent of a text-to-speech system for numbers which can at least infer the relevant properties of its pronunciation and encode them in its lexical representation. This is a rather nontrivial add-on to the basic morphological analyzer.

2.6.3 Foreign Words

Texts contain foreign common words and proper names which when used as constituents in a sentence have to be properly inflected. However, the morphophonological processes again proceed according to the pronunciation of the foreign word in its original language: one sees sequences like ... *serverlar ve clientlar* ... (... servers and clients) where the vowel harmony in the inflected words is fine when the pronunciations of *server* and *client* in English are considered, but not when their written forms (on which the analyzer crucially depends) are considered.[12] Another example is the sports page headline from many years ago:

    Bordeaux'yu yendik (We beat Bordeaux (in soccer)).

Here the poor reader who would not know the pronunciation of *Bordeaux* in French would be puzzled by the selection of the *+yu* accusative case morpheme.

    This is a tough problem to solve in a principled way. One needs again something akin to a full text-to-speech component that would extract enough information from

---

[12]Users of such words have the bizarre presumption that readers know how to pronounce those words in English!

the foreign root to condition the morphology. However this is beyond the scope of a morphological processor for Turkish. It may be better to incorporate a solution involving lenient morphology (Oflazer 2003) which can ignore the violation of a very small number of morphographemic rules but only across root-morpheme boundaries.

2.6.4 Unknown Words

When large amounts of text are processed there will always be unknown words that cannot be recognized and analyzed by a morphological processor. Such words can be misspelled words, or to a lesser extent words whose roots are missing from the root lexicons of the morphological analyzer. In a morphological processing pipeline, words found to be misspelled can be corrected with a spelling corrector and an analysis can be reattempted. However for words with roots missing from the lexicon, one can probably do better by analyzing any suffix sequence (assuming the affixes are correct) and then based on those, infer what the roots can be. For instance, when a word sequence such as . . . *showları zapladım* . . . (I zapped through the shows) is encountered, assuming both are unknown by the morphological analyzer, one can either posit that both words are singular nouns with no possessive agreement and with nominative case or attempt to processes with a version the morphological analyzer whose noun and verb root lexicons have been replaced by a single entry that always matches the regular expression, $?^+$, which matches one or more occurrence of any symbol in the morphological analyzer's surface alphabet. Such an analyzer will skip over any prefix of the unknown words positing the skipped portion as the root, provided the rest of the unknown token can be parsed properly as a valid sequence of morphemes in Turkish. For example, the first word above could be segmented as *showlar+ı* or *show+lar+ı* positing roots *showlar* and *show* respectively and then emitting features corresponding to recognized morphemes. Similarly, the second word can be segmented as *zapla+dı+m* and *zapla* can be posited as a verb root since the remaining morphemes can only attach to verbs.

2.7 Multiword Processing

Multiword expression recognition is an important component in lexical processing that aims to identify segments of input text where the syntactic structure and the semantics of a sequence of words (possibly not contiguous) are usually not compositional. Idiomatic forms, support verbs, verbs with specific particle or pre/postposition uses, morphological derivations via partial or full word duplications are some examples of multiword expressions. Further, constructs such as time-date expressions which can be described with simple (usually finite state) grammars or named-entities and whose internal structure is of no real importance to the overall analysis of the sentence can also be considered under this heading. Marking multiword expressions in text usually reduces (though not significantly) the number of actual tokens that further processing modules use as input, although this reduction may depend on the domain the text comes from.

Turkish presents some interesting issues for multiword expression processing as it makes substantial use of support verbs with lexicalized direct or oblique objects, subject to various morphological constraints. It also uses partial and full reduplication of forms of various parts-of-speech, across their whole domain to form what we call *non-lexicalized* collocations, where it is the duplication and contrast of certain morphological patterns that signal a collocation rather than the specific root words used. This can be considered another example of morphological derivation involving a sequence of words.

Turkish employs multi-word expressions in essentially four different forms:

1. *Lexicalized Collocations* where all components of the collocations are fixed,
2. *Semi-lexicalized Collocations* where some components of the collocation are fixed and some can vary via inflectional and derivational morphology processes and the (lexical) semantics of the collocation is not compositional,
3. *Non-lexicalized Collocations* where the collocation is mediated by a morphosyntactic pattern of duplicated and/or contrasting components—hence the name *non-lexicalized*, and
4. *Multi-word named-entities* which are multi-word proper names for persons, organizations, places, etc.

2.7.1 Lexicalized Collocations

Under the notion of lexicalized collocations, we consider the usual fixed multiword expressions whose resulting syntactic function and semantics are not readily predictable from the structure and the morphological properties of the constituents.

Here are some examples of the multi-word expressions that we consider under this grouping[13]:

- *hiç olmazsa*
  - hiç+Adverb
    ol+Verb+Neg+Aor+Cond+A3sg
  - hiç_olmazsa+Adverb
    "at least" (literally "if it never is")

---

[13]In every group we first list the morphological features of all the tokens, one on every line and then provide the morphological features of the multiword construct followed by a gloss and a literal meaning.

- *ipe sapa gelmez*

    - ip+Noun+A3sg+Pnon+Dat
      sap+Noun+A3sg+Pnon+Dat
      gel+Verb+Neg+Aor+A3sg
    - ipe_sapa_gelmez+Adj
      "worthless" (literally "(he) does not come to rope and handle")

2.7.2 Semi-lexicalized Collocations

Multiword expressions that are considered under this heading are compound and
support verb formations where there are two or more lexical items the last of
which is a verb or is a derivation involving a verb. These are formed by a lexically
adjacent, direct or oblique object, and a verb, which for the purposes of syntactic
analysis, may be considered as single lexical item: e.g., *devam et-* (literally *to
make continuation*—to continue), *kafayı ye-* (literally *to eat the head*—to get
mentally deranged), etc.[14] Even though the other components can themselves be
inflected, they can be assumed to be fixed for the purposes of the collocation, and
the collocation assumes its morphosyntactic features from the last verb which itself
may undergo any morphological derivation or inflection process. For instance in

- *kafayı ye-*   "get mentally deranged" (literally "eat the head")

    - kafa+Noun+A3sg+Pnon+Acc ye+Verb...

the first part of the collocation, the accusative marked singular noun *kafayı*, is the
fixed part and the part starting with the verb *ye-* is the variable part which may be
inflected and/or derived in myriads of ways. With multiword processing, these can
be combined into one form

- kafayı_ye+Verb...

    For example, the following are some possible forms of the collocation:

- kafayı ye*dim* "I got mentally deranged"
- kafayı yi*yeceklerdi* "they would have become mentally deranged"
- kafayı yi*yenler* "those who got mentally deranged"
- kafayı ye*diği* "the fact that (s/he) got mentally deranged"
- kafayı ye*dirdi* "(he) caused (us) to get mentally deranged"

---

[14]Here we just show the roots of the verb with - denoting the rest of the suffixes for any inflectional
and derivational markers.

Under certain circumstances, the "fixed" part may actually vary in a rather controlled manner subject to certain morphosyntactic constraints, as in the idiomatic verb:

- *kafa(yı) çek-*   "get drunk" (but literally "to pull the head")

    – kafa+Noun+A3sg+Pnon+Acc çek+Verb...

    which can be replaced by

    – kafa_çek+Verb...

- *kafaları çek-*

    – kafa+Noun+**A3pl**+Pnon+Acc çek+Verb...
    – kafa_çek+Verb...

    which can also be replaced by

    – kafa_çek+Verb...

In these examples, the fixed part has to have the root *kafa* but can be in the nominative or the accusative case, and if it is in the accusative case, it may be marked plural, in which case the verb has to have some kind of plural agreement (i.e., first, second, or third person plural), *but no possessive agreement markers are allowed.*

In their simplest forms, it is sufficient to recognize a sequence of tokens one of whose morphological analyses matches the corresponding pattern, and then coalesce these into a single multiword expression representation. However, some or all variants of these and similar semi-lexicalized collocations present further complications brought about by the relative freeness of the constituent order in Turkish, and by the interaction of various clitics with such collocations.[15]

When such multiword expressions are coalesced into a single morphological entity, the ambiguity in morphological interpretation could be reduced as we see in the following example:

- *devam etti*   "(he) continued" (literally "made a continuation")

    – devam

        · devam+Noun+A3sg+Pnon+Nom "continuation"
        · deva+Noun+A3sg+P1sg+Nom "my therapy"

---

[15]The question and the emphasis clitics which are written as separate tokens can occasionally intervene between the components of a semi-lexicalized collocation. We omit the details of these.

– etti

  · `et+Verb+Pos+Past+A3sg` "made"
  · `et+Noun+A3sg+Pnon+Nom^DB+Verb+Past+A3sg` "it was meat"

– `devam_et+Verb+Pos+Past+A3sg`
  "(he) continued" (literally "made a continuation")

Here, when this semi-lexicalized collocation is recognized, other morphological interpretations of the components (the second in each group) above) can safely be removed, contributing to overall morphological ambiguity reduction.

2.7.3 Non-lexicalized Collocations

Turkish employs quite a number of non-lexicalized collocations where the sentential role of the collocation has (almost) nothing to do with the parts-of-speech and the morphological features of the individual forms involved. Almost all of these collocations involve partial or full duplications of the forms involved and can actually be viewed as morphological derivational processes mediated by reduplication across multiple tokens.

The morphological feature representations of such multiword expressions follow one of the patterns:

* $\omega \, \omega$
* $\omega \, Z \, \omega$
* $\omega + X \, \omega + Y$
* $\omega_1 + X \, \omega_2 + X$

where ω is the duplicated string comprising the root, its part-of-speech and possibly some additional morphological features encoded by any suffixes. X and Y are further duplicated or contrasted morphological patterns and Z is a certain clitic token. In the last pattern, it is possible that ω_1 is different from ω_2.

Below we present list of the more interesting non-lexicalized expressions along with some examples and issues.

* When a noun appears in duplicate following the first pattern above, the collocation behaves like a manner adverb, modifying a verb usually to the right. Although this pattern does not necessarily occur with every possible noun, it may occur with many (countable) nouns without much of a further semantic restriction. Such a sequence has to be coalesced into a representation indicating this derivational process as we see below.

  – *ev ev* ($\omega \, \omega$)   "house by house" (literally "house house")

    · `ev+Noun+A3sg+Pnon+Nom`
      `ev+Noun+A3sg+Pnon+Nom`

are combined into

- `ev+Noun+A3sg+Pnon+Nom^DB+Adverb+By`

- When an adjective appears in duplicate, the collocation behaves like a manner adverb (with the semantics of *-ly* adverbs in English), modifying a verb usually to the right. Thus such a sequence has to be coalesced into a representation indicating this derivational process.

  - *yavaş yavaş* (ω ω)   "slowly" (literally "slow slow")

    - `yavaş+Adj`
      `yavaş+Adj`

    are combined into

    - `yavaş+Adj^DB+Adverb+Ly`

- This kind of duplication can also occur when the adjective is a derived adjective as in

  - *hızlı hızlı* (ω ω)   "rapidly" (literally "with-speed with-speed")

    - `hız+Noun+A3sg+Pnon+Nom^DB+Adj+With`
      `hız+Noun+A3sg+Pnon+Nom^DB+Adj+With`

    being replaced by

    - `hız+Noun+A3sg+Pnon+Nom`
          `^DB+Adj+With^DB+Adverb+Ly`

- Turkish has a fairly large set of onomatopoeic words which always appear in duplicate as a sequence and function as manner adverbs. The words by themselves have no other use and literal meaning, and mildly resemble sounds produced by natural or artificial objects. In these cases, the first word can be duplicated but need not be, but both words should be of the part-of-speech category +Dup that we use to mark such roots.

  - *harıl hurul* ($\omega_1 + X$ $\omega_2 + X$)   "making rough noises" (no literal meaning)

    - `harıl+Dup`
      `hurul+Dup`

    gets combined into
  - `harıl_hurul+Adverb+Resemble`

- Duplicated verbs with optative mood and third person singular agreement function, as manner adverbs, indicating that another verb is executed in a manner indicated by the duplicated verb:

  – *koşa koşa* (ω ω)

    · ```
      koş+Verb+Pos+Opt+A3sg
      koş+Verb+Pos+Opt+A3sg
      ```

    gets combined into
  – ```
    koş+Verb+Pos^DB+Adverb+ByDoingSo
    ```
    "by running" (literally "let him run let him run")

- Duplicated verbs in aorist mood with third person agreement and first with positive then negative polarity, function as temporal adverbs with the semantics "as soon as one has *verb*ed"

  – *uyur uyumaz* (ω + X ω + Y)
  – · ```
      uyu+Verb+Pos+Aor+A3sg
      uyu+Verb+Neg+Aor+A3sg
      ```
      gets combined into

    · ```
      uyu+Verb+Pos^DB+Adverb+AsSoonAs
      ```

    "as soon as (he) sleeps" (literally "(he) sleeps (he) does not sleep")

It should be noted that for most of the non-lexicalized collocations involving verbs (like the last two above), the verbal stem before the inflectional marker for mood can have additional derivational markers and all such markers have to duplicate. For example:

- *sağlamlaştırır sağlamlaştırmaz* "as soon as (he) fortifies (causes to become strong)" (ω + X ω + Y)

    · ```
      sağlam+Adj^DB+Verb+Become
      ^DB+Verb+Caus^DB+Verb+Pos+Aor+A3sg
      sağlam+Adj^DB+Verb+Become
      ^DB+Verb+Caus^DB+Verb+Neg+Aor+A3sg
      ```

    which gets combined into

    · ```
      sağlam+Adj^DB+Verb+Become+
          ^DB+Verb+Caus+Pos
          ^DB+Adverb+AsSoonAs
      ```

An interesting point is that non-lexicalized collocations can interact with semi-lexicalized collocations since they both usually involve verbs. For instance, below we have an example of the verb of a semi-lexicalized collocation being repeated in a non-lexicalized collocation:

– *kafaları çeker çekmez*

  In this case, first the non-lexicalized collocation has to be combined into

– *kafaları* çek+Verb+Pos^DB+Adverb+AsSoonAs

and then the semi-lexicalized collocation is handled, to give

– kafa_çek+Verb+Pos^DB+Adverb+AsSoonAs

to get an idiomatic case combined with duplication meaning "as soon as (we/you/they) get drunk."

- Finally, the following non-lexicalized collocation involving adjectival forms involving duplication and a question clitic is an example of the last type of non-lexicalized collocation.

  – *güzel mi güzel* (ω Z ω)   "very beautiful" (literally "beautiful (is it?) beautiful")

    · güzel+Adj
      mi+Ques
      güzel+Adj

    which gets combined into
– güzel+Adj+Superlative

Oflazer et al. (2004) describe a post-processing system that implemented the multi-word processing scheme described above for Turkish.

2.8 Conclusions

This chapter has presented an overview of Turkish morphology and the architecture of a state-of-the-art wide coverage morphological analyzer for Turkish implemented to be used in a variety of natural language processing downstream applications. We also touched upon issues that one encounters when processing real text such as numeric tokens, acronyms, foreign words, unknown words, etc. Finally we gave an overview of the multiwords that one needs to deal after morphological processing but before any further additional processing is attempted.

Appendix: Turkish Morphological Features

In this appendix we present an overview of the morphological features that the morphological analyzer produces. The general format of an analysis is as given in Sect. 2.4.1: any derivations are indicated by ^DB. The first symbol following a ^DB is the part-of-speech of the derived form and the next feature symbol is usually a semantic marker that indicates the semantic nature of the derivation. If the second symbol is +Zero that indicates a implied covert derivation without any overt morphemes.

1. **Major Root Parts of Speech**: These mark the part-of-speech category of the root word. This is not necessarily the part-of-speech of the final word if the word involves one or more derivations.

Feature	Indicates	Feature	Indicates
+Noun	Noun	+Adj	Adjective/modifier
+Adverb	Adverb	+Verb	Verb
+Pron	Pronoun	+Postp	Postposition
+Num	Number	+Conj	Conjunction
+Det	Determiner	+Interj	Interjection
+Ques	Question clitic	+Punc	Punctuation
+Dup	Onomatopoeia words		

2. **Minor Parts of Speech**: These follow one of the part-of-speech category symbols above and either denotes a further subdivision that is morphosyntactically relevant or a semantic marker that indicates the nature of the derivation.

   (a) After +Noun

Feature	Indicates	Example
+Prop	Proper noun	Çağla, Mahkemesi'nde

   (b) After +Pron

Feature	Indicates	Example
+Demons	Demonstrative pronoun	bu "this"
+Ques	Interrogative pronoun	kim "who"
+Reflex	Reflexive pronoun	kendim "myself"
+Pers	Personal pronoun	biz "we"
+Quant	Quantifying pronoun	hepimiz "all of us"

(c) After +Num

Feature	Indicates	Example
+Card	Cardinal number	iki "two"
+Ord	Ordinal number	ikinci "second"
+Dist	Distributive number	ikişer "two each"

(d) After ^DB+Noun

Feature	Indicates	Example
+Inf1	Infinitive	gitmek "to go"
+Inf2	Infinitive	gitme "going" , gitmem "my going"
+Inf3	Infinitive	gidiş (going)
+PastPart	Past participle	gittiği (the fact that (he) went)
+FutPart	Future participle	gideceği "the fact that he will go"
+FeelLike	"the state of feeling like"	gidesim ((the state of) me feeling like going)

(e) After ^DB+Adj: These are markers that indicate the equivalent of subject, object, or adjunct extracted relative clauses.

Feature	Indicates	Example
+PastPart	Past participle	gittiğim [yer] "[the place] I am going"
+FutPart	Future participle	gideceğim [yer] "[the place] I will be going"
+PresPart	Present participle	giden [adam] "[the man] who is going"
+NarrPart	Evidential participle	gitmiş [adam] "[the man] who (is rumored) to have gone"
+AorPart	Aorist participle	geçer [not] "passing [grade]" , dayanılmaz [sıcak] "unbearable [heat]"

3. Nominal forms (Nouns, Derived Nouns, Derived Nominal and Pronouns) get the
   following inflectional markers. Not all combinations are valid in all cases:

   (a) Number/Person Agreement

Feature	Indicates	Example
+A1sg	1st person singular	ben "I"
+A2sg	2nd person singular	sen "you"
+A3sg	3rd person singular	o "he/she/it", all singular nouns
+A1pl	1st person plural	biz "we"
+A2pl	2nd person plural	siz "you"
+A3pl	3rd person plural	onlar "they", all plural nouns

   (b) Possessive Agreement

Feature	Indicates	Example
+P1sg	1st person singular possessive	kalemim "my pencil"
+P2sg	2nd person singular possessive	kalemin "your pencil"
+P3sg	3rd person singular possessive	kalemi "his/her/its pencil"
+P1pl	1st person plural possessive	kalemimiz "our pencil"
+P2pl	2nd person plural possessive	kaleminiz "your pencil"
+P3pl	3rd person plural possessive	kalemleri "their pencil(s)"
+Pnon	No possessive	kalem "pencil"

   (c) Case

Feature	Indicates	Example
+Nom	Nominative	çocuk "child"
+Acc	Accusative	çocuğu "child as definite object"
+Dat	Dative	çocuğa "to the child"
+Abl	Ablative	çocuktan "from the child"
+Loc	Locative	çocukta "on the child"
+Gen	Genitive	çocuğun "of the child"
+Ins	Instrumental/	kalemle "with a pencil"
	accompanier	çocukla "with the child"
+Equ	Equative (by object)	bizce "by us"

4. Adjectives do not take any inflectional markers. However, the cases ^DB+Adj-
   +PastPart and ^DB+Adj+FutPart will have a possessive marker "one of
   the first six of the seven above" to mark subject agreement with the verb that
   is derived into the modifier participle. For example, *gittiğim [yer]* "[the place]

(that) **I** went" will have ... ^DB+Adj+PastPart+P1sg, *gittiğimiz [yer]* "[the place] (that) **we** went" will have ... ^DB+Adj+PastPart+P1pl.

5. Verbs will have multiple classes of markers

(a) Valency changing voice suffixes are treated as derivations. These voice markers follow ^DB+Verb. A verb may have multiple causative markers.

Feature	Indicates	Example
+Pass	Passive	yıkandı "it was washed"
+Caus	Causative	yıkattı "he had it washed"
+Reflex	Reflexive	yıkandı "he washed himself"
+Recip	Reciprocal/	selamlaştık "we greeted each other"
	Collective	gülüştük "we all giggled"

(b) The following markers marking compounding and/or modality are treated as deriving new verbs with a semantic twist. These markers also follow ^DB+Verb. All except the first have quite limited applicability.

Feature	Indicates	Example
+Able	Able to *verb*	okuyabilir
		"[s/he] can read"
+Repeat	*verb* repeatedly	yapadurdum
		"I kept on doing [it]"
+Hastily	*verb* hastily	siliverdim
		"I quickly wiped [it]"
+EverSince	have been *verb*ing ever since	bilegeldiğimiz
		"that we knew ever since"
+Almost	Almost *verb*ed but did not	düşeyazdım
		"I almost fell"
+Stay	Stayed/frozen while *verb*ing	uyuyakaldılar
		"they fell asleep"
+Start	Start *verb*ing immediately	pişirekoydum
		"I got on cooking [it]"

(c) Verbal polarity attaches to a verb (or the last verbal derivation (if any), unless last verbal derivation is from a +Noun or +Adj is a zero derivation).

Feature	Indicates	Example
+Pos	Positive polarity	okudum "I read"
+Neg	Negative polarity	okumadım "I did not read"

(d) Verbs may have one or two tense, aspect or mood markers. However not all combinations are possible.

Feature	Indicates	Example
+Past	Past tense	okudum "I read"
+Narr	Evidential past tense	okumuşum
		"it is rumored that I read"
+Fut	Future tense	okuyacağım "I will read"
+Prog1	Present continuous tense—process	okuyorum "I am reading"
+Prog2	Present continuous tense—state	okumaktayım
		"I am in a state of reading"
+Aor	Aorist mood	okur "he reads"
+Desr	Desiderative mood	okusam "wish I could read"
+Cond	Conditional aspect	okuyorsam "if I am reading"
+Neces	Necessitative aspect	okumalı "he must read"
+Opt	Optative aspect	okuyalım "let's read"
+Imp	Imperative aspect	oku "read!"

(e) Verbs also have Person/Number Agreement markers. See above. Occasionally finite verbs with have a copula +Cop marker.

6. Semantic markers for derivations

(a) The following markers mark adverbial derivations from a *verb*—hence they appear after $^{\wedge}$DB+Adverb.

Feature	Indicates	Example
+AfterDoingSo	After having *verb*ed	okuyup "after having read"
+SinceDoingSo	Since having *verb*ed	okuyalı "since having read"
+As	As ... *verb*s	okudukça "as he reads"
+When	When ... is done *verb*ing	okuyunca
		"when he is done reading"
+ByDoingSo	By *verb*ing	okuyarak "by reading"
+AsIf	As if *verb*ing	okurcasına
		"as if he is reading"
+WithoutHaving- DoneSo	Without having *verb*ed	okumadan
		"without having read"
		okumaksızın
		"without reading"

(b) +Ly marks manner adverbs derived from an adjective: *yavaş* (slow) derives *yavaşça* "slowly".

(c) +Since marks temporal adverbs derived from a temporal noun: *aylar* "months" derives *aylardır* "since/for months."

(d) +With and +Without mark modifiers derived from nouns: *renk* "color" derives *renkli* "with color" and *renksiz* "without color."

(e) +Ness marks a noun derived from an adjective with semantics akin to *-ness* in English: *kırmızı* "red" derives *kırmızılık* "redness," *uzun* "long" derives *uzunluk* "length."

(f) +Become and +Acquire mark verbs productively derived from nouns with the semantics of becoming like the noun or acquiring the noun: *taş* "stone" derives the verb stem *taşlaş* "become a stone/petrify"; *para* "money" derives the verb stem *paralan* "acquire money."

(g) +Dim marks derives a diminutive form a noun: *kitap* "book" derives *kitapçık* "little book/booklet".

(h) +Agt marks a noun derived from another noun involved in someway with the original noun; the actual additional semantics is not predictable in general but depends on the stem noun: *kitap* derives *kitapçı* "bookseller," *gazete* "newspaper" derives *gazeteci* "journalist," *fotoğraf* derives *fotoğrafçı* "photographer."

7. The following will follow a postposition to indicate the case of the preceding nominal it will subcategorize for. This is not morphologically marked but is generated to help with parsing or morphological disambiguation. Their only use is to disambiguate the case of the preceding noun if it has multiple morphological interpretations.

- +PCAbl
- +PCAcc
- +PCDat
- +PCGen
- +PCIn
- +PCNom

References

Beesley KR, Karttunen L (2003) Finite state morphology. CSLI Publications, Stanford University, Stanford, CA

Clements GN, Sezer E (1982) Vowel and consonant disharmony in Turkish. In: van der Hulst H, Smith N (eds) The structure of phonological representations. Foris, Dordrecht, pp 213–255

Karttunen L (1993) Finite-state lexicon compiler. Technical report, Xerox PARC, Palo Alto, CA

Karttunen L, Beesley KR (1992) Two-level rule compiler. Technical report, Xerox PARC, Palo Alto, CA

Karttunen L, Chanod JP, Grefenstette G, Schiller A (1996) Regular expressions for language engineering. Nat Lang Eng 2(4):305–328

Kornfilt J (1997) Turkish. Routledge, London

Koskenniemi K (1983) Two-level morphology: a general computational model for word-form recognition and production. PhD thesis, University of Helsinki, Helsinki

Oflazer K (1994) Two-level description of Turkish morphology. Lit Linguist Comput 9(2):137–148

Oflazer K (2003) Lenient morphological analysis. Nat Lang Eng 9:87–99

Oflazer K, Inkelas S (2006) The architecture and the implementation of a finite state pronunciation lexicon for Turkish. Comput Speech Lang 20:80–106

Oflazer K, Çetinoğlu Ö, Say B (2004) Integrating morphology with multi-word expression processing in Turkish. In: Proceedings of the ACL workshop on multiword expressions: integrating processing, Barcelona, pp 64–71

Sproat RW (1992) Morphology and computation. MIT Press, Cambridge, MA

van der Hulst H, van de Weijer J (1991) Topics in Turkish phonology. In: Boeschoten H, Verhoeven L (eds) Turkish linguistics today. Brill, Leiden

Chapter 3
Morphological Disambiguation for Turkish

Dilek Zeynep Hakkani-Tür, Murat Saraçlar, Gökhan Tür, Kemal Oflazer, and Deniz Yuret

Abstract Morphological disambiguation is the task of determining the contextually correct morphological parses of tokens in a sentence. A morphological disambiguator takes in a set of morphological parses for each token, generated by a morphological analyzer, and then selects a morphological parse for each, considering statistical and/or linguistic contextual information. This task can be seen as a generalization of the part-of-speech (POS) tagging problem, for morphologically rich languages. The disambiguated morphological analysis is usually crucial for further processing steps such as dependency parsing. In this chapter, we review the morphological disambiguation problem for Turkish and discuss approaches for solving this problem as they have evolved from manually crafted constraint-based rule systems to systems employing machine learning.

3.1 Introduction

Part-of-speech (POS) tagging is one of the most basic steps of natural language processing and aims to assign each word token the correct POS tag, given all the possible tags. The set of POS tags belongs to a pre-defined vocabulary. Typical

D. Z. Hakkani-Tür · G. Tür
Google Inc., Mountain View, CA, USA
e-mail: dilek@ieee.org; gokhan.tur@ieee.org

M. Saraçlar
Boğaziçi University, Istanbul, Turkey
e-mail: murat.saraclar@boun.edu.tr

K. Oflazer (✉)
Carnegie Mellon University Qatar, Doha-Education City, Qatar
e-mail: ko@cs.cmu.edu

D. Yuret
Koç University, Istanbul, Turkey
e-mail: dyuret@ku.edu.tr

© Springer International Publishing AG, part of Springer Nature 2018
K. Oflazer, M. Saraçlar (eds.), *Turkish Natural Language Processing*,
Theory and Applications of Natural Language Processing,
https://doi.org/10.1007/978-3-319-90165-7_3

English tag sets include less than a hundred tags, for example, the Penn Treebank tag set contains 36 POS tags and 12 other tags for punctuation and currency symbols (Marcus et al. 1993).

On the other hand, as we have seen in Chap. 2, the agglutinative morphology of Turkish can allow a very large vocabulary owing mainly to very productive derivational processes. Without considering any context, many word forms can be morphologically ambiguous for many reasons:

- The root word can be ambiguous with respect to its part-of-speech.
- Homograph morphemes can be interpreted in multiple ways
- Word forms can be segmented in multiple different ways, either due to having different root forms and/or morphographemic processes for interacting morphemes leading to the same surface structures.

For example, the following are the possible morphological parses or interpretations for the word *evin*[1]:

ev+Noun+A3sg+P2sg+Nom	*(your) house*
ev+Noun+A3sg+Pnon+Gen	*of the house*
evin+Noun+A3sg+Pnon+Nom	*wheat germ*

Here, the lexical morphemes for the first two parses are different, but they surface the same due to morphographemic processes, while the third one in a example of a root segmentation ambiguity.

As we mentioned earlier correct parse of each word depends on the context given by the sentence in which the word appears. The first one is the correct parse when it appears in the following sentence:

Senin evin nerede? (*Where is your house?*)

while the second one is the correct parse when it appears in this sentence:

Evin kapısını kilitlemeyi unutma.
(*Do not forget to lock the door of the house.*)

The following is another example of a surface word that has two ambiguous parses with the same set of morphological features, but different roots: two parses of the word *takası* are the same except their roots (taka vs takas):

taka+Noun+A3sg+P3sg+Nom	*his/her boat*
takas+Noun+A3sg+P3sg+Nom	*his/her exchange*

Note that we are including the roots as a parts of morphological parses.

Morphological disambiguation is a generalization of POS tagging but in general it is more complex and usually has a lower accuracy. For some applications however,

---

[1] See Chap. 2 for many additional examples of morphological ambiguity.

it might be sufficient to determine the main POS tag for each word or for its root; these can be seen as special cases of full morphological disambiguation.

In this chapter, we summarize the challenges for morphological disambiguation of Turkish and review the previous research. We then discuss the available annotated data sets, evaluation methods, and experimental comparisons where applicable, as well as possible future directions.

3.2 Challenges

The two main challenges for morphological disambiguation in Turkish are the following:

- The number of possible distinct word forms and hence the complex morphological parses to capture all possible morphological distinctions is very large—in the tens of thousands as hinted by Fig. 1.4 (page 7) and Table 1.3 (page 6) in Chap. 1, compared to few tens to few hundreds of tags for heavily studied languages like English.
- The number of possible morphological interpretations per word, especially when very high-frequency words are excluded, is relatively large These are typically caused by root part-of-speech ambiguity, segmentation ambiguity, morphographemic processes, homograph morphemes, but can occasionally be caused by the original source of a (borrowed) word. Furthermore lexicalized derivations inject additional ambiguities when the lexicalized word has gained a totally different meaning and life of its own.

Similar to morphological analysis, explained in Chap. 2, morphological disambiguation of real text requires dealing with acronyms, numbers, foreign and unknown words. Finally, multi-word or idiomatic constructs such as *koşa koşa* and *gelir gelmez* (see Chap. 2) may necessitate some a multi-word preprocessing step between morphological analysis and disambiguation.

3.3 Previous Work

At a high level, studies on morphological disambiguation of Turkish can be categorized as rule-based and machine learning-based methods. Rules are usually manually written in the form of constraints, but they can also be learned. In the rest of this section, we start with reviewing the early rule-based methods, and continue with methods for rule-learning, and finally summarize the statistical methods.

3.3.1 Rule-Based Methods

3.3.1.1 Constraint-Based Morphological Disambiguation

The first work on morphological disambiguation of Turkish is the constraint-based approach of Oflazer and Kuruöz (1994). After all the word tokens in a sentence are morphologically analyzed, this approach uses a set of manually written rules for multi-word and idiomatic expression recognition and morphological disambiguation. These rules are formed of a set of constraint and action pairs $C_i : A_i$:

$$C_1 : A_1; C_2 : A_2; \ldots; C_n : A_n$$

where the actions A_i are executed only when the constraints are satisfied. The constraints specify any available morphological feature (such as derivation type, case, and agreement) or positional feature (such as absolute or relative position of the word token in the sentence) associated with the word token. Actions could be one of the four types:

1. null action (the matching parse is untouched),
2. delete (the matching parse is deleted from the set of morphological parses associated with the word token, if there are more than one associated morphological parses with that word),
3. output (assigns the matching parse to the word token, and removes all other parses for that word), and
4. compose (forms a new parse from various matching parses).

For example, the following rule composes a single token for multi-word expressions which actually functions as a temporal adverb in a sentence by combining two verbs with the same root (indicated by _R1), in aorist form with opposing polarities, matching the constraint in parentheses (e.g., *gelir gelmez*).

```
((Lex = _W1, Root=_R1, Cat=V, Aspect=AOR, Agr=3SG,
  Polarity=POS),
 (Lex=_W2, Root=_R1, Cat=V, Aspect=AOR, Agr=3SG,
  Polarity=NEG):
Compose=((*CAT* ADV)(*R* "_W1 _W2 (_R1)")(*SUB* TEMP)).
```

The following example matches the sentence final (indicated SP=END) adjectival readings derived from verbs, and deletes them from the set of morphological parses of the corresponding word tokens:

```
(Cat=V, Finalcat=ADJ, SP=END) : Delete
```

The authors created about 250 rules that were used for disambiguation with minimal human supervision. They showed that the resulting disambiguated text resulted in halving the ambiguity and parsing time of a downstream lexical functional grammar (LFG) parser developed for Turkish (Güngördü and Oflazer 1995).

3.3.1.2 Constraints with Voting

An issue with the rule-cased approach is that changes in order of the rules may result in different morphological tag sequences, and ordering the rules for optimum tagging accuracy was challenging. Hence, Oflazer and Tür (1997) proposed an approach where votes were assigned to constraints that they cast on matching parses of a given word token. In this approach, instead of immediately applying the actions of matching constraints, all tokens were disambiguated in parallel once all applicable rules were applied to each sentence. Their rules were of the form:

$$R = (C_1, C_2, \ldots, C_n; V)$$

where $C_1, C_2, \ldots, C_n$ are the constraints, and V is the vote assigned to the matching morphological parse.

Votes were assigned to constraints based their static properties: more specific rules (i.e., rules that have higher number of constraints or higher number of features in their constraints) had higher votes. Then the vote for the rule is computed as the sum of the votes of its constraints and during disambiguation, the votes of all the parses matching all the constraints were incremented by the vote of the rule. After all the rules have been applied to all the word tokens, morphological parses with the highest votes were selected.

3.3.2 Learning the Rules

Oflazer and Tür (1996) extended the constraint-based disambiguation of Oflazer and Kuruöz (1994) by incorporating an unsupervised rule learning component based on learned rules in addition to linguistically motivated hand-crafted rules. The hand-crafted rules were tuned to improve precision without sacrificing recall. The unsupervised learning process produces two sets of rules: (1) choose rules which choose morphological parses of a lexical item satisfying constraint effectively discarding other parses, and (2) delete rules, which delete parses satisfying a constraint. Coupled with secondary morphological processor for unknown word processing, this system surpassed the accuracy of the previous rule-based system of Oflazer and Kuruöz (1994).

While coming up with a set of rules for disambiguation is an intuitive approach, coming up with the right/optimum set of rules can be non-trivial. Instead, learning of the rules and their associated weights could be performed jointly. Yuret and Türe (2006) took such an approach that would learn the rules and their weights.

Their approach was based on supervised training of a separate model for every possible morphological feature that can be generated by the morphological analyzer, such as +Acc for an accusative case marker, using text with manually annotated morphological parses. In the case of Yuret and Türe (2006), that required training a model for 126 features. For each unique feature f, they took all word tokens in the manually annotated training set that include f in one of their possible morphological parses. This set was then split into positive and negative examples depending on whether the correct (i.e., manually annotated) parse includes the feature f or not. These examples were then used to learn rules using a novel decision list learning algorithm, namely, Greedy Prepend Algorithm (GPA), that is robust to irrelevant and redundant features.

A decision list is an ordered list of rules, where each rule consists of a pattern and a classification (Rivest 1987). The pattern specifies surface attributes of the target word and words in its context (i.e., surrounding words), and the classification indicates the presence or absence of a morphological feature for the target word. Table 3.1 shows an example decision list for +Det feature, taken from Yuret and Türe (2006).

To learn a decision list from a given set of training examples, the Greedy Prepend Algorithm (GPA) (Yuret and de la Maza 2005) starts with a default rule that matches all instances and classifies them using the most common class in the training data. The algorithm then prepends the rule with the maximum gain in front of the decision list, and repeats this until no rule with a positive gain can be found. The gain of a rule is defined as the increase in the number of correctly classified instances in the training set as a result of prepending the rule to the existing decision list. To find the next rule with the maximum gain, GPA uses a heuristic search algorithm that only considers candidate rules that add a single new attribute to the pattern of an existing rule in the decision list. To use the resulting decision list the algorithm tries to match a new test instance to each rule in order (the last added rule is tried first), and predicts the class given by the first matching rule.

Table 3.1 A five-rule decision list for +Det from Yuret and Türe (2006)

Pattern	Class
W==çok AND R1=+DA	1
L1==pek	1
W=+AzI	0
W==çok	0
–	1

W denotes the target word, L1 denotes the word to the left and R1 denotes the word to the right, == denotes case-insensitive match of the word, and $\mp$ is used to check if the corresponding word contains the feature listed. The class is 1 if the target word should have the +Det feature, and is 0 otherwise

When new data is presented to their tagger, first a morphological analyzer is used to generate all possible parses (i.e., candidates) for all the word tokens. The decision lists are used to predict the presence or absence of each of the features in these candidates. The rules in each decision list are tried in the order, and the first rule with a matching pattern is used to predict the classification of the target word. The final tags are determined by combining the decision of each model by weights assigned to the decision lists, according to their accuracy.

3.3.3 Models Based on Inflectional Group n-Grams

For part-of-speech tagging of English, Hidden Markov Models (HMMs) have been the dominant method for a long time (Charniak et al. 1993). For Turkish, Hakkani-Tür et al. (2002) used a HMM that seeks for the sequence of morphological parses $\hat{T}$ that maximizes the posterior probability, $P(T|W)$, for the sequence of words W, presented formally as:

$$\hat{T} = \operatorname*{argmax}_{T} P(T|W) = \operatorname*{argmax}_{T} P(T) \times P(W|T)$$

Note that the probability $P(W)$ is constant for the sequence of words and can be omitted for evaluating `argmax`.

In part-of speech tagging, such an equation is commonly simplified into the following using two assumptions (Charniak et al. 1993):

$$\hat{T} = \operatorname*{argmax}_{T} \prod_{i=1}^{n} P(t_i|t_{i-2}, t_{i-1}) \times P(w_i|t_i)$$

The first assumption is that the words are independent of each other, given their part-of-speech tags and a word's identity only depends on its tag and not on any other word's tag:

$$P(W|T) = \prod_{i=1}^{n} P\left(w_i|t_1^n\right) \approx \prod_{i=1}^{n} P(w_i|t_i)$$

$P(T)$ can be expanded by using the chain rule and then approximated with a trigram tag model:

$$P(T) = \prod_{i=1}^{n} P(t_i|t_1, \ldots t_{i-1}) \approx \prod_{i=1}^{n} P(t_i|t_{i-2}, t_{i-1})$$

In Turkish, a morphological parse which includes the root word uniquely determines the surface word $P(w_i|t_i) = 1$, simplifying the problem to:

$$\hat{T} = \underset{T}{\operatorname{argmax}} \prod_{i=1}^{n} P(t_i|t_{i-2}, t_{i-1})$$

Since the number of distinct morphological tags in Turkish is very large (and even theoretically infinite), this results in data sparsity when estimating the tag trigram probabilities. Many of the possible tags and tag n-grams remain unseen, even with very large corpora. A practical solution to this problem that has mostly been used in language modeling to use sub-word units (Kneissler and Klakow 2001). In the extreme case, one could use letters, instead of whole words, resulting in a vocabulary that is equal to the alphabet size. However, in that case, the context would be too small, hurting the performance of the modeling. Previous work on language modeling has investigated automatic methods to estimate optimum size units for language modeling (Creutz and Lagus 2005).

Hakkani-Tür et al. (2002) instead followed a linguistically motivated approach and used inflectional groups as basic units in modeling of tag sequences. Each tag, t_i, was represented as a sequence of root r_i and inflectional groups $G_{i,m}$, $m = 1, \ldots, n_i$, and n_i is the number of inflectional groups in the corresponding word w_i:

$$t_i = (r_i, G_{i,1}, \ldots, G_{i,n_i})$$

Then, we can re-write the probabilities and factor them as follows:

$$P(t_i|t_{i-2}, t_{i-1}) = P((r_i, G_{i,1}, \ldots, G_{i,n_i})|(r_i, G_{i,1}, \ldots, G_{i,n_i}), (r_i, G_{i,1}, \ldots, G_{i,n_i}))$$
$$= P(r_i|(r_i, G_{i,1}, \ldots, G_{i,n_i}), (r_i, G_{i,1}, \ldots, G_{i,n_i}))$$
$$\times P(G_{i,1}|(r_i, G_{i,1}, \ldots, G_{i,n_i}), (r_i, G_{i,1}, \ldots, G_{i,n_i}))$$
$$\times \ldots \times P(G_{i,n_i}|(r_i, G_{i,1}, \ldots, G_{i,n_i}), (r_i, G_{i,1}, \ldots, G_{i,n_i}))$$

To further deal with data sparseness problem, one can make further simplifying assumptions related to root and inflectional group dependencies. Hakkani-Tür et al experimented with various models that assume the roots depend only on the preceding roots, and inflectional groups depend only on the final inflectional groups of the previous words. This second assumption was motivated by the property of dependency relationships in Turkish: When a word is considered to be a sequence of inflectional groups, syntactic relation links only emanate from the last inflectional group of a (dependent) word and land on one of the inflectional groups of the (head) word on the right (Oflazer 2003) (see, e.g., Fig. 1.6 on page 10). Later on, such feature-level or sub-word dependencies in language modeling were also investigated in the framework of factored language models (Bilmes and Kirchhoff

2003; Kirchhoff and Yang 2005), but have not been applied to morphological disambiguation.

This approach is also supervised, and a large corpus is used to estimate root and inflectional group trigram probabilities. These are then used to estimate probabilities of candidate morphological parses for a new sequence of words, and the most probable sequence is output.

3.3.4 Discriminative Methods for Disambiguation

Another approach to solve the disambiguation problem aims to determine the best morphological analyses of the words in a sentence in a discriminative fashion. Discriminative training methods proposed by Collins (2002) were applied to morphological disambiguation by Sak et al. (2011). Their discriminative morphological disambiguator based on the linear modeling framework uses the averaged perceptron algorithm for parameter estimation. In the linear modeling framework, the morphological analyses are represented as feature vectors whose inner products with the model parameter vector yield the scores used to rank the analyses. The feature vector representation is very flexible and allows different types of features to be easily incorporated into the model. Sak et al. (2007) first use the baseline trigram-based model of Hakkani-Tür et al. (2002) to enumerate n-best candidates of alternative morphological parses of a sentence, and then rerank these with the discriminative model, whereas Sak et al. (2008, 2011) consider all of the ambiguous morphological parses of the words in a sentence.

The discriminative linear modeling framework is adapted to the disambiguation problem as follows:

- The training examples are the pairs (W_k, T_k) for $k = 1 \ldots N$, where N is the number of training sentences. For the kth sentence, W_k is the word sequence $w_{[1:n_k]}^k$ and T_k is the correct tag sequence $t_{[1:n_k]}^k$, where n_k is the number of words in the sentence.
- The morphological analyzer is represented by the function **GEN**(W) that maps the input sentence W to the candidate parse sequences.
- The representation $\mathbf{\Phi}(W, T) \in \mathfrak{R}^d$ is a d-dimensional feature vector. Each component $\Phi_j(W, T)$ for $j = 1 \ldots d$ is the count of a feature for this sentence W and its candidate analysis T.
- The score for the pair (W, T) is given by the inner product

$$\mathbf{\Phi}(W, T) \cdot \overline{\alpha} = \sum_{j=1}^{d} \Phi_j(W, T)\overline{\alpha}_j$$

where $\overline{\alpha}_j$ is the jth component of the parameter vector $\overline{\alpha}$.
- The best morphological analysis is found as $\widehat{T} = \underset{T \in \mathbf{GEN}(W)}{\text{argmax}} \ \mathbf{\Phi}(W, T) \cdot \overline{\alpha}.$

Fig. 3.1 A variant of the
perceptron algorithm
by Collins (2002) (reprinted
with permission)

Inputs: Training examples $(W_k, T_k), k = 1 \ldots N$
Initialization: Set $\alpha = 0, \overline{\alpha} = 0$
Algorithm:
  for $u = 1 \ldots U, k = 1 \ldots N$
    Calculate $\widehat{T}_k = \underset{T \in \mathbf{GEN}(W_k)}{\mathrm{argmax}} \ \Phi(W_k, T) \cdot \alpha$
    if $(\widehat{T}_k \neq T_k)$ **then** $\alpha = \alpha + \Phi(W_k, T_k) - \Phi(W_k, \widehat{T}_k)$
    $\overline{\alpha} = \overline{\alpha} + \alpha$
Output: Parameters $\overline{\alpha} = \overline{\alpha}/(UN)$

For discriminative training of the model, Sak et al. (2011) used a variant of
the perceptron algorithm given in Fig. 3.1. The algorithm makes multiple passes
(denoted by $u = 1 \ldots U$) over the training examples $(W_k, T_k), k = 1 \ldots N$ to
estimate the parameter vector $\overline{\alpha}$. For each example, it finds the highest scoring
candidate among all candidates using the current parameter values α. If the highest
scoring candidate is not the correct parse, it updates the parameter vector α by
increasing the parameter values for features in the correct candidate and decreasing
the parameter values for features in the competitor. For the final model, the
algorithm computes the "averaged parameters" $\overline{\alpha}$ since they are more robust to noisy
or inseparable data (Collins 2002).

For Turkish, the features are based on the output of the morphological analyzer.
Unlike earlier studies (Hakkani-Tür et al. 2002; Sak et al. 2007) where only
derivational boundaries are used to split the morphological analysis of a word
into chunks called inflectional groups, Sak et al. (2011) use both inflectional and
derivational boundaries to obtain the morphemes. A morphosyntactic tag t_i, which
is a morphological analysis of a word w_i, is split into a root tag r_i and a morpheme
tag m_i. The morpheme tag m_i is the concatenation of the morphosyntactic tags of
morphemes $m_{i,j}$ for $j = 1 \ldots n_i$, where n_i is the number of morphemes in t_i:

$$t_i = r_i m_i = r_i + m_{i,1} + m_{i,2} + \ldots + m_{i,n_i}$$

Thus for example, the morphological analysis of the word $w_i = ula\c{s}mad\i\u{g}\i$ (*that
fact that it did not arrive*)

$$t_i = ula\c{s}[\text{Verb}] + mA[\text{Neg}] + DHk[\text{Noun} + \text{PastPart}] + [\text{A3sg}] + sH[\text{P3sg}] + [\text{Nom}]$$

is represented as its root tag and morpheme tags as follows (Sak et al. 2011):

$$r_i = ula\c{s}[\text{Verb}]$$

$$m_{i,1} = +mA[\text{Neg}]$$

$$m_{i,2} = +DHk[\text{Noun} + \text{PastPart}] + [\text{A3sg}]$$

$$m_{i,3} = +sH[\text{P3sg}] + [\text{Nom}]$$

Table 3.2 Feature templates used for morphological disambiguation

Gloss	Feature
Morphological parse trigram	(1) $t_{i-2}t_{i-1}t_i$
Morphological parse bigram	(2) $t_{i-2}t_i$ and (3) $t_{i-1}t_i$
Morphological parse unigram	(4) t_i
Morpheme tag with previous tag	(5) $t_{i-1}m_i$
Morpheme tag with second to previous tag	(6) $t_{i-2}m_i$
Root trigram	(7) $r_{i-2}r_{i-1}r_i$
Root bigram	(8) $r_{i-2}r_i$ and (9) $r_{i-1}r_i$
Root unigram	(10) r_i
Morpheme tag trigram	(11) $m_{i-2}m_{i-1}m_i$
Morpheme tag bigram	(12) $m_{i-2}m_i$ and (13) $m_{i-1}m_i$
Morpheme tag unigram	(14) m_i
Individual morpheme tags	(15) $m_{i,j}$ for $j = 1 \ldots n_i$
Individual morpheme tags with position	(16) $jm_{i,j}$ for $j = 1 \ldots n_i$
Number of morpheme tags	(17) n_i

The feature set takes into account the current morphosyntactic tag t_i, the previous tag t_{i-1}, and the tag before that t_{i-2}. The feature templates are given in Table 3.2.

Another discriminative approach to the morphological disambiguation problem makes use of Conditional Random Fields (CRFs) (Lafferty et al. 2001) which directly computes the conditional probability of the tag sequence T given the word sequence W using a feature representation $\mathbf{\Phi}(W, T)$:

$$P(T|W) = \frac{e^{\mathbf{\Phi}(W,T) \cdot \overline{\alpha}}}{\sum_{T' \in \mathbf{GEN}(W)} e^{\mathbf{\Phi}(W,T') \cdot \overline{\alpha}}}.$$

Arslan (2009) introduced a method based on conditional random fields (CRF) to disambiguate the alternative morphological analyses for Turkish and Ehsani et al. (2012) applied CRFs to the task of disambiguating the main POS tags which is similar to but simpler than the full disambiguation task. While these approaches used similar feature sets, the models and especially the training procedures are more complicated than the linear modeling framework. In order to deal with the high computational cost, Ehsani et al. (2012) proposed splitting long sentences into sub-sentences and trimming unlikely tags using marginal probabilities.

A classification based approach to the morphological disambiguation problem was proposed by Görgün and Yıldız (2011) where the disambiguation task was redefined as identifying the correct parse from the possible parses *excluding* the root. If choosing among each set of alternative parses is considered as a different classification problem, a 1M word disambiguated corpus has approximately 400K distinct problems. However, when the root is excluded from the morphological parse of a word, there are only 9320 distinct problems. Among the classification algorithms applied to this task, decision trees performed the best.

3.4 Discussion

In this section, we give an overview the data sets used for the morphological disambiguation of Turkish and summarize the results reported in the literature.

3.4.1 Data Sets

Most of the studies on Turkish morphological disambiguation make use of the data prepared by Hakkani-Tür et al. (2002) collected from the web resources of a Turkish daily newspaper. The tokens were analyzed using a morphological analyzer. The training data of Hakkani-Tür et al. (2002) consists of the unambiguous sequences of about 650K tokens in a corpus of 1 million tokens, and two manually disambiguated texts of 12,000 and 20,000 tokens. Their test data consists of 2763 tokens of which 935 ($\approx$34%) have more than one morphological analysis. Yuret and Türe (2006) used the same semi-automatically disambiguated training data and reported that it contained 111,467 unique tokens, 11,084 unique tags, and 2440 unique features. The final evaluation of their model was performed on a manually annotated test data set of 958 instances. Sak et al. (2007, 2008, 2011) used the same training and test data.

More recently, the METU-Sabancı Turkish Treebank (Oflazer et al. 2003) and the ITU validation set (Eryiğit and Pamay 2014) were used for evaluating the performance of the morphological disambiguation with the primary purpose of observing its impact on dependency parsing of Turkish (Eryiğit 2012; Çetinoğlu 2014). The METU-Sabancı Treebank consists of 5635 sentences and 56K words whereas the ITU validation set has 300 sentences and 3.7K words.

3.4.2 Experimental Results

The main evaluation measure for morphological disambiguation is accuracy based on the *exact* match of the chosen morphological analysis with the manually labeled reference. Of course, in real applications the performance will depend on the morphological analyzer that produces the candidate morphological analyses. The coverage of the morphological analyzer constitutes an upper bound on the success of the morphological disambiguation process. In simplified tasks such as main part-of-speech tag disambiguation, only the main part-of-speech tag is taken into account when computing the accuracy. For analysis purposes, accuracy at the IG level is also used.

Table 3.3 The accuracy of the morphological disambiguators trained on the same semi-automatically disambiguated training data

Disambiguator	Manual test	METU-Sabancı Treebank	ITU validation set
Hakkani-Tür et al. (2002)	95.48%	–	–
Yuret and Türe (2006)	95.82%	78.76%	87.67%
Sak et al. (2011)	96.45%	78.23%	87.84%

Table 3.4 The accuracy of the morphological disambiguator trained and tested on matched and mismatched data

Training set	Manual test	METU-Sabancı Treebank	ITU validation set
MD train set	96.29%	87.64%	88.41%
METU-Sabancı Treebank	88.17%	90.19%	89.87%

Following the original papers as well as the results reported by Eryiğit (2012), we give the accuracy of the morphological disambiguators. Hakkani-Tür et al. (2002), Yuret and Türe (2006), Sak et al. (2008, 2011) trained on the same semi-automatically disambiguated training data in Table 3.3 where Manual Test Set stands for the manually annotated test set of Yuret and Türe (2006), and METU-Sabancı Treebank results are tenfold cross-validation results.

Given the decrease of performance on the METU-Sabancı Treebank and the ITU Validation Set, Çetinoğlu (2014) also trained a version of the morphological disambiguator by Sak et al. (2008) on the treebank data and tested on the same data. The results are summarized in Table 3.4 where MD Train Set refers to the original semi-automatically disambiguated training data and METU-Sabancı Treebank test results are tenfold cross-validation results.

3.5 Conclusions

This chapter presented an overview of the morphological disambiguation problem for Turkish and the approaches that have been used in the last 25 years. The approaches have spanned a range from earlier ruled-based methods with first hand-crafted and the machine-learned rules, to various statistical and machine learning approaches based on HMMs, perceptron-based discriminative method and CRFs.

In addition to the challenges that we presented earlier, these approaches, especially those using various forms of supervised learning suffer from the lack of a large scale training corpus with gold-standard annotations.

References

Arslan BB (2009) An approach to the morphological disambiguation problem using conditional random fields. Master's thesis, Sabancı University, Istanbul

Bilmes JA, Kirchhoff K (2003) Factored language models and generalized parallel backoff. In: Proceedings of NAACL-HLT, Edmonton, pp 4–6

Çetinoğlu Ö (2014) Turkish treebank as a gold standard for morphological disambiguation and its influence on parsing. In: Proceedings of LREC, Reykjavík, pp 3360–3365

Charniak E, Hendrickson C, Jacobson N, Perkowitz M (1993) Equations for part-of-speech tagging. In: Proceedings of AAAI, Washington, DC, pp 784–789

Collins M (2002) Discriminative training methods for Hidden Markov Models: theory and experiments with perceptron algorithms. In: Proceedings of EMNLP, Philadelphia, PA, pp 1–8

Creutz M, Lagus K (2005) Unsupervised morpheme segmentation and morphology induction from text corpora using Morfessor 1.0. Publications in Computer and Information Science Report A81, Helsinki University of Technology, Helsinki

Ehsani R, Alper ME, Eryiğit G, Adalı E (2012) Disambiguating main POS tags for Turkish. In: Proceedings of the 24th conference on computational linguistics and speech processing, Chung-Li

Eryiğit G (2012) The impact of automatic morphological analysis and disambiguation on dependency parsing of Turkish. In: Proceedings of LREC, Istanbul

Eryiğit G, Pamay T (2014) ITU validation set. Türkiye Bilişim Vakfı Bilgisayar Bilimleri ve Mühendisliği Dergisi 7(1):103–106

Görgün O, Yıldız OT (2011) A novel approach to morphological disambiguation for Turkish. In: Proceedings of ISCIS, London, pp 77–83

Güngördü Z, Oflazer K (1995) Parsing Turkish using the Lexical-Functional Grammar formalism. Mach Transl 10(4):515–544

Hakkani-Tür DZ, Oflazer K, Tür G (2002) Statistical morphological disambiguation for agglutinative languages. Comput Hum 36(4):381–410

Kirchhoff K, Yang M (2005) Improved language modeling for statistical machine translation. In: Proceedings of the workshop on building and using parallel texts, Ann Arbor, MI, pp 125–128

Kneissler J, Klakow D (2001) Speech recognition for huge vocabularies by using optimized sub-word units. In: Proceedings of INTERSPEECH, Aalborg, pp 69–72

Lafferty JD, McCallum A, Pereira F (2001) Conditional random fields: probabilistic models for segmenting and labeling sequence data. In: Proceedings of ICML, Williams, MA, pp 282–289

Marcus M, Marcinkiewicz M, Santorini B (1993) Building a large annotated corpus of English: the Penn Treebank. Comput Linguist 19(2):313–330

Oflazer K (2003) Dependency parsing with an extended finite-state approach. Comput Linguist 29(4):515–544

Oflazer K, Kuruöz İ (1994) Tagging and morphological disambiguation of Turkish text. In: Proceedings of ANLP, Stuttgart, pp 144–149

Oflazer K, Tür G (1996) Combining hand-crafted rules and unsupervised learning in constraint-based morphological disambiguation. In: Proceedings of EMNLP-VLC, Philadelphia, PA

Oflazer K, Tür G (1997) Morphological disambiguation by voting constraints. In: Proceedings of ACL-EACL, Madrid, pp 222–229

Oflazer K, Say B, Hakkani-Tür DZ, Tür G (2003) Building a Turkish Treebank. In: Treebanks: building and using parsed corpora. Kluwer Academic Publishers, Berlin

Rivest R (1987) Learning decision lists. Mach Learn 2(3):229–246

Sak H, Güngör T, Saraçlar M (2007) Morphological disambiguation of Turkish text with perceptron algorithm. In: Proceedings of CICLING, Mexico City, pp 107–118

Sak H, Güngör T, Saraçlar M (2008) Turkish language resources: morphological parser, morphological disambiguator and web corpus. In: Proceedings of the 6th GoTAL conference, Gothenburg, pp 417–427

Sak H, Güngör T, Saraçlar M (2011) Resources for Turkish morphological processing. Lang Resour Eval 45(2):249–261

Yuret D, de la Maza M (2005) The greedy prepend algorithm for decision list induction. In: Proceedings of ISCIS, Istanbul

Yuret D, Türe F (2006) Learning morphological disambiguation rules for Turkish. In: Proceedings of NAACL-HLT, New York, NY, pp 328–334

Chapter 4
Language Modeling for Turkish Text and Speech Processing

Ebru Arısoy and Murat Saraçlar

Abstract This chapter presents an overview of language modeling followed by a discussion of the challenges in Turkish language modeling. Sub-lexical units are commonly used to reduce the high out-of-vocabulary (OOV) rates of morphologically rich languages. These units are either obtained by morphological analysis or by unsupervised statistical techniques. For Turkish, the morphological analysis yields word segmentations both at the lexical and surface forms which can be used as sub-lexical language modeling units. Discriminative language models, which outperform generative models for various tasks, allow for easy integration of morphological and syntactic features into language modeling. The chapter provides a review of both generative and discriminative approaches for Turkish language modeling.

4.1 Introduction

A statistical language model assigns a probability distribution over all possible word strings in a language. The ultimate goal in statistical language modeling is to find probability estimates for word strings that are as close as possible to their true distribution. In the last couple of decades, a number of statistical techniques have been proposed to appropriately model natural languages. These techniques employ large amounts of text data to robustly estimate model parameters which are then used to estimate probabilities of unseen text.

Statistical language models are used in many natural language applications such as speech recognition, statistical machine translation, handwriting recognition, and

E. Arısoy
MEF University, Istanbul, Turkey
e-mail: ebruarisoy.saraclar@mef.edu.tr

M. Saraçlar (✉)
Boğaziçi University, Istanbul, Turkey
e-mail: murat.saraclar@boun.edu.tr

© Springer International Publishing AG, part of Springer Nature 2018   69
K. Oflazer, M. Saraçlar (eds.), *Turkish Natural Language Processing*,
Theory and Applications of Natural Language Processing,
https://doi.org/10.1007/978-3-319-90165-7_4

spelling correction, as a crucial component to improve the performance of these applications. In these and other similar applications, statistical language models provide prior probability estimates and play the role of the source model in communication theory inspired source-channel formulations of such applications. A typical formulation of these applications allows language models to be used as a predictor that can assign a probability estimate to the next word given the contextual history. Some applications employ more complex language models in reranking scenarios where alternative hypotheses generated by a simpler system are rescored or reranked using additional information. A typical example is the feature-based discriminative language model where model parameters associated with many overlapping features are used to define a cost or conditional probability of the word sequences. Such a model then enables the selection of the best hypothesis among the alternatives based on the scores assigned by the model.

This chapter focuses on language modeling mainly for Turkish text and speech processing applications. First we introduce the foundations of language modeling and describe the popular approaches to language modeling, then we explain the challenges that Turkish presents for language modeling. After reviewing various techniques proposed for morphologically rich languages including Turkish, we summarize the approaches used for Turkish language modeling.

4.2 Language Modeling

Statistical language models assign a prior probability, $P(W)$, to every word string $W = w_1 \, w_2 \, \ldots \, w_N$ in a language. Using the chain rule, the prior probability of a word string can be decomposed into the following form:

$$P(W) = P(w_1 \, w_2 \, \ldots \, w_N) = \prod_{k=1}^{N} P(w_k | w_1 \, \ldots \, w_{k-1}). \tag{4.1}$$

Here the prior probability is calculated in terms of the dependencies of words to a group of preceding words, $w_1 \, \ldots \, w_{k-1}$, called the "history." These conditional probabilities need to be estimated in order to determine $P(W)$. It is, however, not practical to obtain the prior probability as given in Eq. (4.1) for two main reasons. First, if the history is too long, it is not possible to robustly estimate the conditional probabilities, $P(w_k | w_1 \, \ldots \, w_{k-1})$. Second, it is not entirely true that the probability of a word depends on *all* the words in its entire history. It is more practical and realistic to assign histories to equivalence classes $\Psi(w_1 \ldots w_{k-1})$ (Jelinek 1997). Equivalence classes change Eq. (4.1) into the following form:

$$P(W) = P(w_1 \, w_2 \, \ldots \, w_N) = \prod_{k=1}^{N} P(w_k | \Psi(w_1 \, \ldots \, w_{k-1})) \tag{4.2}$$

While the equivalence classes can be based on any classification of the words in the history, or their syntactic and semantic information, the most common approach is based on a very simple equivalence classification which utilizes only the $n - 1$ preceding words as the history. This approach results in the widely used n-gram language models, and $P(W)$ is approximated as

$$P(W) = P(w_1 \, w_2 \, \ldots \, w_N) \approx \prod_{k=1}^{N} P(w_k | w_{k-n+1} \ldots \, w_{k-1}) \qquad (4.3)$$

The n-gram language model probabilities are estimated from a text corpus related to the application domain with Maximum Likelihood Estimation (MLE). In other words, n-gram probabilities are estimated by counting the occurrences of a particular n-gram in the text data and dividing this count by the number of occurrences of all n-grams that start with the same sequence of $n - 1$ words:

$$P(w_k | w_{k-n+1} \, \ldots \, w_{k-1}) = \frac{C(w_{k-n+1} \, \ldots \, w_{k-1} \, w_k)}{C(w_{k-n+1} \, \ldots \, w_{k-1})} \qquad (4.4)$$

where $C(\cdot)$ represents the number of occurrences of the word string given in parentheses in the text data.

If the language model vocabulary contains $|V|$ words, then there may be up to $|V|^n$ n-gram probabilities to be calculated—thus higher order n-grams need a much larger set of language model parameters. Robust estimation of n-gram probabilities with MLE critically depends on the availability of large amounts of text data. However experience with many applications has shown that 3/4/5-gram models are quite satisfactory and higher order models do not provide any further benefits.

The quality of the statistical language models can be best evaluated using the performance of the applications they are used in—for example, speech recognition or statistical machine translation. An alternative approach without including the overall system into the evaluation is to rely on *perplexity* to gauge the generalization capacity of the proposed language model on a separate text that is not seen during model training. Formally, perplexity is defined as:

$$PP(w_1, w_2, \cdots, w_N) = 2^{-\frac{1}{N} \log_2 P(w_1, w_2, \cdots, w_N)} \qquad (4.5)$$

In other words, perplexity shows us how well a language model trained on a text data does on an unseen text data. Minimizing the perplexity corresponds to maximizing the probability of the test data. Even though a lower perplexity usually means a better language model with more accurate prediction performance, perplexity may not always be directly correlated with application performance.

One of the problems in n-gram language modeling is data sparseness. Any finite training corpus contains only a subset of all possible n-grams. So, MLE will assign zero probability to all *unseen* n-grams. A test sentence containing such n-grams not seen in the training corpus will also be assigned zero probability according to

Eq. (4.3). In order to prevent this, a technique known as *smoothing* is employed to reserve some of the probability mass to unseen n-grams so that no n-gram gets zero probability. This also means that this mass comes from the probabilities of the observed n-grams leading to slight reductions in their probabilities. Smoothing techniques thus lead to better language model estimates for unseen data.

Interpolation and back-off smoothing are the most common smoothing methods. In interpolation, higher and lower order n-gram models are linearly interpolated. In back-off smoothing, when a higher order n-gram model assigns zero probability to a particular n-gram, the model backs off to a lower order n-gram model. Good-Turing, Katz, and Kneser-Ney are some examples of popular smoothing algorithms. See Chen and Goodman (1999) for a survey of smoothing approaches for statistical language models.

In addition to these smoothing techniques, class-based n-gram language models (Brown et al. 1992) and continuous space language models (Bengio et al. 2003; Schwenk 2007) have been used to estimate unseen event probabilities more robustly. These approaches try to make more reasonable predictions for the unseen histories by assuming that they are similar to the histories that have been seen in the training data. Class-based language models group words into classes, while continuous space language models project words into a higher dimensional continuous space, with the expectation that words that are semantically or grammatically related will be grouped into the same class or mapped to similar locations in the continuous space. The main goal of these models is to generalize well to unseen n-grams.

One drawback of the conventional n-gram language models is their reliance on only the last $n-1$ words in the history. However, there are many additional sources of information, such as morphology, syntax, and semantics, that can be useful while predicting the probability of the next word. Such additional linguistic information can be either incorporated into the history of the n-gram models or encoded as a set of features to be utilized in feature-based language models.

Structured language models (Chelba and Jelinek 2000), probabilistic top-down parsing in language modeling (Roark 2001), and Super ARV language models (Wang and Harper 2002) are some example approaches that incorporate syntactic information into the n-gram history. The factored language model (Bilmes and Kirchhoff 2003) is another example that incorporates syntactic as well as morphological information into the n-gram history.

Feature-based models allow for easy integration of arbitrary knowledge sources into language modeling by encoding relevant information as a set of features. The maximum entropy language model (Rosenfeld 1994) is a popular example of this type, where the conditional probabilities are calculated with an exponential model,

$$P(w|h) = \frac{1}{Z(h)} e^{\sum_i \alpha_i \Phi_i(h,w)}. \tag{4.6}$$

Here, $Z(h)$ is a normalization term and $\Phi_i(h, w)$'s are arbitrary features which are functions of the word w and the history h. The whole sentence maximum entropy model (Rosenfeld et al. 2001) assigns a probability to the whole sentence using the features $\Phi_i(W)$ with a constant normalization term Z:

$$P(W) = \frac{1}{Z} e^{\sum_i \alpha_i \Phi_i(W)}. \tag{4.7}$$

Discriminative language models (DLMs) (Roark et al. 2007) have been proposed as a complementary approach to the state-of-the-art n-gram language modeling. There are mainly two advantages of DLMs over n-grams. The first advantage is improved parameter estimation with discriminative training, since DLMs utilize both positive and negative examples to optimize an objective function that is directly related with the system performance. In training a DLM, positive examples are the correct or meaningful sentences in a language while negative examples are word sequences that are not legitimate or meaningful sentences in the language.

The second advantage is the ease of incorporating many information sources such as morphology, syntax, and semantics into language modeling. As a result, DLMs have been demonstrated to outperform generative n-gram language models. Linear and log-linear models have been successfully applied to discriminative language modeling for speech recognition (Roark et al. 2004, 2007; Collins et al. 2005). In DLMs based on linear models, model parameters are used to define a cost, $F(W)$, on the word sequence

$$F(W) = \sum_i \alpha_i \Phi_i(W). \tag{4.8}$$

In DLMs based on log-linear models, the cost $F(W)$ has the same form as the log of the probability given by the whole sentence maximum entropy model

$$F(W) = \sum_i \alpha_i \Phi_i(W) - \log Z, \tag{4.9}$$

where Z is approximated by summing over the alternative hypotheses. The details of the DLM framework will be given in Sect. 4.6.

4.3   Challenges in Statistical Language Modeling for Turkish

In the context of language modeling, two aspects of Turkish, very productive agglutinative morphology leading to a very large vocabulary, and free constituent order make statistical language modeling rather challenging, especially for applications such as automatic speech recognition (ASR) and statistical machine translation (SMT).

State-of-the-art ASR and SMT systems utilize predetermined and finite vocabularies that contain the most frequent words related to the application domain. The words that do not occur in the vocabulary but are encountered by the ASR or SMT

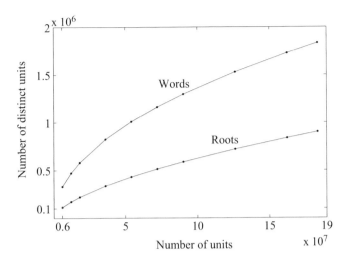

Fig. 4.1 Vocabulary growth curves for words and roots

system are called Out-Of-Vocabulary (OOV) words. Existence of OOV words is one of the causes of degradation in system performance. For instance in an ASR system, if a word is not in the vocabulary and it is uttered by a speaker, it has no chance to be recognized correctly. Hetherington (1995) estimates that as a rule of thumb an OOV word leads to on the average 1.5 recognition errors.

As described in earlier chapters, the very productive morphology of Turkish yields many unique word forms, making it difficult to have a fixed vocabulary covering all these words. Figure 4.1 illustrates the growth for unique Turkish words and roots as a function of the number of tokens in a text corpus of 182.3M word tokens (units) and 1.8M word types (distinct units). It can be observed that the increase in the number of distinct units with the increasing amount of data is much higher for words compared to roots which is an expected result for Turkish. From the morphological analysis of these Turkish words, we have also observed that on the average each root generates 204 words and each word is composed of on the average 1.7 morphemes including the root.[1] The verb *etmek* "to do" accounts for 3348 unique words—the maximum number for any of the roots. The word form *ruhsat+lan+dır+ıl+ama+ma+sı+nda+ki* is an example with the maximum number of morphemes but only occurs once in the corpus.

This significant word vocabulary growth results in high OOV rates even for vocabulary sizes that would be considered as large for English. Figure 4.2 shows the OOV rates calculated on a test data of 23K words, for different vocabulary sizes. For instance, around 9% OOV rate is achieved with a vocabulary size of 60K words.

---

[1]But as noted in Chap. 2, most high-frequency words have a single morpheme so most likely inflected words have more than 1.7 morphemes.

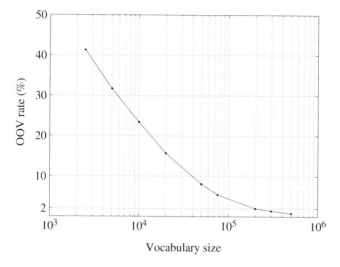

Fig. 4.2 OOV rates for Turkish with different vocabulary sizes

However, with an optimized 60K word lexicon for English, the OOV rate is less than 1% for North American business news text (Rosenfeld 1995). Other morphologically rich languages such as Finnish, Estonian, Hungarian, and Czech also suffer from high OOV rates: 15% OOV with a 69K lexicon for Finnish (Hirsimäki et al. 2006), 10% OOV with a 60K lexicon for Estonian (Kurimo et al. 2006), 15% OOV with a 20K lexicon for Hungarian (Mihajlik et al. 2007), and 8.3% OOV with a 60K lexicon for Czech (Podvesky and Machek 2005). Even though these numbers are not directly comparable with each other, they indicate that high OOV rates are a major problem for morphologically rich languages. Therefore, addressing the OOV problem is crucial for the performance of downstream applications systems that make use of statistical language models.

The free word order is another challenge for statistical language modeling. The relatively free word order contributes to the sparseness data and this can lead to non-robust n-gram language model estimates. However this is more of a problem for speech recognition applications or processing of informal texts—in formal text such as news the dominant constituent order is subject-object-verb but there are no reliable statistics on the distribution of different constituent order in large Turkish corpora. We will not be addressing this issue in the rest of this chapter.

4.4   Sub-lexical Units for Statistical Language Modeling

A commonly proposed solution for reducing high OOV rates for morphologically rich languages is to use sub-lexical units for language modeling. In sub-lexical language modeling, the vocabulary is composed of sub-lexical units instead of words. These could be letters, syllables, morphemes or combination of morphemes or arbitrary word segments. In order to address the OOV problem, the sub-lexical unit vocabulary should be capable of covering most of the words of a language, and clearly these sub-lexical units should be meaningful for prediction using language models. They should have limited confusion and avoid over-generation. For instance, if the letters are used as sub-lexical units, only a vocabulary of 29 letters of the Turkish alphabet will cover all the words in the language. However, letters are not logical sub-lexical unit choices since they require very long histories for accurate language model predictions and they allow more confusable choices in, for instance, speech recognition. Note also that the perplexities of language models based on different units are not directly comparable due to each model having different OOV rates and different number of tokens for evaluation. Assuming no OOVs, perplexity of sub-lexical language models need to be normalized by the number of word tokens for a fair comparison. However, a better way of comparing sub-lexical and word language models is directly measuring the task performance.

Sub-lexical units can be classified as being linguistic or statistical, based on the underlying algorithm utilized in segmenting words into sub-lexical units. Linguistic sub-lexical units are obtained with rule-based morphological analyzers while statistical sub-lexical units are obtained with statistical segmentation approaches that rely on unsupervised model of word segmentation.

4.4.1 Linguistic Sub-lexical Units

In agglutinative languages like Turkish, words are formed as a concatenation of stems and affixes. Therefore, linguistic units such as stems, affixes, or their groupings can be considered as natural choices of sub-lexical units for language modeling. In language modeling with linguistic sub-lexical units, the words are split into morphemes using morphological analyzers, and then a vocabulary composed of chosen morphological units is built for language modeling. However, there is a trade off between using long and short units: long units, e.g., full words will result in OOV problem while shorter units (e.g., morphemes) will require larger n-grams for prediction and risk assigning probabilities to non-words because of over-generation. Since morphemes might be very short, as short as a single letter, Kanevsky et al. (1998) have suggested using stems and endings as vocabulary units as a compromise between words and morphemes, where an ending is what is left after removing the root from the word.

Morphemes, stems, and endings are examples of commonly used linguistic sub-lexical units in language modeling of agglutinative languages like Turkish, Korean, Finnish, Estonian, and Hungarian and highly inflectional languages like Czech, Slovenian, and Arabic. Morpheme-based language models were utilized for language modeling of Korean, another agglutinative language, and to deal with the coarticulation problem rising from very short morphemes, frequent and short morpheme pairs were merged before modeling (Kwon and Park 2003).

Morpheme-based language models were also investigated for language modeling of Finnish (Hirsimäki et al. 2006), Estonian (Alumäe 2006), and Hungarian (Mihajlik et al. 2007), all also agglutinative. These researchers also compared linguistic sub-lexical units with their statistical counterparts for ASR. Kirchhoff et al. (2006) and Choueiter et al. (2006) applied morphology-based language modeling to Arabic ASR and reported better recognition results than words. Rotovnik et al. (2007) used stem and endings for Slovenian language modeling for ASR. Additional constraints to the ASR decoder, such as restricting the correct stem and ending order, and limiting the number of endings for an individual stem were found to reduce over-generation.

The main disadvantage of linguistic sub-lexical units is the need for expert knowledge of the language for building the morphological analyzers. Thus they are not applicable to languages lacking such morphological tools. Additionally, even if a morphological analyzer is available, usually a fixed limited root vocabulary may not necessarily help with the OOV problem. For instance, a Turkish morphological analyzer (Sak et al. 2011) with 54.3K roots can analyze only 52.2% of the 2.2M unique words in a text corpus of 212M words. However, the words that the morphological analyzer cannot parse are usually rare words and only account for about 3% of the word tokens in the text corpus. Hence, this limitation may not necessarily have much impact on the statistical language model. A more important concern is the need for morphological disambiguation of multiple analyses of words.

4.4.2   Statistical Sub-lexical Units

Statistical sub-lexical units are morpheme-like units or segments obtained by data driven approaches, usually in an unsupervised manner. The main advantage of statistical sub-lexical units is that they do not rely on a manually constructed morphological analyzer. These segments do not necessarily match with the linguistic morphemes, however, they are "meaningful" units in terms of language modeling.

One of the earliest works in this area, Harris (1967) posited morpheme boundaries in a word by using letter transition frequencies with the assumption that the predictability of a letter will decrease at the morpheme boundaries.

The last 15 years have seen a surge in data-driven algorithms for unsupervised morpheme discovery based on probabilistic models as well as some heuristics. One of the algorithms with publicly available software is Linguistica (Goldsmith 2001) that utilizes the minimum description length (MDL) principle to learn morphological segmentations in an unsupervised way, aiming to find the segmentations as close as possible to the true morphemes. Whittaker and Woodland (2000), motivated by the productive morphology of Russian, aim to obtain sub-lexical units (called particles) that maximize the likelihood of the training data using a bigram particle language model. In contrast to Linguistica, their algorithm does not aim to find the true morphological segmentations but instead searches for meaningful units for language modeling. Creutz and Lagus (2002, 2005) present

Morfessor, another algorithm for unsupervised discovery of morphemes. Morfessor was inspired by earlier work by Brent (1999) that explored word discovery during language acquisition of young children. Brent (1999) proposed a probabilistic model based on the MDL principle to recover word boundaries in a natural raw text from which they have been removed. The Morfessor algorithm also utilizes the MDL principle while learning a representation of the language in the data, as well as the most accurate segmentations. It is better suited for highly inflectional and agglutinative languages than Linguistica as it is designed to deal with languages with concatenative morphology. The annual Morpho Challenge competitions,[2] held since 2005, have helped the development of new algorithms for sub-lexical units. The Morfessor algorithm itself has been used as the baseline statistical sub-lexical approach in Morpho Challenge tracks and several different algorithms have competed against it.

Statistical sub-lexical units have been explored in language modeling of highly inflected and agglutinative languages. Hirsimäki et al. (2006), Kurimo et al. (2006), Siivola et al. (2003) applied Morfessor to Finnish, while Kurimo et al. (2006) applied it to Estonian and Mihajlik et al. (2007) to Hungarian. The performance of morpheme-based language models was compared with the language models built with Morfessor segmentations for Finnish (Hirsimäki et al. 2006) and Hungarian (Mihajlik et al. 2007) in the context of ASR. In Finnish ASR experiments, statistical units outperformed linguistic morphemes in news reading task where the number of foreign words that could not be handled by the morphological analyzer was quite high. In Hungarian ASR experiments for spontaneous speech, the best result was obtained with statistical segmentations. Hirsimäki (2009) describes the advances in building efficient speech recognition systems with Morfessor based segmentations. Kneissler and Klakow (2001) used an optimized sub-lexical approach for Finnish dictation and German street names recognition tasks. Pellegrini and Lamel (2007, 2009) modified the Morfessor algorithm to incorporate basic phonetic knowledge and explored its use for ASR of Amharic, a highly inflectional language.

4.5   Statistical Language Modeling for Turkish

This section reviews statistical language modeling units explored in Turkish text and speech processing systems. Figure 4.3 illustrates segmentations of the same Turkish phrase using different sub-lexical units. When applicable, the examples also show the lexical and surface form representations and "morphs" denote statistical sub-lexical units. In the rest of this section we will describe the details of Turkish language models based on these units.

---

[2] Aalto University, Finland. Department of Computer Science. "Morpho Challenge": morpho.aalto. fi/events/morphochallenge/ (Accessed Sept. 14, 2017).

Words:	derneklerinin öncülüğünde

Syllables:	der −nek −le −ri −nin ön −cü −lü −ğün −de

Morphemes
Lexical:	dernek +lArH +nHn öncü +lHk +sH +nDA
Surface:	dernek +leri +nin öncü +lüğ +ü +nde

Stem+Endings
Lexical:	dernek +lArH+nHn öncü +lHk+sH+nDA
Surface:	dernek +lerinin öncü +lüğünde

Morphs:	dernek +lerinin öncü +lüğü +nde

Fig. 4.3 A Turkish phrase segmented into linguistic and statistical sub-lexical units

4.5.1 Language Modeling with Linguistic Sub-lexical Units

Over the last 15 years, various linguistic sub-lexical units for Turkish language modeling have been explored in the literature. Here we first review some of the earlier work and then summarize our work using such units.

Çarkı et al. (2000) were the first to investigate sub-lexical language models for Turkish. Due to the ambiguity in morphological analyses, they utilized syllables instead of morphemes as language modeling units and syllables were merged to obtain longer units with word-positioned syllable classes. While this approach addressed the serious OOV problem, it did not yield any improvements over the word-based language model built with a 30K vocabulary. Hakkani-Tür (2000) proposed groupings of morphemes, called inflectional groups as language modeling units. Mengüşoğlu and Deroo (2001) explored an extension of inflectional groups to *n*-gram language modeling as well as utilizing stem+ending models for Turkish. Dutağacı (2002) presented a comparative study of morpheme, stem+ending, and syllable language models in terms of generalization capacity of language models and OOV handling. ASR experiment results were also reported for these sub-lexical units, however, for a small vocabulary isolated word recognition task. This work was extended to continuous speech recognition by Arısoy (2004), and Arısoy et al. (2006) with a new model utilizing words, stem+endings and morphemes together in the same model vocabulary. Such a hybrid vocabulary combined model slightly outperformed the word model in terms of recognition accuracy when 10K units were used in combined and word bigram language models. Çiloğlu et al. (2004) compared bigram stem+ending model with a bigram stem model in terms of recognition accuracy in a small vocabulary ASR task and found that the stem model outperformed the stem+ending model when the language models were trained on a very small text corpus (less than 1M words). However, the stem+ending model was shown to outperform stem model when the text corpus size was increased to approximately 6M words (Bayer et al. 2006).

Erdoğan et al. (2005) was one of the most comprehensive previous research on language modeling for Turkish ASR. The acoustic and language models in this work were trained on much larger amounts of data (34 h of speech corpus and 81M words text corpus). They investigated words, stem+endings and syllables as language modeling units and compared their performances on an ASR task and reported that the stem+ending model outperformed word and syllable models in recognition accuracy. This work also dealt with the over-generation problem of sub-lexical units by a post-processing approach imposing phonological constraints of Turkish and achieved further improvements over the best scoring stem+ending model.

Arısoy and Saraçlar (2009) presented another approach for dealing with the over-generation problem of sub-lexical units, especially for statistical sub-lexical units. This work along with Arısoy et al. (2009a) used a 200 million word text corpus collected from the web. The Turkish morphological parser described in Sak et al. (2011) was used to decompose words into morphemes and the Turkish morphological disambiguation tool developed by Sak et al. (2007) was used to disambiguate multiple morphological parses. Both the lexical and surface form representations of morphemes, stems and endings were used as linguistic sub-lexical units for Turkish. The details of these units are given in the following sections.

4.5.1.1 Surface Form Stem+Ending Model

Instead of using words as vocabulary items as in the word-based model, the surface form stem+ending model uses a vocabulary comprising surface form stem and endings and the words in the text data are split into their stems and endings. This is done by first extracting the stem from morphological analyses and taking the remaining part of the word as the ending.

In this approach, no morphological disambiguation was done. Instead Arısoy et al. (2009a) investigated building language models with all the ambiguous parses, with the parses with the smallest number of morphemes, and with randomly selected parses for each word token and type. They found no significant difference between the first two methods and these fared better than random choice of a parse. Sak et al. (2010) showed that utilizing the parse with the smallest number of morphemes performed slightly better than using the disambiguated parse in Turkish ASR. The method of selecting the parse with the smallest number of morphemes is not only extremely simple but also avoids more complex and error-prone approaches such as morphological disambiguation.

4.5.1.2   Lexical Form Stem+Ending Model

Morpholexical language models are trained as standard n-gram language models over morpholexical units. The one important advantage of using morpholexical units is that they allow conflating different surface forms of morphemes to one underlying form thereby alleviating the sparseness problem. For instance, the plural in Turkish is indicated by surface morphemes *+ler* or *+lar*, depending on the phonological (and not morphological) context. Thus representing these morphemes with a single lexical morpheme *+lAr* allows counts to be combined leading to more robust parameter estimation. Combining lexical morphemes also naturally leads to lexical stem+ending models (Arısoy et al. 2007).

Morpholexical language models have the advantage that they give probability estimates for sequences consisting of only valid words, that is they do not over-generate like the other sub-lexical models. Sak et al. (2012) have demonstrated the importance of both morphotactics and morphological disambiguation when producing the morpholexical units used for language modeling.

4.5.2   Statistical Sub-lexical Units: Morphs

As discussed earlier, statistical sub-lexical units obtained via unsupervised word segmentation algorithms have been used as an alternative to linguistic sub-lexical units. In fact, Turkish has been a part of the Morpho Challenge since 2007.[3]

Hacıoğlu et al. (2003) were the first to model Turkish with statistical sub-lexical units obtained with the Morfessor algorithm and showed that they outperform a word-based model with 60K word vocabulary, even though the language models were built on a text corpus containing only 2M words. Arısoy et al. (2009a) used statistical sub-lexical units for extensive experimentation using large corpora for Turkish ASR.

Arısoy et al. (2009b) proposed an enhanced Morfessor algorithm with phonetic features for Turkish. The main idea in this work was to incorporate simple phonetic knowledge of Turkish into Morfessor in order to improve the segmentations. Two main modifications were made to enhance Morfessor: a phone-based feature, called "DF" for distinctive feature, and a constraint called 'Cc' for confusion constraint. DF was directly incorporated into Morfessor's probability estimates and Cc was indirectly incorporated into Morfessor as a yes/no decision in accepting candidate splits. Both of these modifications aimed at reducing the number of confusable morphs in the segmentations by taking phonetic and syllable confusability into account.

---

[3] Aalto University, Finland. Department of Computer Science. "Morpho Challenge: Results": morpho.aalto.fi/events/morphochallenge/results-tur.html (Accessed Sept. 14, 2017).

4.6 Discriminative Language Modeling for Turkish

Recent ASR and MT systems utilize discriminative training methods on top of traditional generative models. The advantage of discriminative parameter estimation over generative parameter estimation is that discriminative training takes alternative (negative) examples into account as well as the correct (positive) examples. While generative training estimates a model that can generate the positive examples, discriminative training estimates model parameters that discriminate the positive examples from the negative ones. In ASR and MT tasks, positive examples are the correct transcriptions or translations and negative examples are the erroneous candidate transcriptions or translations. Discriminative models utilize these examples to optimize an objective function that is directly related to the system performance. Discriminative acoustic model training for ASR utilizes objective functions like Maximum Mutual Information (MMI) (Povey and Woodland 2000; Bahl et al. 1986) and Minimum Phone Error (MPE) (Povey and Woodland 2002) to estimate the acoustic model parameters that represent the discrimination between alternative classes. Discriminative language model (DLM) training for ASR aims to optimize the WER while learning the model parameters that discriminate the correct transcription of an utterance from the other candidate transcriptions. Another advantage of DLM is that discriminative language modeling is a feature-based approach, like conditional random fields (CRFs) (Lafferty et al. 2001) and maximum entropy models (Berger et al. 1996), therefore, it allows for easy integration of relevant knowledge sources, such as morphology, syntax, and semantics, into language modeling. As a result of improved parameter estimation with discriminative training and ease of incorporating overlapping features, discriminatively trained language models have been demonstrated to consistently outperform generative language modeling approaches (Roark et al. 2007, 2004; Collins et al. 2005; Shafran and Hall 2006).

In this section we will briefly explain the DLMs and the linguistically and statistically motivated features extracted at lexical and sub-lexical levels for Turkish DLMs.

4.6.1 Discriminative Language Model

This section describes the framework for discriminatively trained language models used for ASR. The definitions and notations given in Roark et al. (2007) are modified to match the definitions and notations of this chapter. The main components of DLMs are as follows:

1. **Training Examples:** These are the pairs (X_i, W_i) for $i = 1 \ldots N$. Here, X_i are the utterances and W_i are the corresponding reference transcriptions.

2. **GEN(X)**: For each utterance X, this function enumerates a finite set of alternative hypotheses, represented as a lattice or N-best list output of the baseline ASR system of that utterance.
3. **$\Phi(X, W)$**: A d-dimensional real-valued feature vector ($\Phi(X, W) \in \Re^d$). The representation Φ defines the mapping from the (X, W) pair to the feature vector $\Phi(X, W)$. When the feature depends only on W, we simplify the notation to $\Phi(W)$ to match the notation used for other feature-based language models.
4. **$\bar{\alpha}$**: A vector of discriminatively learned feature parameters ($\bar{\alpha} \in \Re^d$).

Like many other supervised learning approaches, DLM requires labeled input:output pairs as the training examples. Utterances with the reference transcriptions are utilized as the training examples, $(X_1, W_1) \ldots (X_N, W_N)$. These utterances are decoded with the baseline acoustic and language models in order to obtain the lattices or the N-best lists, in other words, the output of the $GEN(X)$ function. Since speech data with transcriptions are limited compared to the text data, it may not be possible to train the baseline acoustic and in-domain language models, and the DLM on separate corpora. Therefore, DLM training data is generated by breaking the acoustic training data into k-folds, and recognizing the utterances in each fold using the baseline acoustic model (trained on all of the utterances) and an n-gram language model trained on the other $k - 1$-folds to alleviate over-training of the language models. Acoustic model training is more expensive and less prone to over-training than n-gram language model training (Roark et al. 2007), so it is not typically controlled in the same manner.

Discriminative language modeling is a feature-based sequence modeling approach, where each element of the feature vector, $\Phi_0(X, W) \ldots \Phi_{d-1}(X, W)$, corresponds to a different feature. Each candidate hypothesis of an utterance has a score from the baseline acoustic and language models. This score is used as the first element of the feature vector, $\Phi_0(X, W)$. This feature is defined as the "log-probability of W in the lattice produced by the baseline recognizer for utterance X." In the scope of this chapter, the rest of the features depend only on W and will be denoted by $\Phi(W)$. The basic approach for the other DLM features is to use n-grams in defining features. The n-gram features are defined as the number of times a particular n-gram is seen in the candidate hypothesis. The details of the features used in Turkish DLMs will be explained in Sect. 4.6.2.

Each DLM feature has an associated parameter, i.e., α_i for $\Phi_i(X, W)$. The best hypothesis under the $\bar{\alpha}$ model, W^*, maximizes the inner product of the feature and the parameter vectors, as given in Eq. (4.10). The values of $\bar{\alpha}$ are learned in training and the best hypothesis under this model is searched for in decoding.

$$W^* = \operatorname*{argmax}_{W \in GEN(X)} \langle \bar{\alpha}, \Phi(X, W) \rangle$$

$$= \operatorname*{argmax}_{W \in GEN(X)} (\alpha_0 \Phi_0(X, W) + \alpha_1 \Phi_1(W) + \cdots + \alpha_{d-1} \Phi_{d-1}(W)) \quad (4.10)$$

Fig. 4.4 A variant of the perceptron algorithm given in Roark et al. (2007). $\bar{\alpha}_t^i$ represents the feature parameters after the tth pass on the ith example. R_i is the gold-standard hypothesis

Inputs: Training examples (X_i, R_i) for $i = 1 \ldots N$
Initialization: $\bar{\alpha}_0^N = (\alpha_0, 0, \ldots, 0)$
Algorithm:
  For $t = 1 \ldots T$
    $\bar{\alpha}_t^0 = \bar{\alpha}_{t-1}^N$
    For $i = 1 \ldots N$
      $W^* = \underset{W \in GEN(X_i)}{\mathrm{argmax}} \; \langle \bar{\alpha}_t^{i-1}, \bar{\Phi}(X_i, W) \rangle$
      $\bar{\alpha}_t^i = \bar{\alpha}_t^{i-1} + \Phi(X_i, R_i) - \Phi(X_i, W^*)$
Output: Averaged parameters $\bar{\alpha} = \sum_{i,t} \bar{\alpha}_t^i / NT$

In basic DLM training, the parameters are estimated using a variant of the perceptron algorithm (shown in Fig. 4.4). The main idea in this algorithm is to penalize features associated with the current 1-best hypothesis, and to reward features associated with the gold-standard hypothesis (reference or lowest-WER hypothesis). It has been found that the perceptron model trained with the reference transcription as the gold-standard hypothesis is much more sensitive to the value of the α_0 constant (Roark et al. 2007). Therefore, we use the lowest-WER hypothesis (oracle) as the gold-standard hypothesis. Averaged parameters, $\bar{\alpha}_{AVG}$, are utilized in decoding held-out and test sets, since averaged parameters have been shown to outperform regular perceptron parameters in tagging tasks and also give much greater stability of the tagger (Collins 2002). See Roark et al. (2007) for the details of the training algorithm.

4.6.2 Feature Sets for Turkish DLM

This section describes the feature sets utilized in Turkish DLM experiments in the context of ASR (Arısoy et al. 2012; Sak et al. 2012). In order to generate the negative examples, we used a baseline Turkish ASR system to decode the DLM training set utterances yielding an N-best list for each training utterance. We then extracted the features from the correct transcriptions of the utterances together with the N-best list outputs of the baseline ASR system. In this section we investigate linguistically and statistically motivated features in addition to the basic n-gram features extracted from the word and sub-lexical ASR hypotheses.

4.6.2.1 Basic n-Gram Features

The basic n-gram features consist of word n-gram features extracted from word ASR hypotheses and sub-lexical n-gram features extracted from sub-lexical ASR hypotheses. Consider the Turkish phrase "derneklerinin öncülüğünde" given in Fig. 4.3. The unigram and bigram word features extracted from this phrase are as follows:

$\Phi_i(W)$ = number of times "derneklerinin" is seen in W
$\Phi_j(W)$ = number of times "öncülüğünde" is seen in W
$\Phi_k(W)$ = number of times "derneklerinin öncülüğünde" is seen in W

We use a statistical morph-based ASR system to obtain the sub-lexical ASR hypotheses from which we extract the basic sub-lexical n-gram features. Some examples of the morph unigram and bigram features for the phrase in Fig. 4.3 are given as follows:

$\Phi_i(W)$ = number of times "dernek" is seen in W
$\Phi_j(W)$ = number of times "+lerinin" is seen in W
$\Phi_k(W)$ = number of times "öncü +lüğü" is seen in W

where the non-initial morphs were marked with "+" in order to find the word boundaries easily after recognition.

4.6.2.2 Linguistically Motivated Features

This section describes the morphological and syntactic features utilized in Turkish DLM. The rich morphological structure of Turkish introduces challenges for ASR systems (see Sect. 4.3). We aim to turn this challenging structure into a useful information source when reranking N-best word hypotheses with DLMs. Therefore, we utilize information extracted from morphological decompositions as DLM features. In our work, we have used root and stem+ending n-grams as the morphological features. In order to obtain the features, we first morphologically analyzed and disambiguated all the words in the hypothesis sentences using a morphological parser (Sak et al. 2011). The words that cannot be analyzed with the parser are left unparsed and represented as nominal nouns.

In order to obtain the root n-gram features, we first replace the words in the hypothesis sentences with their roots using the morphological decompositions. Then we generate the n-gram features as before, treating the roots as words. The root unigram and bigram features, with examples from Fig. 4.3, are listed below:

$\Phi_i(W)$ = number of times "dernek" is seen in W
$\Phi_j(W)$ = number of times "öncü" is seen in W
$\Phi_k(W)$ = number of times "dernek öncü" is seen in W

For the stem+ending n-gram features, we first extract the stem from the morphological decomposition and take the remaining part of the word as the ending. If there is no ending in the word, we use a special symbol to represent the empty ending. After converting the hypothesis sentences to stem and ending sequences, we generate the n-gram features in the same way with words as if stems and endings were words. The stem+ending unigram and bigram features, with examples from

Fig. 4.3, are listed below:

$\Phi_j(W)$ = number of times "+lerinin" is seen in W
$\Phi_k(W)$ = number of times "öncü +lüğünde" is seen in W

Syntax is an important information source for language modeling due to its role in sentence formation. Syntactic information has been incorporated into conventional generative language models using left-to-right parsers to capture long distance dependencies in addition to $n-1$ previous words (Chelba and Jelinek 2000; Roark 2001). Feature-based reranking approaches (Collins et al. 2005; Rosenfeld et al. 2001; Khudanpur and Wu 2000) also make use of syntactic information. The success of these approaches lead us to investigate syntactic features for Turkish DLMs.

For the syntactic DLM features, we explored feature definitions similar to Collins et al. (2005). We used part-of-speech tag n-grams and head-to-head (H2H) dependency relations between lexical items or their part-of-speech tags as the syntactic features. Part-of-speech tag features were utilized in an effort to obtain class-based generalizations that may capture well-formedness tendencies. H2H dependency relations were utilized since the presence of a word or morpheme can depend on the presence of another word or morpheme in the same sentence and this information is represented in the dependency relations.

The syntactic features will be explained with the dependency analysis given in Fig. 4.5 for a Turkish sentence, which translates as "Patrol services will also be increased throughout the city." The incoming and outgoing arrows in the figure show the dependency relations between the head and the dependent words with the type of the dependency. The words with English glosses, part-of-speech tags associated with the words are also given in the example. The dependency parser by Eryiğit et al. (2008) was used for the dependency analysis.

To obtain the syntactic features from the training examples, we first generated the dependency analyses of hypothesis sentences. Then we extracted the part-of-speech tag and H2H features from these dependency analyses. Here, it is important to note that hypothesis sentences contain recognition errors and the parser generates

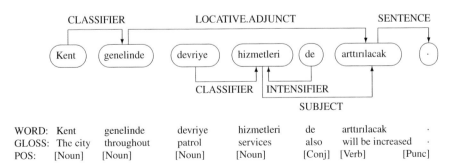

Fig. 4.5 Example dependency analysis for syntactic features

the best possible dependency relations even for incorrect hypotheses. The syntactic features are listed below with examples from Fig. 4.5.

- Part-of-speech tag n-gram features:
  Example for the word 'Kent':
  $\Phi_k(W)$ = number of times " [Noun] " is seen in W
  Example for the words 'hizmetleri de':
  $\Phi_k(W)$ = number of times " [Noun] [Conj] " is seen in W

- Head-to-Head (H2H) dependencies:
  Examples for the words 'Kent genelinde':
  - dependencies between lexical items:
    $\Phi_k(W)$ = number of times "CLASSIFIER Kent genelinde" is seen in W
  - dependencies between a single lexical item and the part-of-speech of another item:
    $\Phi_k(W)$ = number of times "CLASSIFIER Kent [Noun]" is seen in W
    $\Phi_l(W)$ = number of times "CLASSIFIER [Noun] genelinde" is seen in W
  - dependencies between part-of-speech tags of lexical items:
    $\Phi_k(W)$ = number of times "CLASSIFIER [Noun] [Noun]" is seen in W

4.6.2.3 Statistically Motivated Features

The advantage of statistical sub-lexical units compared to their linguistic counterparts is that they do not require linguistic knowledge for word segmentation. As a result, statistical morphs do not convey explicit linguistic information like morphemes and obtaining linguistic information from morph sequences is not obvious. One way of information extraction from morphs is to convert them into word-like units and to apply the same procedure with words. However, this indirect approach tends to fail when concatenation of morph sequences does not generate morphologically correct words. In addition, this approach contradicts with the main idea of statistical morphs—obtaining sub-lexical units without any linguistic tools. Therefore, we focused on exploring representative features of implicit morpho-syntactic information in morph sequences. We explored morph-based features similar to part-of-speech tag and H2H dependency features using data driven approaches.

The first feature set is obtained by clustering morphs. We applied two hierarchical clustering approaches on morphs to obtain meaningful categories. The first one is the algorithm by Brown et al. (1992) which aims to cluster words into semantically-based or syntactically-based groupings by maximizing the average mutual information of adjacent classes. Brown et al.'s algorithm is proposed for class-based n-gram language models and the optimization criterion in clustering is directly related to the n-gram language model quality. Utilizing n-gram features in DLMs makes this clustering an attractive approach for our investigation. The second approach utilizes minimum edit distance (MED) as the similarity function in bottom-up clustering. The motivation in this algorithm is to capture the syntactic similarity of morphs using their graphemic similarities, since a non-initial morph can cover a linguistic morpheme, a group of morphemes or pieces of morphemes. In our application, we modify MED to softly penalize the variations in the lexical and surface forms of morphemes. Note that this clustering is only meaningful for non-initial morphs since graphemic similarity of initial morphs does not reveal any linguistic information. Therefore, we only cluster the non-initial morphs and all the initial morphs are assigned to the same cluster with MED-based clustering approach.

Clustering is applied to morph sequences and each morph is assigned to one of the predetermined number of classes. The class associated with a particular morph is considered as the tag of that morph and utilized in defining DLM features. Clustering-based features are defined in a similar way with part-of-speech tag n-gram features, the class labels of morphs playing the role of the part-of-speech tags of words.

The second feature set is obtained with the triggering information obtained from morph sequences. These features are motivated by the H2H dependency features in words. Considering initial morphs as stems and non-initial morphs as suffixes, we assume that the existence of a morph can trigger another morph in the same sentence. The morphs in trigger pairs are believed to co-occur for a syntactic function, like the syntactic dependencies of words, and these pairs are utilized to define the long distance morph trigger features. Long distance morph trigger features are similar to the trigger features proposed in Rosenfeld (1994) and Singh-Miller and Collins (2007). We only consider sentence level trigger pairs to capture the syntactic-level dependencies instead of discourse-level information. The candidate morph trigger pairs are extracted from the hypothesis sentences (1-best and oracle) to obtain also the negative examples for DLMs. An example morph hypothesis sentence with the candidate trigger pairs is given in Fig. 4.6. Among the possible candidates, we try to select only the pairs where morphs are occurring together for a special function. This is formulated with hypothesis testing where the null hypothesis (H_0) represents the independence and the alternative hypothesis

Morph hypothesis:
```
dernek +lerinin öncü +lüğü +nde
```
Candidate trigger pairs:
```
dernek +lerinin   dernek öncü   dernek +lüğü   dernek +nde
        +lerinin öncü   +lerinin +lüğü   +lerinin +nde
                 öncü +lüğü   öncü +nde
                        +lüğü +nde
```

Fig. 4.6 A morph hypothesis sentence and the candidate trigger pairs extracted from this hypothesis

(H_1) represents the dependence assumptions of morphs in the pairs (Manning and Schütze 1999). The pairs with higher likelihood ratios (log $\frac{L(H_1)}{L(H_0)}$) are assumed to be the morph triggers and utilized as features. The number of times a morph trigger pair is seen in the candidate hypothesis is defined as long-distance trigger features. For instance, if among the candidate trigger pairs, given in Fig. 4.6, "öncü +lüğü" is selected as a trigger pair, the feature for this pair is defined as follows:

$\Phi_k(W)$ = number of times "öncü +lüğü" is seen in W

4.7 Conclusions

In this chapter, we summarized the language modeling research for Turkish text and speech processing applications. The agglutinative nature of Turkish results in high OOV rates which can be alleviated by using sub-lexical units for language modeling. Knowledge-based linguistic methods and data-driven unsupervised statistical methods have both been used for segmenting words into sub-lexical units. Language models based on these units have advantages of those based on words and often result in improved performance. After many years of research, n-gram language models are still the most popular language modeling technique. However, in certain applications such as ASR, discriminative language models have been shown to improve the task performance. The ASR performance of the language models presented in this chapter is provided in Chap. 5. Despite significant progress in the recent years, language modeling for morphologically rich languages such as Turkish remains an active field of research.

References

Alumäe T (2006) Methods for Estonian large vocabulary speech recognition. PhD thesis, Tallinn University of Technology, Tallinn

Arısoy E (2004) Turkish dictation system for radiology and broadcast news applications. Master's thesis, Boğaziçi University, Istanbul

Arısoy E, Saraçlar M (2009) Lattice extension and vocabulary adaptation for Turkish LVCSR. IEEE Trans Audio Speech Lang Process 17(1):163–173

Arısoy E, Dutağacı H, Arslan LM (2006) A unified language model for large vocabulary continuous speech recognition of Turkish. Signal Process 86(10):2844–2862

Arısoy E, Sak H, Saraçlar M (2007) Language modeling for automatic Turkish broadcast news transcription. In: Proceedings of INTERSPEECH, Antwerp, pp 2381–2384

Arısoy E, Can D, Parlak S, Sak H, Saraçlar M (2009a) Turkish broadcast news transcription and retrieval. IEEE Trans Audio Speech Lang Process 17(5):874–883

Arısoy E, Pellegrini T, Saraçlar M, Lamel L (2009b) Enhanced Morfessor algorithm with phonetic features: application to Turkish. In: Proceedings of SPECOM, St. Petersburg

Arısoy E, Saraçlar M, Roark B, Shafran I (2012) Discriminative language modeling with linguistic and statistically derived features. IEEE Trans Audio Speech Lang Process 20(2):540–550

Bahl L, Brown P, deSouza P, Mercer R (1986) Maximum mutual information estimation of Hidden Markov Model parameters for speech recognition. In: Proceedings of ICASSP, Tokyo, pp 49–52

Bayer AO, Çiloğlu T, Yöndem MT (2006) Investigation of different language models for Turkish speech recognition. In: Proceedings of IEEE signal processing and communications applications conference, Antalya, pp 1–4

Bengio Y, Ducharme R, Vincent P, Jauvin C (2003) A neural probabilistic language model. J Mach Learn Res 3:1137–1155

Berger AL, Della Pietra SD, Della Pietra VJD (1996) A maximum entropy approach to natural language processing. Comput Linguist 22(1):39–71

Bilmes JA, Kirchhoff K (2003) Factored language models and generalized parallel backoff. In: Proceedings of NAACL-HLT, Edmonton, pp 4–6

Brent MR (1999) An efficient, probabilistically sound algorithm for segmentation and word discovery. Mach Learn 34:71–105

Brown PF, Pietra VJD, deSouza PV, Lai JC, Mercer RL (1992) Class-based n-gram models of natural language. Comput Linguist 18(4):467–479

Çarkı K, Geutner P, Schultz T (2000) Turkish LVCSR: towards better speech recognition for agglutinative languages. In: Proceedings of ICASSP, Istanbul, pp 1563–1566

Chelba C, Jelinek F (2000) Structured language modeling. Comput Speech Lang 14(4):283–332

Chen SF, Goodman J (1999) An empirical study of smoothing techniques for language modeling. Comput Speech Lang 13(4):359–394

Choueiter G, Povey D, Chen SF, Zweig G (2006) Morpheme-based language modeling for Arabic. In: Proceedings of ICASSP, Toulouse, pp 1052–1056

Çiloğlu T, Çömez M, Şahin S (2004) Language modeling for Turkish as an agglutinative language. In: Proceedings of IEEE signal processing and communications applications conference, Kuşadası, pp 461–462

Collins M (2002) Discriminative training methods for Hidden Markov Models: theory and experiments with perceptron algorithms. In: Proceedings of EMNLP, Philadelphia, PA, pp 1–8

Collins M, Saraçlar M, Roark B (2005) Discriminative syntactic language modeling for speech recognition. In: Proceedings of ACL, Ann Arbor, MI, pp 507–514

Creutz M, Lagus K (2002) Unsupervised discovery of morphemes. In: Proceedings of the workshop on morphological and phonological learning, Philadelphia, PA, pp 21–30

Creutz M, Lagus K (2005) Unsupervised morpheme segmentation and morphology induction from text corpora using Morfessor 1.0. Publications in Computer and Information Science Report A81, Helsinki University of Technology, Helsinki

Dutağacı H (2002) Statistical language models for large vocabulary continuous speech recognition of Turkish. Master's thesis, Boğaziçi University, Istanbul

Erdoğan H, Büyük O, Oflazer K (2005) Incorporating language constraints in sub-word based speech recognition. In: Proceedings of ASRU, San Juan, PR, pp 98–103

Eryiğit G, Nivre J, Oflazer K (2008) Dependency parsing of Turkish. Comput Linguist 34(3):357–389

Goldsmith J (2001) Unsupervised learning of the morphology of a natural language. Comput Linguist 27(2):153–198

Hacıoğlu K, Pellom B, Çiloğlu T, Öztürk Ö, Kurimo M, Creutz M (2003) On lexicon creation for Turkish LVCSR. In: Proceedings of EUROSPEECH, Geneva, pp 1165–1168

Hakkani-Tür DZ (2000) Statistical language modeling for agglutinative languages. PhD thesis, Bilkent University, Ankara

Harris Z (1967) Morpheme boundaries within words: report on a computer test. In: Transformations and discourse analysis papers, vol 73. University of Pennsylvania, Philadelphia, PA

Hetherington IL (1995) A characterization of the problem of new, out-of-vocabulary words in continuous-speech recognition and understanding. PhD thesis, Massachusetts Institute of Technology, Cambridge, MA

Hirsimäki T (2009) Advances in unlimited vocabulary speech recognition for morphologically rich languages. PhD thesis, Helsinki University of Technology, Espoo

Hirsimäki T, Creutz M, Siivola V, Kurimo M, Virpioja S, Pylkkönen J (2006) Unlimited vocabulary speech recognition with morph language models applied to Finnish. Comput Speech Lang 20(4):515–541

Jelinek F (1997) Statistical methods for speech recognition. The MIT Press, Cambridge, MA

Kanevsky D, Roukos S, Sedivy J (1998) Statistical language model for inflected languages. US patent No: 5,835,888

Khudanpur S, Wu J (2000) Maximum entropy techniques for exploiting syntactic, semantic and collocational dependencies in language modeling. Comput Speech Lang 14:355–372

Kirchhoff K, Vergyri D, Bilmes J, Duh K, Stolcke A (2006) Morphology-based language modeling for conversational Arabic speech recognition. Comput Speech Lang 20(4):589–608

Kneissler J, Klakow D (2001) Speech recognition for huge vocabularies by using optimized sub-word units. In: Proceedings of INTERSPEECH, Aalborg, pp 69–72

Kurimo M, Puurula A, Arısoy E, Siivola V, Hirsimäki T, Pylkkönen J, Alumäe T, Saraçlar M (2006) Unlimited vocabulary speech recognition for agglutinative languages. In: Proceedings of NAACL-HLT, New York, NY, pp 487–494

Kwon OW, Park J (2003) Korean large vocabulary continuous speech recognition with morpheme-based recognition units. Speech Comm 39:287–300

Lafferty JD, McCallum A, Pereira F (2001) Conditional random fields: probabilistic models for segmenting and labeling sequence data. In: Proceedings of ICML, Williams, MA, pp 282–289

Manning C, Schütze H (1999) Foundations of statistical natural language processing. MIT Press, Cambridge, MA

Mengüşoğlu E, Deroo O (2001) Turkish LVCSR: database preparation and language modeling for an agglutinative language. In: Proceedings of ICASSP, Salt Lake City, UT, pp 4018–4021

Mihajlik P, Fegyò T, Tüske Z, Ircing P (2007) A morpho-graphemic approach for the recognition of spontaneous speech in agglutinative languages like Hungarian. In: Proceedings of INTER-SPEECH, Antwerp, pp 1497–1500

Pellegrini T, Lamel L (2007) Using phonetic features in unsupervised word decompounding for ASR with application to a less-represented language. In: Proceedings of INTERSPEECH, Antwerp, pp 1797–1800

Pellegrini T, Lamel L (2009) Automatic word decompounding for ASR in a morphologically rich language: application to amharic. IEEE Trans Audio Speech Lang Process 17(5):863–873

Podvesky P, Machek P (2005) Speech recognition of Czech – inclusion of rare words helps. In: Proceedings of the ACL student research workshop, Ann Arbor, MI, pp 121–126

Povey D, Woodland PC (2000) Large-scale MMIE training for conversational telephone speech recognition. In: Proceedings of NIST speech transcription workshop, College Park, MD

Povey D, Woodland PC (2002) Minimum phone error and i-smoothing for improved discriminative training. In: Proceedings of ICASSP, Orlando, FL, pp 105–108

Roark B (2001) Probabilistic top-down parsing and language modeling. Comput Linguist 27(2):249–276

Roark B, Saraçlar M, Collins MJ, Johnson M (2004) Discriminative language modeling with conditional random fields and the perceptron algorithm. In: Proceedings of ACL, Barcelona, pp 47–54

Roark B, Saraçlar M, Collins M (2007) Discriminative n-gram language modeling. Comput Speech Lang 21(2):373–392

Rosenfeld R (1994) Adaptive statistical language modeling: a maximum entropy approach. PhD thesis, Carnegie Mellon University, Pittsburgh, PA

Rosenfeld R (1995) Optimizing lexical and n-gram coverage via judicious use of linguistic data. In: Proceedings of EUROSPEECH, Madrid, pp 1763–1766

Rosenfeld R, Chen SF, Zhu X (2001) Whole-sentence exponential language models: a vehicle for linguistic-statistical integration. Comput Speech Lang 15(1):55–73

Rotovnik T, Maučec MS, Kačic Z (2007) Large vocabulary continuous speech recognition of an inflected language using stems and endings. Speech Commun 49(6):437–452

Sak H, Güngör T, Saraçlar M (2007) Morphological disambiguation of Turkish text with perceptron algorithm. In: Proceedings of CICLING, Mexico City, pp 107–118

Sak H, Saraçlar M, Güngör T (2010) Morphology-based and sub-word language modeling for Turkish speech recognition. In: Proceedings of ICASSP, Dallas, TX, pp 5402–5405

Sak H, Güngör T, Saraçlar M (2011) Resources for Turkish morphological processing. Lang Resour Eval 45(2):249–261

Sak H, Saraçlar M, Güngör T (2012) Morpholexical and discriminative language models for Turkish automatic speech recognition. IEEE Trans Audio Speech Lang Process 20(8):2341–2351

Schwenk H (2007) Continuous space language models. Comput Speech Lang 21(3):492–518

Shafran I, Hall K (2006) Corrective models for speech recognition of inflected languages. In: Proceedings of EMNLP, Sydney, pp 390–398

Siivola V, Hirsimäki T Teemu, Creutz M, Kurimo M (2003) Unlimited vocabulary speech recognition based on morphs discovered in an unsupervised manner. In: Proceedings of EUROSPEECH, Geneva, pp 2293–2296

Singh-Miller N, Collins M (2007) Trigger-based language modeling using a loss-sensitive perceptron algorithm. In: Proceedings of ICASSP, Honolulu, HI, pp 25–28

Wang W, Harper MP (2002) The SuperARV language model: investigating the effectiveness of tightly integrating multiple knowledge sources. In: Proceedings of EMNLP, Philadelphia, PA, pp 238–247

Whittaker E, Woodland P (2000) Particle-based language modelling. In: Proceedings of ICSLP, Beijing, vol 1, pp 170–173

Chapter 5
Turkish Speech Recognition

Ebru Arısoy and Murat Saraçlar

Abstract Automatic speech recognition (ASR) is one of the most important applications of speech and language processing, as it forms the bridge between spoken and written language processing. This chapter presents an overview of the foundations of ASR, followed by a summary of Turkish language resources for ASR and a review of various Turkish ASR systems. Language resources include acoustic and text corpora as well as linguistic tools such as morphological parsers, morphological disambiguators, and dependency parsers, discussed in more detail in other chapters. Turkish ASR systems vary in the type and amount of data used for building the models. The focus of most of the research for Turkish ASR is the language modeling component covered in Chap. 4.

5.1 Introduction

Automatic Speech Recognition (ASR) aims to produce a transcription of a given speech input by finding the most likely word sequence corresponding to a spoken utterance. There are various ASR tasks, such as digit and command recognition, dictation, and transcription of telephone conversations or broadcast news. As smartphones have become popular, ASR has started playing an important role as an input method for mobile devices. The difficulty of these tasks depends on the vocabulary size (small or large), speaking mode (isolated or continuous), speaking style (planned or spontaneous speech), acoustic environment (clean or noisy), and recording conditions (telephone, close-talking microphone, far-field microphone). For instance, large vocabulary, continuous and spontaneous telephone

E. Arısoy
MEF University, Istanbul, Turkey
e-mail: ebruarisoy.saraclar@mef.edu.tr

M. Saraçlar (✉)
Boğaziçi University, Istanbul, Turkey
e-mail: murat.saraclar@boun.edu.tr

© Springer International Publishing AG, part of Springer Nature 2018
K. Oflazer, M. Saraçlar (eds.), *Turkish Natural Language Processing*,
Theory and Applications of Natural Language Processing,
https://doi.org/10.1007/978-3-319-90165-7_5

speech recognition under noisy acoustic conditions is a much more difficult task than small vocabulary isolated word recognition of speech recorded with a close-talking microphone in a quiet room.

A recent and popular application of ASR is voice search, where a speech query is used to search the internet, as in Google Voice Search (Schalkwyk et al. 2010), or to answer a question, as in intelligent personal assistant applications such as Apple's Siri or Microsoft's Cortana. ASR is also one of the main components in spoken document retrieval systems which aim to retrieve audio clips related to a query, in spoken term detection systems, which aim to locate occurrences of a term in a spoken archive, and in spoken dialogue systems, which aim to accomplish a task using spoken language interaction. Additionally, speech-to-speech translation systems also require ASR technology to obtain accurate transcriptions for the input to the machine translation system.

In this chapter, we first explain the basics of an ASR system. Then, we review available Turkish speech and text resources required for building a Turkish ASR system. Finally, we summarize the research on Turkish ASR systems developed for transcribing telephone conversations, dictating read speech, and transcribing television broadcast news.

5.2 Foundations of Automatic Speech Recognition

The state-of-the-art approach for ASR can be explained with the source-channel model (Jelinek 1997). This model was inspired by communication theory and illustrates the path from the source to the target. Here, the source is the speaker's mind and the target is the hypothesized word sequence $\hat{W}$ corresponding to the word sequence W uttered by the speaker. First, the speaker decides on a word string to utter and utters it. Then, this word string passes through a noisy acoustic channel which is composed of the human speech production mechanism, the transmission channel, and the acoustic processor (front-end) parts. The speech produced by the speaker is converted into the acoustic observations, A, by the front-end. The linguistic decoder picks the most likely word string corresponding to the speech input using statistical acoustic and language models. It is important to note that although the model is inspired by communication theory, since there is no way of modifying the encoder, an ASR system only consists of a pre-processor and the decoder.

The words in the word sequence W are taken from the vocabulary of a language. For ASR purposes, the vocabulary, V, is assumed to be fixed and finite. The vocabulary size, $|V|$, can be as small as two words for recognition of yes/no answers to questions or 10 words for digit recognition and as large as thousands of words for dictation systems or millions of words for voice-enabled search applications. Note that the decoder can only recognize the words in its vocabulary. Therefore, the vocabulary of the ASR system should match with the words that the speaker utters. However, this is not easy and not always possible, since the speakers' word choices for a dictation system for a medical domain are completely different than the choices for a spoken dialog system developed for flight information. If the speaker utters a word which is not in the recognition vocabulary, this word is called an Out-of-Vocabulary (OOV) word and in the ASR output it will be substituted with one of the words in the recognition vocabulary.

The **acoustic processor (front-end)** is one of the main components of an ASR system and is required to convert the speech signal into a form that can be processed by digital computers. The basic approach in ASR is to use Mel Frequency Cepstral Coefficients (MFCC) as the features. Perceptual Linear Prediction (PLP) based features are also commonly used as they provide better robustness. The details of front-end design can be found in Jurafsky and Martin (2000) and Huang et al. (2001). With the recent developments in deep learning, the boundary between the front-end and the acoustic model has been blurred. When deep neural networks are used as the acoustic model, the input can be as simple as the log-spectral features or even the raw waveform (Tuske et al. 2014).

The ASR problem is stated as finding the most probable word string among all possible word strings given the input speech and mathematically formulated as follows:

$$\widehat{W} = \underset{W}{\mathrm{argmax}} \, P(W|A) \tag{5.1}$$

Using the Bayes' formula, Eq. (5.1) can be rewritten as

$$\widehat{W} = \underset{W}{\mathrm{argmax}} \, \frac{P(A|W)P(W)}{P(A)} = \underset{W}{\mathrm{argmax}} \, P(A|W)P(W) \tag{5.2}$$

which is known as "the fundamental equation of speech recognition." Note that Eq. (5.2) is simplified by ignoring $P(A)$, since it is fixed and has no effect when finding the word string maximizing the product of the acoustic model probability, $P(A|W)$, and the language model probability, $P(W)$.

In ASR systems, a simple approach for **acoustic modeling** is to model each phone in a language using a 3-state left-to-right Hidden Markov Model (HMM). Figure 5.1 shows an example phone model for the phone "æ". The composite HMM for a word is the concatenation of the phone models. The basic tool for mapping words to phones is a pronunciation lexicon consisting of the correct pronunciations of words. Figure 5.2 shows a composite HMM for the word "cat" pronounced as "k æ t". The HMM parameters are trained using a large set of acoustic training data, A, and the corresponding transcriptions, W. Most ASR systems use context-dependent phone (typically triphone) models instead of context-independent phone

Fig. 5.1 3-state HMM model for phone "æ"

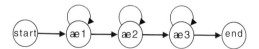

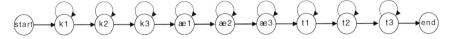

Fig. 5.2 A composite HMM model for the word "cat" pronounced as "k æ t"

models to take the right and the left context of the phones into account. For instance, the possible triphones for the word "cat" are as follows: "ε-k+æ", "k-æ+t", "æ-t+ε" where "-" sign represents the left context, "+" sign represents the right context, and ε represents the word boundary. In order to handle the data sparsity due to context-dependency, the HMM states are clustered using decision trees. Traditionally, the output distribution of each HMM state is modeled by Gaussian Mixture Models (GMMs), but the GMMs are being replaced or complemented by deep neural networks (DNNs), following the recent developments in deep learning. See Jelinek (1997), Jurafsky and Martin (2000), Huang et al. (2001), Rabiner (1989) for details of HMMs and acoustic models and Hinton et al. (2012) for application of DNNs to ASR.

The most common approach for **language modeling** is to use n-gram language models that provide a probability estimate for the current word given the previous $n-1$ words in the history. These models assign probabilities to all the word strings in a language and their parameters are learned from a large text corpus related to the ASR domain. Traditionally, these models are trained using maximum likelihood estimation followed by smoothing. However, neural network language models (Bengio et al. 2003; Schwenk 2007; Mikolov et al. 2010) are also gaining popularity (Arısoy et al. 2012). More details on the fundamentals of language modeling can be found in Chap. 4.

The **decoder**, corresponding to the argmax in Eq. (5.2), is responsible for finding the most likely word string among all possible word strings. If there are N words in a word string and if the ASR vocabulary is composed of $|V|$ words, then there will be totally $|V|^N$ possible word strings that need to be searched by the decoder to find the most probable word string. Fortunately, the Markov assumptions in modeling allow the search space to be represented as a weighted finite state transducer (WFST). WFSTs provide a unified framework for representing different knowledge sources in ASR systems (Mohri et al. 2002). In this framework, the speech recognition problem is treated as a transduction from input speech signal to a word sequence. A typical set of knowledge sources consists of a HMM (H) mapping state ID sequences to context-dependent phones, a context-dependency network (C) transducing context-dependent phones to context-independent phones, a lexicon (L) mapping context-independent phone sequences to words, and a language model (G) assigning probabilities to word sequences. The composition of these models $H \circ C \circ L \circ G$ results in an all-in-one search network that directly maps HMM state ID sequences to weighted word sequences. The search network can be optimized by WFST determinization and minimization algorithms. For improved efficiency, the decoder prunes the low probability paths from the search space with the help of the acoustic and language models. Efficient decoding algorithms make the search tractable even for very large vocabulary sizes.

In Eq. (5.2), $\widehat{W}$ is the most likely word string (1-best hypothesis) obtained from the decoder with the current acoustic and language models. The most likely hypothesis may not be the same with the word string that is uttered by the speaker. The possible sources of errors include all components of the ASR system: the limited recognition vocabulary, the imperfect statistical language models, the front-end not being robust to noisy acoustic conditions, the conditional independence assumptions of HMMs in acoustic modeling and heuristic pruning algorithms in the decoder to deal with the astronomically large number of word strings in the search space, among others. The ratio of the total number of errors in the hypothesis strings to the total number of words in the reference strings, called word error rate (WER), defines the performance of the speech recognizer and is formulated as:

$$\text{WER} (\%) = \frac{\#D + \#S + \#I}{\#\text{reference words}} \times 100, \tag{5.3}$$

where $\#D$, $\#S$, and $\#I$, respectively, represent the minimum number of deletion, substitution, and insertion operations required to transform the hypothesis strings to the reference strings. The number of errors in each hypothesis string is calculated with the minimum edit distance algorithm (Levenshtein 1966).

In addition to the 1-best hypothesis, the decoder can output alternative hypotheses in the form of a lattice or N-best list. A word lattice is an efficient representation of the possible word sequences in the form of a directed graph. Figure 5.3 shows an example lattice output from the decoder where the arcs are labeled with words and their probabilities. The most likely hypothesis obtained from the lattice is "haberler sundu" which has an error rate of 50%. However, the path with the lowest WER is "haberleri sundu," which has no errors. Oracle error rate or lattice word error rate is the error rate of the path in the lattice with the lowest word error. This is the lower bound on the error rate that can be achieved given a lattice. In this example, the oracle error rate is 0% since the hypothesis with the lowest WER exactly matches with the reference string.

An N-best list contains the most probable N hypotheses. For the lattice given in Fig. 5.3, there are six hypotheses, ordered by their posterior probabilities:

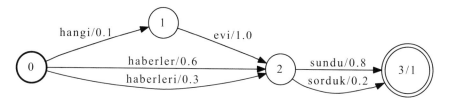

Fig. 5.3 Lattice example for the reference sentence "haberleri sundu"

```
haberler  sundu 0.48
haberleri sundu 0.24
haberler  sorduk 0.12
hangi evi sundu 0.08
haberleri sorduk 0.06
hangi evi sorduk 0.02
```

Note that the correct hypothesis is the second best hypothesis in the list.

In many ASR systems, the initial decoder output consisting of the alternative hypotheses is rescored by more sophisticated models and reranked to get improved performance. This multi-pass strategy also allows for adapting the models or normalizing the acoustic features for a particular speaker. Speaker adaptation or normalization has been shown to be effective even for small amounts of data from a particular speaker.

5.3 Turkish Language Resources for ASR

In this section, we review the available Turkish acoustic and text data as well as the linguistic tools utilized in building Turkish ASR systems. Existence of shared language resources is crucial in ASR research for system training and development. Robust estimation of the acoustic and language models requires large amounts of acoustic and text data. Linguistic Data Consortium (LDC)[1] and European Language Resources Association (ELRA)[2] provide language resources in different languages. These resources include read, broadcast, and conversational speech together with reference transcripts and pronunciation lexicons for acoustic modeling, as well as text corpora for language modeling.

5.3.1 Turkish Acoustic and Text Data

The lack of standard and publicly available speech and text corpora has been a major problem for Turkish ASR research. Over the last two decades, various researchers have collected application specific Turkish acoustic and text data. However, the amount of the data collected with individual efforts was quite limited and the research results were not comparable with each other because of not having a standard training and test corpus.

The GlobalPhone project (Schultz and Waibel 2001; Schultz 2002) was one of the early efforts in collecting a standard multilingual audio and text corpus and aimed to

---

[1]University of Pennsylvania, PA, USA. Linguistic Data Consortium: www.ldc.upenn.edu (Accessed Sept. 14, 2017).

[2]European Language Resources Association. Catalogue of Language Resources: catalog.elra.info (Accessed Sept. 14, 2017).

generate a high quality read speech and text database suitable for the development of language independent and language adaptive speech recognition as well as for language identification tasks. Currently, the GlobalPhone speech and text corpus contains 20 widespread languages in the world, including Turkish. The entire corpus contains around 450 h of speech spoken by more than 1900 native adult speakers. For each language, there are recordings from around 100 native speakers reading about 100 sentences selected from national newspapers. The speech data, recorded with a close-speaking microphone, is provided in 16 bit, 16 kHz mono quality. The Turkish portion of GlobalPhone speech and text corpus offers approximately 17 h of Turkish speech with reference transcriptions. The recordings are taken from 100 speakers (28 males, 72 females) with a balanced age distribution (30 speakers are below 19, 30 speakers are between 20 and 29, 23 speakers are between 30 and 39, 14 speakers are between 40 and 49, and 3 speakers are over 50). The GlobalPhone speech and text corpus can be obtained through ELRA and Appen Butler Hill Pty Ltd. In addition to the speech and text corpora, the GlobalPhone project also provides pronunciation dictionaries covering the vocabulary of the transcripts and n-gram language models. The Turkish vocabulary contains 34K words seen in the transcriptions of the acoustic data and the dictionary contains the pronunciation of these 34K words. A Turkish n-gram language model was built using a text corpus containing approximately 7M words collected from the web. The statistics about the databases and the details of speech recognition systems built with these databases can be found in Schultz et al. (2013). The GlobalPhone dictionaries are distributed by ELRA and GlobalPhone n-gram language models are freely available for download.[3] A Turkish system was integrated into the multilingual recognition engine of the GlobalPhone project. This system used the GlobalPhone speech corpus for acoustic modeling and an approximately 15M word text corpus, collected from the web, for language modeling. In order to handle the OOV problem caused by the agglutinative nature of Turkish, morpheme-based language modeling units and vocabulary adaptation were investigated. More details can be found in Çarkı et al. (2000).

Mengüşoğlu and Deroo (2001) describe a Turkish read-speech data collection effort with the aim of creating a phonetically balanced corpus where the speakers (10 male and 10 female) were asked to utter 100 isolated words and 215 continuous sentences selected from television programs and newspaper articles. About 25 min of recordings were taken from each speaker. This corpus was used to train a Turkish ASR system and the performance of the system was evaluated in an isolated word recognition task.

Another acoustic and text data collection effort for investigating various language modeling units for Turkish ASR was presented in Dutağacı (2002), Arısoy (2004) and Arısoy et al. (2006). In these, a Turkish ASR system was built using more than 7 h of acoustic data, recorded from around 200 speakers, and around 10M

---

[3]Vu, Ngoc Thang and Schultz, Tanja. GlobalPhone Language Models. University of Bremen, Germany. Cognitive Systems Lab: http://www.csl.uni-bremen.de/GlobalPhone/ (Accessed Sept. 14. 2017).

words text corpus for language modeling. Approximately 95% of this text database contains text from literature (novels, stories, poems, essays), law, politics, social sciences (history, sociology, economy), popular science, information technology, medicine, newspapers, and magazines. The rest of the text data was collected as in-domain data for developing a newspaper content transcription system. The text material consists of articles from various domains such as world news, economics, contemporary news, politics and daily life, collected from a Turkish newspaper over a 1 month period. In order to evaluate the speech recognition performance of the developed system, approximately 1 h of speech was recorded from a female speaker reading sentences from newspaper articles. More details of this newspaper content transcription system will be given in Sect. 5.4.1.

Researchers from the Middle East Technical University (METU) collaborated with the Center for Spoken Language Research (CSLR) of the University of Colorado at Boulder to design a standard phonetically-balanced Turkish microphone speech corpus similar to TIMIT for American English. As the first step, a pho-netically rich sentence set, containing 2.5M word tokens, was collected from the web and triphone frequencies were extracted from this corpus as a representative set of Turkish triphones. Then the first 2000 sentences of the TIMIT corpus were translated into Turkish and 462 new sentences were added in order to cover the most frequent 5000 triphones in the text corpus. These 2462 sentences provide a triphone-balanced sentence set. For speech recordings, a subset of 40 sentences among 2462 sentences were randomly selected for each speaker and speakers uttered each sentence once. Speech from 193 native Turkish speakers (89 female and 104 male) was collected in a quiet office environment using a Sennheiser condenser microphone with 16 kHz sampling rate. The final corpus consists of audio files, associated text transcriptions and phone-level, word-level and HMM-state-level alignments from a phonetic aligner developed by porting CSLR speech recognition tool SONIC (Pellom 2001) to Turkish. More details on this data collection effort and developing speech tools can be found in Salor et al. (2007). This corpus was released by LDC in 2006 as "Middle East Technical University Turkish Microphone Speech v 1.0 (LDC2006S33)". The LDC release contains 500 min of speech from 120 speakers (60 male and 60 female) speaking 40 sentences each (approximately 300 words per speaker). This speech corpus together with a 2M word text corpora was used in building an ASR system for Turkish for comparing morphological and data-driven units in language modeling (Hacıoğlu et al. 2003). In addition, a 6M word text corpus was collected from sports articles of Turkish online newspapers and the ASR performance of morphologically driven units, i.e., stems and endings, were investigated using an ASR system trained on this speech corpus. Moreover, the corpus was used in Turkish text-to-speech synthesis (Salor et al. 2003) as well as voice transformation and development of related speech analysis tools for Turkish (Salor 2005).

Another data collection effort was led by the Sabancı University. Speech recordings were collected from 367 different speakers with equal gender distri-bution. Each speaker read around 120 sentences among about 1000 phonetically balanced sentences, resulting in around 30 h of speech data. A headset microphone

(Plantronics Audio 50) connected to a laptop was used and speech was captured in 16 bit, 16 kHz quality. In addition to the speech data, a text corpus containing 81M words (5.5M different sentences) was collected mostly from internet news web sites and e-books in Turkish. The news data contains daily life, sports news, and op-ed pieces. The speech and text corpora were used to build a Turkish ASR system and various language modeling units such as syllables, stems, and endings and words were investigated in this ASR system. More details can be found in Erdoğan et al. (2005).

In addition to microphone speech corpora, telephone speech recordings were also collected for Turkish. One of the efforts was the collection of a Turkish telephone speech database within the framework of OrienTel project, funded by the European Union as a fifth Framework activity. The aim of the project was collecting a database for the development of speech based, interactive, multilingual communication services for the languages in the Mediterranean countries of Africa, Middle East, Turkey, and Cyprus. Within the project, 21 telephone speech databases in Arabic, Turkish, German (spoken by Turks), Hebrew, Greek, French, and English were collected using GSM and fixed networks. OrienTel Turkish telephone speech database contains recordings of Turkish digits, numbers, time, date, words, and sentences uttered by 1700 speakers (921 males, 779 females), stored in 8-bit 8 kHz A-law format. The database is balanced in gender, age, dialect (seven dialect regions based on seven official geographical regions of Turkey) and calling environment (home/office, public places, vehicles). A pronunciation lexicon with a phonemic transcription in SAMPA is also included in the database. More details on the OrienTel Turkish database can be found in Çiloğlu et al. (2004). This corpus can also be obtained through ELRA.[4]

Another Turkish telephone speech corpus was collected for call center conversations and utilized for developing an LVCSR system for call tracking (Haznedaroğlu et al. 2010; Haznedaroğlu and Arslan 2011, 2014). In contrast to OrienTel Turkish telephone speech corpus, this corpus contains the real call center conversations collected from different domains such as banking, insurance, and telecommunications. The calls were recorded in two channels (stereo), 8 kHz, 8-bit, A-law format. All of the calls were manually transcribed for the agent and customer channels separately. The total amount of the speech data is approximately 1000 h. More details on the call center LVCSR system built using this data will be given in Sect. 5.4.3.

IARPA Babel Turkish Language Pack (IARPA-babel105b-v0.5) was developed by Appen for the IARPA (Intelligence Advanced Research Projects Activity) Babel program which focuses on low resource languages and seeks to develop speech recognition technology that can be rapidly applied to any human language to support keyword search over large amounts of recorded speech. It contains approximately 213 h of Turkish conversational and scripted telephone speech collected in 2012 along with corresponding transcripts. The Turkish speech in this corpus is

---

[4]European Language Resources Association. Catalogue of Language Resources: catalog.elra.info/ (Accessed Sept. 14, 2017).

representative of the seven dialect regions in Turkey. The gender distribution among speakers is approximately equal and the speakers' ages range from 16 years to 70 years. Calls were made using different telephones (e.g., mobile, landline) from a variety of environments including the street, a home or office, a public place, and inside a vehicle. The audio data presented as 8 kHz 8-bit a-law encoded audio in sphere format and the corresponding transcripts encoded in UTF-8 were released through LDC in October 2016 as "IARPA Babel Turkish Language Pack IARPA-babel105b-v0.5 (LDC2016S10)."

There is also a Turkish Broadcast News (BN) corpus collected at Boğaziçi University. This corpus aims to facilitate the Turkish Broadcast News Transcription and Retrieval research. The BN database is composed of the recordings of the Broadcast News programs and their corresponding reference transcriptions. In this database, Broadcast News programs were recorded daily from the radio channel Voice of America (VOA) and four different TV channels (CNN Türk, NTV, TRT1, and TRT2). Then these recordings were segmented, transcribed, and verified. The segmentation process used automatic segment generation and manual correction steps. The transcription guidelines were adapted from Hub4 BN transcription guidelines. The annotation includes topic, speaker and background information for each acoustic segment. The acoustic data were converted to 16 kHz 16-bit PCM format, and segmentation, speaker and text information was converted to the NIST STM format. Out of a total of over 500 h of speech that was recorded and transcribed, approximately 200 h were partitioned into disjoint training (187.9 h), held-out (3.1 h), and test sets (3.3 h) to be used in LVCSR research. The reference transcriptions of the acoustic training data include 1.3M words. Table 5.1 gives the breakdown of the BN data in terms of acoustic conditions. Here classical Hub4 classes are used: (f0) clean speech, (f1) spontaneous speech, (f2) telephone speech, (f3) background music, (f4) degraded acoustic conditions, and (fx) other. A part of this corpus consisting of approximately 130 h of Voice of VOA Turkish radio broadcasts collected between December 2006 and June 2009 and the corresponding transcripts were released through LDC as "Turkish Broadcast News Speech and Transcripts (LDC2012S06)."

In addition to acoustic resources, large text corpora are required for estimating reliable statistical language model parameters. In order to facilitate building such models for general statistical language processing research for Turkish, a large text

Table 5.1 Size of the data for various acoustic conditions (in hours) for the Turkish BN data

Partition	f0	f1	f2	f3	f4	fx	Total
Train	67.2	15.7	8.3	19.8	73.6	3.3	187.9
Held-out	1.1	0.1	0.1	0.5	1.3	0.0	3.1
Test	0.9	0.1	0.1	0.7	1.4	0.1	3.3
AM subset	69.2	15.9	8.5	21.0	76.3	3.4	194.3
Other	112.0	33.6	24.6	39.9	141.7	11.8	363.7
Total	181.2	49.5	33.1	60.9	218.0	15.2	558.0

corpus containing 423M words was collected by crawling Turkish web sites. Around 184M words of this corpus were collected from three major newspapers portals (Milliyet, Ntvmsnbc, and Radikal) and the rest of the corpus contains text obtained from general sampling of Turkish web pages. In order to build this corpus, around 8M web pages were downloaded and the text in these web pages were cleaned using a morphological parser and some heuristics. More details about this Turkish corpora can be found in Sak et al. (2011) and Sak (2011). The part of this corpora containing text from news portals and BN acoustic data were used to build a BN transcription system for Turkish. More details on that BN system will be given in Sect. 5.4.2.

5.3.2 Linguistic Tools Used in Turkish ASR

This section reviews some Turkish linguistic tools such as morphological parsers, a morphological disambiguator, and a dependency parser that are used in building Turkish LVCSR systems. These tools are useful for language modeling, especially for investigating linguistic units in n-gram language modeling and for integrating linguistic information into language modeling.

Chapter 4 covers various linguistically driven language modeling units for Turkish and explains how to use morphological information in feature-based language models, specifically in discriminative language modeling. Estimating a language model based on morphological units and extracting morphological information for feature-based language modeling require a morphological parser. There are finite-state-based morphological parsers (Oflazer 1994; Sak et al. 2011) and Prolog-based morphological parsers (Çetinoğlu 2000; Dutağacı 2002) developed for Turkish. Each parser can yield different number of segmentations for the same word and there may be differences in the parser outputs. These may be the result of the differences in the lexicon and the morphotactics in each parser implementation. Additionally, parsers can output either lexical or surface form representations of morphemes. Lexical form representations are especially important in n-gram morpholexical language modeling and in feature-based language modeling as they yield richer information. Lexical form representations are converted to surface form representations using a morphophonemic transducer.

The morphological parsing of a word may result in multiple interpretations of that word due to complex morphology of Turkish. This ambiguity can be resolved using morphological disambiguation tools developed for Turkish (Hakkani-Tür et al. 2002; Yuret and Türe 2006; Sak et al. 2007). Sak et al. (2012) showed that morphological disambiguation improves the prediction power of morpholexical language model. See Fig. 5.4 in Sect. 5.4.2 for a detailed analysis of morphological disambiguation on speech recognition performance. In the absence of morphological disambiguation, choosing the morphological segmentation with the smallest number of morphemes has been shown to perform better than choosing one of the segmentations randomly in speech recognition with surface form stem and ending units (Arısoy et al. 2009).

Syntax, the rules of sentence formation, is another useful information source in language modeling. Syntax determines the organization of the sentences either with the constituent structure representation or using the syntactic dependencies. Constituent structure refers to the hierarchical organization of the constituents of a sentence and syntactic dependencies specify the lexical dependencies in the sentence, for instance the presence of a word or a morpheme can depend on the presence of another word or a morpheme in the same sentence (Fromkin et al. 2003). Dependency parsers try to find the underlying dependency tree that shows the dependency links between head and dependent words in a sentence. A Turkish dependency parser (Eryiğit et al. 2008) was used to obtain the dependency relations between the head and the dependent words with the type of the dependency, for ASR output hypotheses. The output of this parser was encoded as features and utilized in discriminative reranking of ASR hypotheses. Chapter 4 details the use of the dependency parser output in discriminative language modeling.

5.4 Turkish ASR Systems

This section reviews some Turkish ASR systems that are noteworthy in terms of amount of the data used in acoustic and language modeling and the techniques used especially in building the language models. They also provide examples of various applications based on ASR.

5.4.1 Newspaper Content Transcription System

A newspaper content transcription system was developed for Turkish to investigate several ways of handling the out-of-vocabulary (OOV) word problem in Turkish ASR (Arısoy and Saraçlar 2009; Arısoy 2004). Acoustic and text databases required for building the acoustic and language models mostly came from the data collected by METU and Sabancı University. The speech recognition performance of the system was evaluated on a test corpus consisting of 1 h of recordings of newspaper sentences (6989 words) read by one female speaker.

The acoustic model of the newspaper content transcription system was built using 17 h of microphone recordings of phonetically balanced sentences read by over 250 male and female native Turkish speakers. Since Turkish is almost a phonetic language, graphemes were used in acoustic modeling instead of phonemes. Decision-tree state clustered cross-word triphone HMMs with approximately 5000 states were used as the acoustic model. Each HMM state had a GMM with six mixture components. The baseline acoustic model was speaker independent. The language model of the newspaper content transcription system was built using a text corpora containing approximately 27M words (675K word types) collected from Turkish web pages. A vocabulary size of 50K words, yielding 11.8% OOV rate, was used to train a 3-gram word-based language model. The language model was built using the SRILM toolkit (Stolcke 2002) with interpolated Kneser-Ney smoothing. Entropy-based pruning (Stolcke 1998) was applied to the language model due to the computational limitations. The word-based language model resulted in 38.8% WER.

A major problem for Turkish LVCSR systems is the high OOV rate and statistical morphs were investigated as one of the ways of eliminating the OOV problem. For building a morph-based language model, the words in the training data were split into their morphs using the Baseline-Morfessor algorithm (Creutz and Lagus 2005). This resulted in around 34K morphs, which could generate all the words in the test data by concatenation. As a result, the OOV rate for morphs is 0%. In order to recover the word sequence from the morph sequence, a special symbol was used to mark the word boundaries in the ASR output of the morph-based model. The ratio of morph tokens to word tokens was calculated as 2.37 including the word boundary symbol which suggests higher order n-gram language models for morphs. A 5-gram morph-based language model built in a similar way with words yielded 33.9% WER.

As another way of handling OOV words, multi-pass lattice extension and vocabulary adaptation approaches were investigated for Turkish using the newspaper content transcription system. Geutner et al. (1998a) introduced a lattice extension method called hypothesis driven lexical adaptation whose main idea is to generate a lattice output from an ASR system and to extend each word in the output lattice with words similar to that word from a fallback vocabulary. Then second-pass recognition is performed on the extended lattice with a language model built with the words in the extended lattice. In the lattice extension framework, the similarity can be morphology-based (Geutner et al. 1998a), where each word is extended with the words having the same stem, or phonetic-distance-based (Geutner et al. 1998b), where the Levenstein distance (minimum edit distance) of phonemes is used. Since the phonetic distance based similarity requires mapping graphemes to phonemes, using graphemes directly in measuring the similarity was proposed in Geutner et al. (1999) for Serbo–Croatian where orthography closely matches its pronunciation. For Turkish, morphology-based and grapheme-based similarities were investigated for lattice extension and a position dependent minimum edit distance was proposed (Arısoy and Saraçlar 2009). Position dependent minimum edit distance penalizes the edit operations at the beginning of the strings more than the operations at the end of the strings. In other words, errors in the stems are penalized more than the errors in the suffixes. Classical and position dependent minimum edit distance did not reveal any significant difference and they all gave better results than morphology-based similarity.

In vocabulary adaptation, the baseline vocabulary is modified after the first pass recognition with the extended lattice words and re-recognition of the input speech is performed using the language model built with the adapted vocabulary. This technique was applied to Czech speech recognition using morphology-based similarity (Ircing and Psutka 2001). In Turkish vocabulary adaptation experiments, the baseline vocabulary is adapted by adding all the words in the extended lattice.

Table 5.2 WER results on newspaper content transcription system

Approach	OOV (%)	WER (%)
Word-based system	11.8	38.8
+ lattice extension/vocabulary adaptation	<2	34.2
Morph-based system	0	33.9
+ lattice extension	>0	32.3

Arısoy and Saraçlar (2009) showed that both lattice extension and vocabulary adaptation approaches reduce the WER compared to the word-based model in Turkish newspaper transcription system. Lattice extension and vocabulary adaptation yielded similar results, 34.2% WER, for larger extended lattice, whereas vocabulary adaptation performed better than lattice extension if the extended lattice was small. The analysis of the improvements obtained by lattice extension and vocabulary adaptation revealed that the gains come from better OOV handling (Arısoy and Saraçlar 2009), which is the main motivation of the proposed techniques. However, it is important to note that morph-based system yielded 0% OOV rate and resulted in better performance than the proposed techniques. See Table 5.2 for a comparison.

The idea of lattice extension was also applied to the morph-based ASR system. In contrast to the lattice extension approach in word-based ASR system where the motivation is to deal with the OOV problem, the motivation of applying lattice extension to morph-based system is to deal with the shortcoming of morphs in ASR. Even though the OOV problem is best handled by using sub-lexical recognition units, statistical morphs, a shortcoming of sub-lexical units is over-generation. A sequence of sub-lexical units is converted into a sequence of words using word boundary markers (symbols inserted into word boundaries or attached to suffixes). However, sub-lexical units can generate non-word recognition sequences after concatenation. In case of linguistic sub-lexical units, over-generation problem can be handled by imposing morphological constraints as in morpholexical units. In contrast to linguistic sub-lexical units, statistical units require more sophisticated approaches and the lattice extension strategy was modified for morphs to solve the over-generation problem. In lattice extension for morphs, the morph lattice is converted to a morph sequence lattice by merging the morphs between two word boundary symbols into a single word. Then this lattice is extended with similar words in the fallback vocabulary and utilized in second-pass recognition. Although lattice extension for morphs brings back the OOV problem, it reduced the WER from 33.9% to 32.3%, yielding 1.6% absolute improvement over the baseline.

5.4.2   Turkish Broadcast News Transcription System

A major effort in Turkish LVCSR research was the development of a broadcast news transcription and retrieval system. As part of this project, a state-of-the-art ASR system based on sub-lexical language modeling units was built and the discriminative language modeling framework was used to incorporate linguistic knowledge into language modeling (Arısoy 2004, 2009; Sak 2011).

The acoustic model of the broadcast news transcription system was built using the Turkish Broadcast News (BN) database introduced in Sect. 5.3.1. Approximately 194 h of speech from the Turkish BN database were used as the acoustic data, partitioned into disjoint training (188 h), held-out (3.1 h), and test sets (3.3 h). As Turkish graphemes are essentially in a one-to-one correspondence with phonemes, graphemes were utilized instead of phonemes in acoustic modeling.[5] Decision-tree state-clustered cross-word triphone models with 10,843 HMM states were used as the acoustic models. Each HMM state has a GMM with 11 mixture components while the silence model has a GMM with 23 mixtures.

Statistical n-gram language models were built with words and sub-lexical units using 184M-word Turkish text collected from online newspapers (see Sect. 5.3.1) and the reference transcriptions of the acoustic training data (1.3M words). In order to reduce the effect of pruning on the recognition accuracy, the first-pass lattice outputs were rescored with unpruned language models. As sub-lexical language modeling units, both linguistically and statistically driven units were investigated. Linguistically driven sub-lexical units include morphemes in lexical form as well as stems and endings in both surface and lexical forms. Statistically driven sub-lexical units include the statistical morphs.

As it is common practice to use longer histories in sub-lexical language models, word level language models use 3-grams while the sub-lexical language models use 4-grams. The vocabulary size for each model is chosen to balance the trade off between the model complexity and OOV rate. For the word model, a vocabulary of 200K words yields 2% OOV rate on the test data. For statistical morphs, the words in the text data is split into their morphs using the Baseline-Morfessor algorithm (Creutz and Lagus 2005). In order to find the word boundaries after recognition, the non-initial morphs are marked with a special symbol and this results in around 76K morph types in the vocabulary. Since these morph types can generate all the words in the test data by concatenation, the OOV rate for morphs is 0%. For the morpholexical language models, the text corpus is morphologically parsed and disambiguated to get the lexical-grammatical morpheme and lexical stem ending representations of corpora. The lexicon of the morphological parser contains about 88K symbols. The OOV rate of the morphological parser on the test set is about 1.3%. The lexical-grammatical morpheme representation results in about 175K symbols. The lexical stem ending representation yields about 200K symbols. For the surface form stem and ending model, the surface form representation of the endings are obtained by removing the stem from the word. In both lexical and surface form stem ending based models, endings are marked with a special symbol in order to find the word boundaries easily after recognition.

---

[5]Phonetic acoustic models together with a finite-state transducer based pronunciation lexicon similar to Oflazer and Inkelas (2006) result in similar overall performance, possibly due to a small number of Turkish words with exceptional pronunciation.

Since Turkish morphology can be represented as a WFST, it is possible to directly integrate it into the WFST framework of an ASR system. The lexical transducer of the morphological parser maps the letter sequences to lexical and grammatical morphemes annotated with morphological features. The lexical transducer can be considered as a computational dynamic lexicon in ASR in contrast to a static lexicon. The computational lexicon has some advantages over a fixed-size word lexicon. It can generate many more words using a relatively smaller number of root words in its lexicon. Therefore it achieves lower OOV rates. In contrast to a static lexicon, even if we have never seen a specific word in the training corpus, the speech decoder has the chance to recognize that word. Compared to some other sub-lexical approaches, this approach only allows valid words, since it incorporates the morphotactics into the decoding process. Another benefit of the computational lexicon is that it outputs the morphological analysis of the word generated. This morphological information is useful for language modeling and further analysis.

Table 5.3 shows the WER and the stem error rate (SER) of the ASR system with word and sub-lexical units. All the sub-lexical approaches yield better results than the word-based model and the lexical form stem and ending model gives the best performance. The analysis of SER in addition to the WER confirms that the lexical stem ending model also improves the recognition accuracy of stems.

Figure 5.4 demonstrates the effect of morphotactics and morphological disambiguation on speech recognition performance using the lexical stem+ending model. Here the best model is the lexical *stem+ending* model with the correct morphotactics and morphological disambiguation. The *stem+ending:no-mt* model represents the experiment where the morphotactics component of the lexical transducer allows any ordering of the morphemes. The *stem+ending:no-disamb* model represents the case where the morphological disambiguation chooses the morphological parse with the least number of morphemes. The final model, *stem+ending:no-mt-no-disamb*, shows the cumulative effect for the absence of morphotactics and morphological disambiguation. These results show that, for the lexical *stem+ending* model, both morphotactics and morphological disambiguation are useful in improving the performance.

As explained in Chap. 4, the discriminative language modeling (DLM) framework provides a feature based mechanism for integrating linguistic knowledge into language modeling. For the broadcast news transcription system, the DLM framework was used with sub-lexical language modeling units as well as various linguistic information sources (Arısoy 2009; Sak 2011). For the DLM experiments,

Table 5.3 WER results on broadcast news transcription system (Sak et al. 2012) (reprinted with permission)

Language model	WER (%)	SER (%)
Word	23.1	20.7
Stem + ending (surface)	21.9	20.1
Morpheme (lexical)	21.8	19.9
Statistical morph	21.7	19.8
Stem + ending (lexical)	21.3	19.5

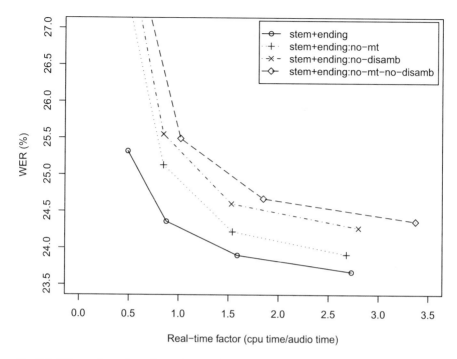

Fig. 5.4 Effects of morphotactics and morphological disambiguation on the lexical stem+ending model (Sak et al. 2012) (reprinted with permission)

N-best lists were extracted from the word-based and morph-based ASR systems. While generating the DLM training data, language model over-training was controlled via 12-fold cross validation. Utterances in each fold were decoded with the baseline acoustic model trained on all the utterances and the fold-specific language model. The fold-specific language model was generated by linearly interpolating the generic language model with the in-domain language model built from the reference transcriptions of the utterances in the other 11 folds. The perceptron algorithm presented in Chap. 4 was used for training the feature parameters.

Table 5.4 shows the results for the DLM experiments with word and sub-lexical units and after integrating linguistically and statistically motivated features with these units. The details of the features experimented in Table 5.4 are given in Sect. 4.6. The unigram features (word unigrams, lexical form stem and ending unigrams and morph unigrams), where a feature is defined as the number of times a unigram is seen in the training data, have been shown to yield improvements for each model. For the word-based system, syntactic features such as part-of-speech tag n-grams and head-to-head dependencies extracted from dependency parser analyses of sentences were explored. The PoS tag unigram and bigram features yielded additive 0.5% improvement on top of the gain obtained with word unigram features, whereas no additional improvement was achieved on top of PoS tag features with head-to-head dependency relations.

Table 5.4 WER results with DLM experiments on broadcast news transcription system (Sak et al. 2012) (reprinted with permission)

Language model	WER (%)	Δ WER (%)
Word	23.4	–
+ word unigrams	23.0	0.4
+ word unigrams + PoS tags	22.5	0.9
+ word unigrams + PoS tags + Head-to-Head dependencies	22.6	0.8
Stem + ending (lexical)	21.6	–
+ stem + ending (lexical) unigrams	20.9	0.7
Statistical morph	22.4	–
+ morph unigrams	21.8	0.6
+ morph unigrams + morph classes	21.3	1.1
+ morph unigrams + morph classes + morph triggers	21.2	1.2

For the morph-based system, statistically motivated features such as automatically induced morph classes and morph trigger pairs were investigated. Automatically induced morph unigram and bigram features gave additional 0.5% improvements on top of the gain obtained with morph unigram features. Long distance morph trigger features were extracted from the oracle and 1-best morph hypotheses in the 50-best lists. Incorporating long distance morph trigger features into morph unigram and morph cluster features yielded no significant additional improvements.

5.4.3 LVCSR System for Call Center Conversations

The aim of developing an LVCSR system for call center conversations is to automatically retrieve statistics about customer satisfaction based on call center conversations. Call tracking can be done manually but it is an expensive process in terms of time and resources. Therefore, automatic retrieval of call center conversations is a cost-effective solution. Existence of a high accuracy LVCSR system is quite important in order to obtain correct statistics with automatic call tracking. The performance of a call ranking system, which labels the calls as "good" or "bad," highly depends on the speech recognition performance and the correlation between human and computer generated scores is highly sensitive to word error rate (Saon et al. 2006). However, recognizing call center conversation with high accuracy or with low word error rate is a challenging task due to telephone channel distortions, external noises on telephone channels, speaking style variability of the customers and also the spontaneous nature of the conversations.

An LVCSR system devoted to transcribing call center conversations for speech mining was developed for Turkish (Haznedaroğlu et al. 2010). The acoustic model was trained with 197 h of speech data, containing 51 h of speech from call center conversations, and their manual transcriptions (see Sect. 5.3.1). Context-dependent triphone models with 12 Gaussian mixtures were trained using the acoustic data. The transcriptions of 27 h of call center recordings were utilized for training word-based bigram language models, as well as statistical morph-based trigram language models. Different language models were built for agent and customer channels. Word-based language models performed better than statistical morph-based models which could be explained with the limited amount of the call center transcriptions used in language modeling.

If the LVCSR system results in high word error rates for the call center conversations, the transcriptions corresponding to these conversations will cause incorrect statistics. Therefore, confidence measures are important to evaluate the reliability of the speech recognition results. Unreliable transcriptions can be filtered out in order to retrieve correct statistics. Haznedaroğlu and Arslan (2011) proposed conversational two-channel (customer and agent channels) confidence measures based on speech overlap ratio and opposite channel energy level. If customer and agent speak simultaneously, this may affect the recognizability of the speech in each channel. Speech overlap ratio metric is based on this premise and is defined as the ratio of duration of the overlapped speech from both channels to the duration of a single channel. Opposite channel energy level shows the average energy of the opposite's party speech. This metric is based on the observation that speakers tend to speak louder when the other channel's voice level is low in telephone conversations which may effect the recognition accuracy. A support vector machine (SVM) was trained to rate the reliability of the call segments. This SVM used speech overlap ratio and opposite channel energy level as features in addition to the features extracted from the decoder output (acoustic and language model scores, posterior probability) and prosodic features extracted from utterances. The performance of the proposed confidence measures was evaluated on an LVCSR system developed for call center conversations. In order to train the system, approximately 200 h of data were collected and manually transcribed (see Sect. 5.3.1). Twelve hours and nine hours of this data were set apart for system evaluation and SVM training purposes, respectively. The rest of the data was used for acoustic model training. A system similar to the one given in Haznedaroğlu et al. (2010) was trained for evaluation. Conversational measures have been shown to improve the rating accuracy of utterances with high and low WERs, and speech overlap ratio became a more discriminative measure than the opposite party energy level. Also, confidence measures were used to select a set of automatically obtained transcriptions for language model adaptation to improve the recognition accuracy (Haznedaroğlu and Arslan 2014). The proposed approach was evaluated in a system similar to the one given in Haznedaroğlu et al. (2010) but using larger amounts of acoustic and text data. In this work, acoustic models were trained with 1000 h of call center acoustic data and the transcriptions of 150 h of recordings were used for language model training. Language model adaptation data consisted of automatically obtained transcriptions of 38,000 h of call-center recordings. A 4% relative WER reduction was achieved with the proposed adaptation set selection approach.

5.5   Conclusions

In recent years, automatic speech recognition has improved so much that practical ASR systems have become a reality and provide a front-end to spoken language technology. Still, there are many language specific challenges, especially for low resource languages. In this chapter, we first summarized the speech and language resources available for building Turkish ASR systems. In addition to speech and text corpora required to train the basic acoustic and language models, various NLP tools such as morphological parsers, morphological disambiguators, and dependency parsers provide the data necessary to enhance these models. Most of the research on Turkish ASR has focused on language modeling discussed in Chap. 4. The high OOV rates caused by the agglutinative nature of Turkish poses a challenge for ASR systems. The most common approach to deal with this challenge is to use sub-lexical units for language modeling. Furthermore, discriminative language models incorporating morphological and syntactic knowledge improve the performance for some ASR tasks.

References

Arısoy E (2004) Turkish dictation system for radiology and broadcast news applications. Master's thesis, Boğaziçi University, Istanbul

Arısoy E (2009) Statistical and discriminative language modeling for Turkish large vocabulary continuous speech recognition. PhD thesis, Boğaziçi University, Istanbul

Arısoy E, Saraçlar M (2009) Lattice extension and vocabulary adaptation for Turkish LVCSR. IEEE Trans Audio Speech Lang Process 17(1):163–173

Arısoy E, Dutağacı H, Arslan LM (2006) A unified language model for large vocabulary continuous speech recognition of Turkish. Signal Process 86(10):2844–2862

Arısoy E, Can D, Parlak S, Sak H, Saraçlar M (2009) Turkish broadcast news transcription and retrieval. IEEE Trans Audio Speech Lang Process 17(5):874–883

Arısoy E, Sainath TN, Kingsbury B, Ramabhadran B (2012) Deep neural network language models. In: Proceedings of the NAACL-HLT workshop: will we ever really replace the *n*-gram model? On the future of language modeling for HLT, Montreal, pp 20–28

Bengio Y, Ducharme R, Vincent P, Jauvin C (2003) A neural probabilistic language model. J Mach Learn Res 3:1137–1155

Çarkı K, Geutner P, Schultz T (2000) Turkish LVCSR: towards better speech recognition for agglutinative languages. In: Proceedings of ICASSP, Istanbul, pp 1563–1566

Çetinoğlu Ö (2000) Prolog-based natural language processing infrastructure for Turkish. Master's thesis, Boğaziçi University, Istanbul

Çiloğlu T, Acar D, Tokatlı A (2004) Orientel Turkish: telephone speech database description and notes on the experience. In: Proceedings of INTERSPEECH, Jeju, pp 2725–2728

Creutz M, Lagus K (2005) Unsupervised morpheme segmentation and morphology induction from text corpora using Morfessor 1.0. Publications in Computer and Information Science Report A81, Helsinki University of Technology, Helsinki

Dutağacı H (2002) Statistical language models for large vocabulary continuous speech recognition of Turkish. Master's thesis, Boğaziçi University, Istanbul

Erdoğan H, Büyük O, Oflazer K (2005) Incorporating language constraints in sub-word based speech recognition. In: Proceedings of ASRU, San Juan, PR, pp 98–103

Eryiğit G, Nivre J, Oflazer K (2008) Dependency parsing of Turkish. Comput Linguist 34(3):357–389

Fromkin V, Rodman R, Hyams N (2003) An introduction to language. Thomson Heinle, Boston, MA

Geutner P, Finke M, Scheytt P, Waibel A, Wactlar H (1998a) Transcribing multilingual broadcast news using hypothesis driven lexical adaptation. In: Proceedings of DARPA broadcast news workshop, Herndon, VA

Geutner P, Finke M, Waibel A (1998b) Phonetic-distance-based hypothesis driven lexical adaptation for transcribing multilingual broadcast news. In: Proceedings of ICSLP, Sydney, pp 2635–2638

Geutner P, Finke M, Waibel A (1999) Selection criteria for hypothesis driven lexical adaptation. In: Proceedings of ICASSP, Phoenix, AZ, pp 617–619

Hacıoğlu K, Pellom B, Çiloğlu T, Öztürk Ö, Kurimo M, Creutz M (2003) On lexicon creation for Turkish LVCSR. In: Proceedings of EUROSPEECH, Geneva, pp 1165–1168

Hakkani-Tür DZ, Oflazer K, Tür G (2002) Statistical morphological disambiguation for agglutinative languages. Comput Hum 36(4):381–410

Haznedaroğlu A, Arslan LM (2011) Confidence measures for Turkish call center conversations. In: Proceedings of INTERSPEECH, Florence, pp 1957–1960

Haznedaroğlu A, Arslan LM (2014) Language model adaptation for automatic call transcription. In: Proceedings of ICASSP, Florence, pp 4102–4106

Haznedaroğlu A, Arslan LM, Büyük O, Eden M (2010) Turkish LVCSR system for call center conversations. In: Proceedings of IEEE signal processing and communications applications conference, Diyarbakır, pp 372–375

Hinton G, Deng L, Yu D, Dahl GE, Mohamed Ar, Jaitly N, Senior A, Vanhoucke V, Nguyen P, Sainath TN, Kingsbury B (2012) Deep neural networks for acoustic modeling in speech recognition: the shared views of four research groups. IEEE Signal Process Mag 29(6):82–97

Huang X, Acero A, Hon HW (2001) Spoken language processing: a guide to theory, algorithm and system development. Prentice Hall, Upper Saddle River, NJ

Ircing P, Psutka J (2001) Two-pass recognition of Czech speech using adaptive vocabulary. In: Proceedings of conference on text, speech and dialogue, Zelezna Ruda, pp 273–277

Jelinek F (1997) Statistical methods for speech recognition. The MIT Press, Cambridge, MA

Jurafsky D, Martin JH (2000) Speech and language processing: an introduction to natural language processing, computational linguistics, and speech recognition. Prentice Hall, Upper Saddle River, NJ

Levenshtein V (1966) Binary codes capable of correcting deletions, insertions and reversals. Sov Phys Dokl 10:707

Mengüşoğlu E, Deroo O (2001) Turkish LVCSR: database preparation and language modeling for an agglutinative language. In: Proceedings of ICASSP, Salt Lake City, UT, pp 4018–4021

Mikolov T, Karafiat M, Burget L, Cernocky J, Khudanpur S (2010) Recurrent neutral network based language model. In: Proceedings of INTERSPEECH, Saint-Malo, pp 1045–1048

Mohri M, Pereira F, Riley M (2002) Weighted finite-state transducers in speech recognition. Comput Speech Lang 16(1):69–88

Oflazer K (1994) Two-level description of Turkish morphology. Lit Linguist Comput 9(2):137–148

Oflazer K, Inkelas S (2006) The architecture and the implementation of a finite state pronunciation lexicon for Turkish. Comput Speech Lang 20:80–106

Pellom BL (2001) Sonic: The University of Colorado continuous speech recognizer. Tech. Rep. TR-CSLR-01, University of Colorado, Boulder, CO

Rabiner L (1989) A tutorial on HMM and selected applications in speech recognition. Proc IEEE 77(2):257–286

Sak H (2011) Integrating morphology into automatic speech recognition: morpholexical and discriminative language models for Turkish. PhD thesis, Boğaziçi University, Istanbul

Sak H, Güngör T, Saraçlar M (2007) Morphological disambiguation of Turkish text with perceptron algorithm. In: Proceedings of CICLING, Mexico City, pp 107–118

Sak H, Güngör T, Saraçlar M (2011) Resources for Turkish morphological processing. Lang Resour Eval 45(2):249–261

Sak H, Saraçlar M, Güngör T (2012) Morpholexical and discriminative language models for Turkish automatic speech recognition. IEEE Trans Audio Speech Lang Process 20(8):2341–2351

Salor Ö (2005) Voice transformation and development of related speech analysis tools for Turkish. PhD thesis, Middle East Technical University, Ankara

Salor Ö, Pellom BL, Demirekler M (2003) Implementation and evaluation of a text-to-speech synthesis system for Turkish. In: Proceedings of EUROSPEECH, Geneva

Salor Ö, Pellom BL, Çiloğlu T, Demirekler M (2007) Turkish speech corpora and recognition tools developed by porting sonic: towards multilingual speech recognition. Comput Speech Lang 21(4):580–593

Saon G, Ramabhadran B, Zweig G (2006) On the effect of word error rate on automated quality monitoring. In: Proceedings of IEEE spoken language technology workshop, Palm Beach, pp 106–109

Schalkwyk J, Beeferman D, Beaufays F, Byrne W, Chelba C, Cohen M, Kamvar M, Strope B (2010) "Your Word is my Command": Google search by voice: a case study. In: Neustein A (ed) Advances in speech recognition: mobile environments, call centers and clinics, Springer, Boston, MA, pp 61–90

Schultz T (2002) Globalphone: a multilingual speech and text database developed at Karlsruhe University. In: Proceedings of ICSLP, Denver, CO

Schultz T, Waibel A (2001) Language-independent and language-adaptive acoustic modeling for speech recognition. Speech Commun 35:31–51

Schultz T, Vu NT, Schlippe T (2013) Globalphone: a multilingual text and speech database in 20 languages. In: Proceedings of ICASSP, Vancouver

Schwenk H (2007) Continuous space language models. Comput Speech Lang 21(3):492–518

Stolcke A (1998) Entropy-based pruning of backoff language models. In: Proceedings of DARPA broadcast news workshop, Herndon, VA, pp 270–274

Stolcke A (2002) SRILM – An extensible language modeling toolkit. In: Proceedings of ICSLP, Denver, CO, vol 2, pp 901–904

Tuske Z, Golik P, Schluter R, Ney H (2014) Acoustic modeling with deep neural networks using raw time signal for LVCSR. In: Proceedings of INTERSPEECH, Singapore, pp 890–894

Yuret D, Türe F (2006) Learning morphological disambiguation rules for Turkish. In: Proceedings of NAACL-HLT, New York, NY, pp 328–334

Chapter 6
Turkish Named-Entity Recognition

Reyyan Yeniterzi, Gökhan Tür, and Kemal Oflazer

Abstract Named-entity recognition is an important task for many other natural language processing tasks and applications such as information extraction, question answering, sentiment analysis, machine translation, etc. Over the last decades named-entity recognition for Turkish has attracted significant attention both in terms of systems development and resource development. After a brief description of the general named-entity recognition task, this chapter presents a comprehensive overview of the work on Turkish named-entity recognition along with the data resources various research efforts have built.

6.1 Introduction

Named-entity recognition (NER) can be defined as the process of identifying and categorizing the named-entities, such as person, location, product and organization names, or date/time and money/percentage expressions in unstructured text. This is an important initial stage for several natural language processing tasks including information extraction, question answering, and sentiment analysis.

Earlier approaches to this task in English relied on handcrafted rule-based systems but over time machine learning became the dominant paradigm (Nadeau and Sekine 2007). State-of-the-art NER systems have been developed for many

R. Yeniterzi
Özyeğin University, Istanbul, Turkey
e-mail: reyyan.yeniterzi@ozyegin.edu.tr

G. Tür
Google Research, Mountain View, CA, USA
e-mail: gokhan.tur@ieee.org

K. Oflazer (✉)
Carnegie Mellon University Qatar, Doha-Education City, Qatar
e-mail: ko@cs.cmu.edu

© Springer International Publishing AG, part of Springer Nature 2018
K. Oflazer, M. Saraçlar (eds.), *Turkish Natural Language Processing*,
Theory and Applications of Natural Language Processing,
https://doi.org/10.1007/978-3-319-90165-7_6

```
"Büyük Şehirlerin Deprem Güvenliği" konulu konferansın ilk günü
ağırlıkla bilimsel deprem senaryolarına ayrılmıştı. Bu çerçevede
<b_enamex Type="LOCATION"> Türkiye <e_enamex>'den
<b_enamexTYPE="ORGANIZATION"> ODTÜ <e_enamex> öğretim üyelerinden
<b_enamex TYPE="PERSON"> Derin Ural<e_enamex>
<n_enamex TYPE="LOCATION"> İzmir <e_enamex> için hazırlanan böyle
bir senaryo çalışmasını aktardı.
```

Fig. 6.1 An example ENAMEX labeled Turkish text

languages and for widely studied languages like English, NER can be considered as a solved problem with an accuracy of around 95%.

Named-entity recognition task was initially introduced by DARPA, and evaluated as an understanding task in both the Sixth and Seventh Message Understanding Conferences (MUC) (Sundheim 1995; Chinchor and Marsh 1998). Later, CoNLL shared tasks (Tjong Kim Sang 2002; Tjong Kim Sang and De Meulder 2003) and Automatic Content Extraction (ACE) program (Doddington et al. 2004) stimulated further research and competition in NER system development.

These conferences defined three basic types of named-entities:

- ENAMEX (person, location, and organization names)
- TIMEX (date and time expressions)
- NUMEX (numerical expressions like money and percentages)

Depending on the application, additional types of entities can also be introduced such as proteins, medicines, etc., in medical text or particle names in quantum physics text. An example Turkish text annotated with ENAMEX entities can be seen in Fig. 6.1.

6.2   NER on Turkish

Initial studies on NER on Turkish texts started in the late 90s. Cucerzan and Yarowsky (1999) proposed a language independent bootstrapping algorithm that uses word-internal and contextual information about entities. They applied this approach to Turkish as well as four other languages. Tür (2000) and Tür et al. (2003) proposed an HMM-based NER system, that was specifically developed for Turkish, together with some other tools for similar information extraction related tasks. They also created the first widely used tagged Turkish newspaper corpora for the NER task. Later, Bayraktar and Temizel (2008) applied a local grammar approach to Turkish financial texts in order to identify person names. Küçük and Yazıcı (2009a,b) developed the first rule-based NER system for Turkish and applied it to Turkish news articles as well as to other domains like children's stories, historical texts, and speech recognition outputs. Dalkılıç et al. (2010) is another rule-based system for Turkish NER.

Recently, NER systems predominantly use machine learning approaches. Küçük and Yazıcı (2010, 2012) extended their rule-based system into a hybrid recognizer in order to perform better when applied to different domains. Yeniterzi (2011) explored the use of morphological features and developed a CRF-based NER system for Turkish. Other CRF-based systems were proposed by Özkaya and Diri (2011) and Şeker and Eryiğit (2012). Şeker and Eryiğit (2012) compared their system with other Turkish NER systems and among the ones that use the same data collection, their system outperformed other systems. Demir and Özgür (2014) developed a neural network based semi-supervised approach, which outperformed Şeker and Eryiğit (2012) over the same dataset (news articles) but without the use of gazetteers. Another notable approach (Tatar and Çiçekli 2011) proposed an automatic rule learner system for Turkish NER.

With the popularity and availability of social media collections, Turkish NER tools that can be used on more informal domains like tweets and forums have recently been developed (Küçük et al. 2014; Küçük and Steinberger 2014; Çelikkaya et al. 2013; Eken and Tantuğ 2015). Küçük et al. (2014) and Küçük and Steinberger (2014) applied rule-based NER systems to tweets. Çelikkaya et al. (2013) applied CRF-based approach of Şeker and Eryiğit (2012) to tweets, forums, and spoken data. Most recently Kısa and Karagöz (2015) applied NLP from Scratch approach to the NER task to propose more generalized models. They tested their system on both formal and informal texts.

6.3 Task Description

6.3.1 Representation

There are several ways to represent the named-entities and the choice of representation can have a big impact on the performance of NER systems. The most basic and simple format is to just use the raw named-entity tags by marking each token of a named-entity with a tag indicating its type. While simple, this has the important problem that it is not possible to annotate two or more consecutive named-entities properly.

The most common representation scheme for named-entities is the IOB2 representation (Tjong Kim Sang 2002) (also known as BIO). It is a variant of the IOB scheme (Ramshaw and Marcus 1995). With this representation, the first token of any named-entity gets the prefix "B-" in the tag type, and the "I-" prefix is used in the rest of the tokens in the named-entity if it involves multiple tokens. Tokens that are not part of a named-entity are tagged with "O".

Table 6.1 An example tagging for Turkish (Raw, IOB2, and BILOU tags)

Token	Raw tag	IOB2 tag	BILOU
Mustafa	PERSON	B-PERSON	B-PERSON
Kemal	PERSON	I-PERSON	I-PERSON
Atatürk	PERSON	I-PERSON	L-PERSON
23	DATE	B-DATE	B-DATE
Nisan	DATE	I-DATE	I-DATE
1920	DATE	I-DATE	L-DATE
'de	O	O	O
Türkiye	ORGANIZATION	B-ORGANIZATION	B-ORGANIZATION
Büyük	ORGANIZATION	I-ORGANIZATION	I-ORGANIZATION
Millet	ORGANIZATION	I-ORGANIZATION	I-ORGANIZATION
Meclisi	ORGANIZATION	I-ORGANIZATION	L-ORGANIZATION
'ni	O	O	O
Ankara	LOCATION	B-LOCATION	U-LOCATION
'da	O	O	O
kurdu	O	O	O
.	O	O	O

Another representation scheme which is not as popular as IOB2 is the BILOU (Ratinov and Roth 2009). In contrast to BIO (IOB2), the BILOU scheme identifies not only the beginning, inside or outside of a named-entity, but also the last token using the "L-" prefix, in addition to identifying the unit length named-entities with a "U-" prefix. This scheme has been shown to significantly outperform the IOB2 representation (Ratinov and Roth 2009).

The named-entities in the following example Turkish sentence

> Mustafa Kemal Atatürk 23 Nisan 1920'de Türkiye Büyük Millet Meclisi'ni Ankara'da kurdu.
> "Mustafa Kemal Atatürk established the Turkish Grand National Assembly on 23 Nisan 1920 in Ankara."

can be represented with all these three formats as shown in Table 6.1.[1]

Among these representations, the IOB2 scheme has been used most commonly by Turkish NER systems (Yeniterzi 2011; Şeker and Eryiğit 2012; Çelikkaya et al. 2013; Önal et al. 2014). Tür (2000) and Tür et al. (2003) used a different representation, which has been shown to reduce the performance compared to IOB2 representation (Şeker and Eryiğit 2012). They showed that using raw labels is also not as effective as the IOB2 representation. Demir and Özgür (2014) seem to be the only ones who have used the BILOU representation in Turkish NER.

---

[1]Note that any suffixes on the last word of a named-entity is split as a separate token.

6.3.2 Evaluating NER Performance

Evaluation of NER performance has used three metrics: (1) MUC, (2) CoNLL, and (3) ACE. For Turkish NER, researchers have used the first two therefore only those will be detailed in this section. Detailed information on all these three evaluation metrics is available in Nadeau and Sekine (2007).

The MUC metric was initially used when NER was part of the understanding task in both the Sixth and Seventh Message Understanding Conferences (Sundheim 1995; Chinchor and Marsh 1998). This metric has two components that evaluate different aspects of NER tasks. MUC TEXT evaluates only the boundaries of the identified entities, and MUC TYPE evaluates whether the identified type of the entity is correct or not. For each of these two criteria, the following values are computed:

- *Correct*: number of named-entities recognized correctly by the system
- *Actual*: number of segments of tokens the system has indicated as named-entities by marking boundaries
- *Possible*: number of named-entities manually annotated in the data.

These values arc used in *Precision* and *Recall* calculations as follows:

$$Precision = \frac{CorrectType + CorrectText}{ActualType + ActualText} \qquad (6.1)$$

$$Recall = \frac{CorrectType + CorrectText}{PossibleType + PossibleText} \qquad (6.2)$$

Recall measures the percentage of actual existing named-entities in a text that a system correctly recognizes, while precision measures the percentage of the named-entities that are correct among all the named-entities recognized by the system. The *f-measure*, the weighted harmonic mean of the *Precision* and *Recall* defined as

$$f\text{-}measure = \frac{2 \times Precision \times Recall}{Precision + Recall} \qquad (6.3)$$

combines these into one quantity.

While MUC evaluates the identification and classification steps of the NER task separately, the CoNLL metric (Tjong Kim Sang 2002) is more strict, and only accepts labelings in which both the boundary (start and end positions) and the type of entity recognized are correct. A named-entity is counted as correct only if it is an exact match of the corresponding entity both in terms of boundary and type. Similar to MUC, CoNLL metric also uses the *f-measure* to report the finalized score.

MUC metric was commonly used in earlier Turkish NER research (Tür 2000; Tür et al. 2003; Bayraktar and Temizel 2008; Şeker and Eryiğit 2012) while more recent studies have preferred the CoNLL metric (Yeniterzi 2011; Şeker and Eryiğit

2012; Demir and Özgür 2014; Önal et al. 2014; Eken and Tantuğ 2015; Kısa and Karagöz 2015).[2]

6.4 Domain and Datasets

This section describes some commonly used data resources used for Turkish NER system development and evaluation. In general there is a distinction between formal and informal texts. While some basic preprocessing schemes like basic tokenization and morphological processing, etc., are usually enough for the formal datasets such as news texts, informal texts such as those found in social media abound with misspelled forms, incomplete or fragment sentences, require many additional preprocessing steps for more accurate NER.

6.4.1 Formal Texts

Formal texts include but are not limited to news articles and books. They can be defined as well-formed texts with correct spellings of words, proper sentence structure, and capitalization.

One of the first datasets for Turkish NER was created by Tür (2000). This dataset consists of news articles from Milliyet newspaper, covering the period between January 1, 1997 and September 12, 1998. This dataset was annotated with ENAMEX type entities and divided into a training set of 492,821 words containing 16,335 person, 11,743 location, and 9199 organization names, for a total of 37,277 named-entities, and a test set of about 28,000 words, containing 924 person, 696 location, and 577 organization names for a total of 2197 named-entities. Parts of this dataset have been widely used in other Turkish NER studies as well (Yeniterzi 2011; Şeker and Eryiğit 2012; Çelikkaya et al. 2013; Demir and Özgür 2014; Eken and Tantuğ 2015; Kısa and Karagöz 2015).[3]

Another newspaper dataset was constructed by Küçük and Yazıcı (2010) using METU Turkish corpus (Say et al. 2004) as the source. A total of 50 news articles were labeled in MUC style with ENAMEX, NUMEX, and TIMEX tags. This dataset contains 101,700 words with 3280 person, 2470 location, 3124 organization names along with 1413 date/time and 919 money/percent expressions. A subset of this dataset with ten news articles has been also used in several other studies (Küçük and Yazıcı 2009a,b; Kısa and Karagöz 2015).

---

[2]The evaluation scripts from the CONLL 2000 shared task can be found at github.com/newsreader/evaluation/tree/master/nerc-evaluation (Accessed on Sept. 14, 2017).

[3]The entity type counts are different in these studies due to either using different subsets or counting multiple token entities as one or not.

A financial news articles dataset was compiled by Küçük and Yazıcı (2010, 2012). This dataset contains 350 annotated financial news articles retrieved from a news provider, Anadolu Agency, with only person and organization names annotated. It comprises 84,300 words and has 5635 named-entities, 1114 are person names and 4521 are organization names.

Another Turkish newspaper text dataset is the TurkIE dataset (Tatar and Çiçekli 2011), which consists of 355 news articles on terrorism, with 54,518 words. The collection includes 1335 person, 2355 location, 1218 organization names, and 373 date and 391 time expressions for a total of 5672 named-entities.

Texts from two books have also been used in several Turkish NER studies (Küçük and Yazıcı 2009a,b, 2010, 2012). The first one consists of two children's stories, with around 19,000 words which contains manually annotated 836 person, 157 location, 6 organization names, and 65 date/time and 20 money/percent expressions. The second dataset comprises of the first three chapters of a book on Turkish history. It contains about 20,100 words and 387 person, 585 location, 122 organization names, and 79 date/time expressions, all manually annotated.

6.4.2 Informal Texts

Following the general trend in NLP, social media text has become a popular domain for the NER research in recent years. Özkaya and Diri (2011) have used an informal email corpora for NER. Çelikkaya et al. (2013) compiled two social media collections, one from an online forum and another from tweets. The first was from a crawl of a popular online forum for hardware product reviews www. donanimhaber.com (Accessed Sept. 14, 2017). With 54,451 words, this collection contains 21 person, 858 organization, 34 location names and 7 date, 2 time, 67 money, 11 percentage expressions (Çelikkaya et al. 2013). Kısa and Karagöz (2015) present some results from this dataset. The tweet dataset includes 54,283 words with around 676 person, 419 organization, 241 location names and 60 date, 23 time, 14 money, 4 percentage expressions. This tweet dataset has been also used in other NER studies (Küçük et al. 2014; Küçük and Steinberger 2014; Eken and Tantuğ 2015; Kısa and Karagöz 2015).

Another Turkish twitter dataset was compiled by Küçük et al. (2014) and Küçük and Steinberger (2014). Tweets posted on June 26, 2013 in between 12:00 and 13:00 GMT were crawled and after removing non-Turkish tweets, the total number of words was 20,752. In addition to the regular ENAMEX, TIMEX, and NUMEX tags, the authors also annotated TV program series, movies, music bands, and products (Küçük et al. 2014). This dataset includes 457 person, 282 location, 241 organization names, 206 date/time, 25 money/percent expressions, and 111 other named-entities. This collection was also used by Kısa and Karagöz (2015).

Finally, Eken and Tantuğ (2015) crawled and tagged around 9358 tweets consisting of 108,743 tokens with 2744 person, 1419 location, 2935 organization names, and 351 date, 86 time, 212 money expressions.

In addition to these social media texts, a spoken language dataset was also complied through a mobile application (Çelikkaya et al. 2013), by recording and converting spoken utterances into written text by using Google Speech Recognition Service. This dataset has 1451 words and contains 79 person, 64 organization, 90 location names and 70 date, 34 time, 27 money, 26 percentage expressions. This collection has been used in other studies as well (Kısa and Karagöz 2015).

Küçük and Yazıcı (2012) constructed two news video transcriptions data collections. The first one includes 35 manually transcribed news videos of around 4 h broadcast by Turkish Radio and Television (TRT). The second video data collection includes 19 videos with a total duration of 1.5 h. Unlike the first one, this video collection has been transcribed automatically using a sliding text recognizer.

6.4.3   Challenges of Informal Texts for NER

The switch from formal domains to these informal ones brings several challenges which cause significant reductions in NER performance (Ritter et al. 2011). As in other similar NLP tasks, the state-of-the-art NLP tools which assume properly constructed input texts may not perform as expected when applied to text in informal domains which contains a lot of misspelled words, ungrammatical constructs and extra-grammatical tokens such as user handles or hashtags.

For instance, Küçük et al. (2014) identified several peculiarities in informal texts especially in tweets. These include but not limited to grammar and spelling errors like incorrect use of capitalization, not using apostrophes to separate suffixes from named-entities, repeating letters for emphasis, using ASCII characters instead of proper Turkish characters. There are also some challenges due to size limitation in tweets leading to lack of useful contextual clues like person titles, professions, or using contracted forms of words or just using single forenames, surnames instead of the full names (Küçük et al. 2014). For instance, wrong use of capitalization and apostrophe makes it harder to recognize proper nouns which are also valid common nouns. Other spelling and grammar errors cause some language analysis tools like morphological analyzers to fail. Therefore, NER systems that depend on significant linguistic analysis of the texts may not perform as expected in such conditions.

In tweets there is also the case of named-entities occurring within a single hashtag but as a single token, for example, *#Istanbuldabahar*, or they can cover the whole hashtag like *#MustafaKemalAtaturk* (Küçük et al. 2014). Clearly these cases impose significant challenges for NER.

6.5   Preprocessing for NER

Depending on the NER system, there can be several data preprocessing steps that come before the identification of named-entities. These are tokenization, morphological analysis, and normalization.

6.5.1   Tokenization

Most NER systems use a word-level tokenizer. The apostrophe symbol that is used in standard formal Turkish orthography to indicate the boundary of the stem and suffixes in proper nouns can be used to split such tokens so that those suffixes appear as a separate token. Other punctuation characters that are not legitimate parts of tokens (e.g., decimal points) are considered as separate tokens (Şeker and Eryiğit 2012). Of course, other tokenization schemes are also possible: Yeniterzi (2011) has considered a morpheme-level tokenization where roots and connected morphemes were considered as separate tokens. The idea was to introduce explicit morphological information to the model, which, while not degrading the performance, did not produce a significant improvement. In her experiments, morpheme-level tokenization outperformed word-level tokenization in identification of person and location named-entities but caused drops for others.

6.5.2   Morphological Analysis

Morphological analysis is among the commonly used preprocessing steps. In order to deal with data sparsity issues, some NER systems use stems or root words in addition to the lexical form of the words. Also, some feature-based systems use inflectional morphemes to identify named-entities. Most Turkish NER systems (Yeniterzi 2011; Şeker and Eryiğit 2012; Eken and Tantuğ 2015) used Oflazer's two-level morphological analyzer (Oflazer 1994) to construct the morphological analysis of the word. A morphological disambiguator (Sak et al. 2011) was also used to resolve the morphological ambiguity.

Küçük and Yazıcı (2009a) also used their own morphological analyzer for their rule-based system. Their analyzer only considers the noun inflections on tokens which exist in the dictionaries and match an existing pattern.

In informal texts, like tweets, morphological analyzers do not work as expected because of spelling errors, capitalization errors, use of nonstandard orthographical forms or not using proper Turkish characters. In order to deal with these, some systems (Çelikkaya et al. 2013; Küçük and Steinberger 2014) have attempted normalizing text as described in the next section. Eken and Tantuğ (2015) proposed using the first and the last four characters instead of the root and inflectional morphemes. Their experiments over tweets showed using such a heuristic provides similar results compared to using a morphological analyzer.

6.5.3 Normalization

As alluded to before, one way to deal with text in informal domains is to tailor the text so that NER systems developed over formal datasets can work with them. Several authors (Çelikkaya et al. 2013; Küçük and Steinberger 2014; Kısa and Karagöz 2015; Eken and Tantuğ 2015) have looked at this as a normalization procedure and applied steps to deal with the following:

- *Slang words*: Slang words are replaced with their more formal usage. For instance, *nbr* is replaced with *ne haber?—what's up?* (Çelikkaya et al. 2013)
- *Repeated characters*: Characters that are repeated for emphasis purposes but lead to a misspelled form are removed. (i.e., *çoooook* for *çok—many*) (Çelikkaya et al. 2013; Küçük and Steinberger 2014)
- *Special tokens*: Hash tags, mentions, smiley icons, and vocatives are replaced with certain tags (Çelikkaya et al. 2013)
- *Emo style writing*: Emo style writing and characters are replaced with their correct characters (i.e., *$eker 4 you* instead of *Seker senin için—Sweety! for you* (Çelikkaya et al. 2013)
- *Capitalization*: All characters are lowercased. (i.e., "aydin" for "Aydin") (Çelikkaya et al. 2013; Kısa and Karagöz 2015)
- *Asciification*: Special Turkish characters *(ç, ğ, ı, ö, ş, ü)* are replaced with equivalent nearest ASCII characters *(c, g, i, o, s, u)*. (Eken and Tantuğ 2015)

Çelikkaya et al. (2013) applied the CRF-based approach of Şeker and Eryiğit (2012) to one formal and three types of informal texts with different subsets of features. While normalization provided observable improvements when applied to tweets, it degraded the performance when applied to formal news dataset, and did not result in an improvement with forum and speech datasets (Çelikkaya et al. 2013). Overall, for informal domains, there is still room for improvement.

Apart from normalizing informal texts like tweets, normalization can also be applied to formal texts to make generalizations. For instance, Demir and Özgür (2014) normalized all numerical expressions into a generic number pattern so that unseen number tokens during testing could be handled properly.

6.6 Approaches Used in Turkish NER

The approaches for NER task can be divided into three main categories: (1) hand-crafted rule-based systems, (2) machine learning based systems, and (3) combination of the first two, hybrid systems. In this section, we review Turkish NER systems and categorize them with respect to these approaches and describe some in detail. Even though it is impossible to make a fair comparison among these systems due to the differences between datasets used, their highest performance scores are nevertheless reported in order to give the reader some idea of the state-of-the-art performance.

6.6.1 Rule-Based Approaches

Küçük and Yazıcı (2009a,b) developed the first rule-based NER system for Turkish. They used two types of information sources: (1) lexical resources and (2) patterns. Lexical resources consists of a gazetteer of person names and lists of well-known people, organizations, and locations. Pattern bases include manually constructed patterns for identifying location names, organization names, and temporal and numerical expressions. Example patterns are as follows:

- Patterns for location names:

  X Sokak/Yolu/Kulesi/Stadyumu/...
  X Street/Road/Tower/Stadium/...

- Patterns for organization names:

  X Grubu/A.Ş./Partisi/Üniversitesi/...
  X Group/Inc./Party/University/...

- Patterns for temporal and numeric expressions:

  X başı/ortası/sonu...
  X start/middle/end...
  'The start/middle/end...of X'

While the authors targeted news text, they also tested their system over different text genres, including children's stories, historical texts, and news video transcriptions. Since not all these (like video transcriptions) have proper capitalization and punctuation, they were not able to exploit these clues for NER. The f-measures for their system were 78.7% on news articles, 69.3% on children's stories, 55.3% on historical texts, and 75.1% on video transcriptions. Even though their results were not even close to the state-of-the-art systems at that time, this study can be considered as a good baseline point for rule-based Turkish NER systems.

This system has been also applied to informal text like tweets with some simple modifications, in order to deal with the peculiarities of the data (Küçük and Steinberger 2014). Due to lack of proper use of capitalization in such texts, the authors initially relaxed the capitalization constraint of the system. They also extended their lexical resources to include both diacritic and non-diacritic variants of the entries. Several tweet normalization techniques were also applied. Experiments over two different tweet collections showed that these modifications were useful

(Küçük and Steinberger 2014). Önal et al. (2014) also applied a rule-based approach inspired by Küçük and Yazıcı (2009a,b) to tweets in order to identify locations.

Küçük et al. (2014) used Europe Media Monitor (EMM) multilingual media analysis and information extraction system (Pouliquen and Steinberger 2009) for Turkish NER. EMM is a language independent rule-based system which uses dictionary lists which contain language-specific words for titles, professions, etc. The EMM system can be adapted to a language by using these lists together with some capitalization related rules. Küçük et al. (2014) identified frequently mentioned person and organization names from news articles and used them to extend the existing resources of the system and applied it to Turkish tweets. On news domain they got an f-measure of 69.2% while on tweets the f-measure was 42.7%.

Dalkılıç et al. (2010) proposed another rule-based system where tokens and morphemes that frequently occur close to person, organization, and location entities can be used to classify other entities. This system was tested over economics, politics, and health domain texts and the best performance was observed in identifying locations with an f-measure of 87.0% on average. Unlike location, person and organization identification performances are lower with f-measures of 80.0% and 81.0%, respectively.

Bayraktar and Temizel (2008) used a system with several manually constructed patterns to identify person named-entities. They applied a local grammar approach (Traboulsi 2006) to recognize person names from Turkish financial texts.[4] Bayraktar and Temizel (2008) initially identified common reporting verbs in Turkish, such as *dedi* (said), *sordu* (asked), then they used these reporting verbs to generate patterns for locating person names. This approach returned an f-measure of 82.0% on news articles.

6.6.2 Hybrid Approaches

The problem with the rule-based systems is that they require the addition of more and more rules and their performance degrades when ported to new domains. In order to overcome this problem, Küçük and Yazıcı (2009a,b) extended their rule-based NER tool into a hybrid recognizer (Küçük and Yazıcı 2010, 2012), so that in a new domain, it can learn from the available annotated data and extend its knowledge resources. They used rote learning (Freitag 2000), which basically groups and stores available named-entities in the training set. When applying this system on different domains, the system starts with the same set of patterns and lexicons, but in the learning stage, it adapts itself to the particular domain by learning

---

[4]A local grammar is "a way of describing the syntactic behavior of groups of individual elements, which are related but whose similarities cannot be easily expressed using phrase structure rules" (Mason 2004).

from the new domain's training data. Küçük and Yazıcı (2009a,b) used their rule-based NER system originally targeted for news texts, and applied it to financial news texts, historical texts, and children's stories. In these experiments, the hybrid entity recognizer outperformed the rule-based system with an f-measure of 86.0% on news, 74.2% on financial news, 85.0% on child stories, and 67.0% on historical texts.[5] These scores were improved further (up to 90.1% on news domain) when they turned on the capitalization feature.

Yavuz et al. (2013) proposed another hybrid approach where they use a Bayesian learning together with the rule-based system by Küçük and Yazıcı (2009a,b).

6.6.3 Machine Learning Approaches

Due to their ability to learn from annotated data and not relying on hand-crafted-rules, and easy adaptability to new domains, machine learning approaches have been used widely in developing NER systems. These approaches however depend on having datasets where named-entities of interest are properly annotated.

The first work on Turkish NER describes a language independent EM-style bootstrapping algorithm that learns from word internal and contextual information of entities (Cuerzan and Yarowsky 1999). The bootstrapping algorithm is a semi-supervised learning algorithm, which starts with a seed set of examples or patterns and iteratively learns new patterns using the clues seeds provide. The authors used hierachically smoothed trie structures for modeling the word internal (morphological) and contextual probabilities. The first set of clues refers to the patterns of prefixes or suffixes which are good indicators of a named-entity. For instance, for Turkish, '-oğlu' (son of) is a strong surname indicator. The contextual patterns either preceding or following a named-entity can also help identify them: for example, "Bey" (Mr.) or "Hanım" (Mrs.) can help identify preceding words as person names. Turkish was one of the five languages evaluated (along with English, Greek, Hindi, and Romanian). With a training size of 5207 tokens and 150 seeds, an f-measure of 53.0% was reported for Turkish.

Tür (2000) and Tür et al. (2003) developed a statistical name tagger system specifically for Turkish which depends on n-gram language models embedded in HMMs. They used four information sources and augmented lexical model with contextual, morphological, and tag models. In their lexical model, which can be considered as a baseline, they only used the lexical forms of the tokens. A word/tag combination HMM was built and trained, where a tag represents whether the word is part of a named-entity and if so its type. In the contextual model, in order to deal with words that do not appear in training data, they built another model with named entities tagged as *unknown*. This model provided useful clues regarding

---

[5]The data collection used in this study is not exactly the same with data used in Küçük and Yazıcı (2009a,b).

Table 6.2 F-measure results from Tür et al. (2003)

Model	Text	Type	F-measure
Lexical	80.87%	91.15%	86.01%
Lexical + Contextual	86.00%	91.72%	88.86%
Lexical + Contextual + Morphological	87.12%	92.20%	89.66%
Lexical + Contextual + Tag	89.54%	92.13%	90.84%
Lexical + Contextual + Morphological + Tag	90.40%	92.73%	91.56%

Table 6.3 F-measure results from Yeniterzi (2011)

Model	Person	Location	Organization	Overall
Lexical	80.88%	77.05%	88.40%	82.60%
Lexical + Root	83.32%	80.00%	90.30%	84.96%
Lexical + Root + POS	84.91%	81.63%	90.18%	85.98%
Lexical + Root + POS + Prop	86.82%	82.66%	90.52%	87.18%
Lexical + Root + POS + Prop + Case	88.58%	84.71%	91.47%	88.71%

the preceding and following tokens inside and around the named entities. Their morphological model captures information related to the morphological analysis of the token. The name tag model ignores the lexical form of the words and only captures the name tag information (like person, location, organization) of the words. Using only the tags and boundary information is useful for identifying multi-token named entities.

Tür (2000) and Tür et al. (2003) used MUC scoring to evaluate these four models and their combinations. The experimental results including both *text* and *type* and the overall *f-measure* scores of these models are summarized in Table 6.2. The baseline lexical model starts with 86.01% f-measure. Using the contextual cues in recognizing unknown words returned improvements up to 5.13% in *text* score. Furthermore, incorporating the tag model increased the *text* score by more than 3% points due to decreasing the improbable tag sequences. Combination of all these four models provided the best performance with 91.6% f-measure.

Conditional Random Fields (CRF) (Lafferty et al. 2001) have been used in several Turkish NER tools. Yeniterzi (2011) built a CRF-based NER tool for Turkish where she used features like stem, part-of-speech, proper noun markers, and case markers, in addition to the lexical form of the token. The individual effects of these features are summarized in Table 6.3. As a morphologically rich language, even adding the root (stem) as a feature to the lexical model improved the system by 2–3%. Other exploited features provided 1–2% improvements to the system individually, which at the end resulted in around 6% improvement in overall f-measure.

In order to see the effects of morphology more clearly, Yeniterzi (2011) also employed a morpheme level tokenization in which a word is represented in several states in the CRF: one state for the root and one state for each morphological feature.

Morpheme-level tokenization model, exploiting the same set of features as in the case of word-level model, improved the overall f-measure to 88.94%.

Özkaya and Diri (2011) applied a CRF-based NER system to emails. They used features like capitalization, punctuation, context, and email field related features like whether the token belongs to *from*, *to*, or other similar fields. The system showed the highest performance in the identification of person named-entities with an 95% f-measure. The authors did not explore the impact of specific features over the results, so it is possible that the field related feature can be the determining component.

Şeker and Eryiğit (2012) also proposed a CRF-based system. Similar to Yeniterzi (2011), they employed lexical and morphological features like stem, part-of-speech, noun case, proper noun markers, various inflectional features. They applied the approach in Sha and Pereira (2003) to these features and manually selected the useful ones. All these added features improved the performance of the system to an f-measure of 91.9%, and outperformed some of the prior work. Çelikkaya et al. (2013) applied a similar approach to tweets, forum, and speech datasets. For training they used the same news dataset used by Şeker and Eryiğit (2012). As expected the CRF model performed at a much worse level when tested on these informal domains with f-measures 6.9% with speech dataset, 5.6% with forum dataset, and 12.2% with tweets. Even though the performance of tweets increased to 19.3% after normalizing them, the performance level was not comparable to that on formal datasets. Önal et al. (2014) also applied this approach to tweets just to recognize locations. Eken and Tantuğ (2015) compared the approach of Şeker and Eryiğit (2012) by using a simpler preprocessing used the first and last four characters of tokens instead of features extracted from morphological analysis of words. Their model exhibited a similar performance to the morphological model. This model which was trained on news articles was tested over tweets with low performance as expected but when training was performed over tweets, the test provided an f-measure 64.03 on tweets.

Tatar and Çiçekli (2011) proposed an automatic rule learning system for NER task. They started with a set of seeds selected from the training set, and then extracted rules over these examples. They generalized the named-entities by using contextual, lexical, morphological, and orthographic features. During this generalization procedure, they used several rule filtering and refinement techniques in order to keep their accuracy high with an f-measure of 91.1%.

Yavuz et al. (2013) were the first to apply the Bayesian Learning approach to Turkish NER. They employed a modified version of the BayesIDF approach (Freitag 2000) with features like case sensitivity, case, token length, etc., which exhibited an f-measure of 88.4%. Two hybrid systems were also constructed by combining this system with a rule-based system (Küçük and Yazıcı 2009a,b). In the first system the training data was used to train the Bayesian learner, and then the rule-based tagged NER data was used as additional training data to update the system. In the second system, the tagged output of the rule-based system was used as an additional feature by the Bayesian learner. Both hybrid systems outperformed the Bayesian learner alone, the first one with an f-measure of 90.0% and the second with 91.4%.

Another semi-supervised approach to Turkish NER was recently proposed by Demir and Özgür (2014). Their neural network based approach had two stages.

In the unsupervised stage, neural networks were used to obtain continuous vector representation of words by using large amounts of unlabeled data. In the supervised stage, these feature vectors and additional language independent features like capitalization patterns of previous tag predictions were used in another neural network to train the NER system. These word representations were also clustered to identify semantically similar words, and cluster ids were used as additional feature. This system has an f-measure of 91.85.

Another recently published semi-supervised approach to NER has also used word embeddings (Kısa and Karagöz 2015). The author have applied NLP from Scratch method (Collobert et al. 2011) to NER on social media texts. Initially a language model and word embeddings were learned from a large unannotated dataset and later these word embeddings were used as features to train a neural network classifier on labeled data. The authors have experimented with different datasets and domains: On formal text their approach outperformed the rule-based system of Küçük and Yazıcı (2009a) but was not better than the CRF-based system by Şeker and Eryiğit (2012) or neural network-based approach of Demir and Özgür (2014). However, when applied to informal texts, this system also outperformed a CRF-based system (Çelikkaya et al. 2013).

6.7   Conclusions

This section presented an overview of Turkish NER systems that have been developed in the last two decades, covering their salient aspects and performance, in addition to pointing out some of the datasets used for developing such systems. It is clear that there is significant room for improvement for Turkish NER systems especially in informal text domains and while performance of these systems is reasonably high on formal texts, further improvements and quick adaptability are the ongoing concerns.

References

Bayraktar Ö, Temizel TT (2008) Person name extraction from Turkish financial news text using local grammar based approach. In: Proceedings of ISCIS, Istanbul

Çelikkaya G, Torunoğlu D, Eryiğit G (2013) Named entity recognition on real data: a preliminary investigation for Turkish. In: Proceedings of the international conference on application of information and communication technologies, Baku

Chinchor N, Marsh E (1998) Appendix D: MUC-7 information extraction task definition (version 5.1). In: Proceedings of MUC, Fairfax, VA

Collobert R, Weston J, Bottou L, Karlen M, Kavukçuoğlu K, Kuksa P (2011) Natural language processing (almost) from scratch. J Mach Learn Res 12:2493–2537

Cucerzan S, Yarowsky D (1999) Language independent named entity recognition combining morphological and contextual evidence. In: Proceedings of EMNLP-VLC, College Park, MD, pp 90–99

Dalkılıç FE, Gelişli S, Diri B (2010) Named entity recognition from Turkish texts. In: Proceedings of IEEE signal processing and communications applications conference, Diyarbakır, pp 918–920

Demir H, Özgür A (2014) Improving named entity recognition for morphologically rich languages using word embeddings. In: Proceedings of the international conference on machine learning and applications, Detroit, MI, pp 117–122

Doddington G, Mitchell A, Przybocki M, Ramshaw L, Strassel S, Weischedel R (2004) The automatic content extraction (ACE) program–tasks, data, and evaluation. In: Proceedings of LREC, Lisbon, pp 837–840

Eken B, Tantuğ C (2015) Recognizing named-entities in Turkish tweets. In: Proceedings of the international conference on software engineering and applications, Dubai

Freitag D (2000) Machine learning for information extraction in informal domains. Mach Learn 39(2–3):169–202

Kısa KD, Karagöz P (2015) Named entity recognition from scratch on social media. In: Proceedings of the international workshop on mining ubiquitous and social environments, Porto

Küçük D, Steinberger R (2014) Experiments to improve named entity recognition on Turkish tweets. Arxiv – computing research repository. arxiv.org/abs/1410.8668. Accessed 14 Sept 2017

Küçük D, Yazıcı A (2009a) Named entity recognition experiments on Turkish texts. In: Proceedings of the international conference on flexible query answering systems, Roskilde, pp 524–535

Küçük D, Yazıcı A (2009b) Rule-based named entity recognition from Turkish texts. In: Proceedings of the international symposium on innovations in intelligent systems and applications, Trabzon

Küçük D, Yazıcı A (2010) A hybrid named entity recognizer for Turkish with applications to different text genres. In: Proceedings of ISCIS, London, pp 113–116

Küçük D, Yazıcı A (2012) A hybrid named entity recognizer for Turkish. Expert Syst Appl 39(3):2733–2742

Küçük D, Jacquet G, Steinberger R (2014) Named entity recognition on Turkish tweets. In: Proceedings of LREC, Reykjavík, pp 450–454

Lafferty JD, McCallum A, Pereira F (2001) Conditional random fields: probabilistic models for segmenting and labeling sequence data. In: Proceedings of ICML, Williams, MA, pp 282–289

Mason O (2004) Automatic processing of local grammar patterns. In: Proceedings of the annual colloquium for the UK special interest group for computational linguistics, Birmingham, pp 166–171

Nadeau D, Sekine S (2007) A survey of named entity recognition and classification. Lingvisticae Investigationes 30(1):3–26

Oflazer K (1994) Two-level description of Turkish morphology. Lit Linguist Comput 9(2):137–148

Önal KD, Karagöz P, Çakıcı R (2014) Toponym recognition on Turkish tweets. In: Proceedings of IEEE signal processing and communications applications conference, Trabzon, pp 1758–1761

Özkaya S, Diri B (2011) Named entity recognition by conditional random fields from Turkish informal texts. In: Proceedings of IEEE signal processing and communications applications conference, Antalya, pp 662–665

Pouliquen B, Steinberger R (2009) Automatic construction of multilingual name dictionaries. In: Goutte C, Cancedda N, Dymetman M, Foster G (eds) Learning machine translation. The MIT Press, Cambridge, MA, pp 266–290

Ramshaw LA, Marcus MP (1995) Text chunking using transformation-based learning. In: Proceedings of the workshop on very large corpora, Cambridge, MA, pp 82–94

Ratinov L, Roth D (2009) Design challenges and misconceptions in named entity recognition. In: Proceedings of CONLL, Boulder, CO, pp 147–155

Ritter A, Clark S, Mausam, Etzioni O (2011) Named entity recognition in tweets: an experimental study. In: Proceedings of EMNLP, Edinburgh, pp 1524–1534

Sak H, Güngör T, Saraçlar M (2011) Resources for Turkish morphological processing. Lang Resour Eval 45(2):249–261

Say B, Zeyrek D, Oflazer K, Özge U (2004) Development of a corpus and a treebank for present-day written Turkish. In: Proceedings of the international conference on Turkish linguistics, Magosa, pp 183–192

Şeker GA, Eryiğit G (2012) Initial explorations on using CRFs for Turkish named entity recognition. In: Proceedings of COLING, Mumbai, pp 2459–2474

Sha F, Pereira F (2003) Shallow parsing with conditional random fields. In: Proceedings of NAACL-HLT, Edmonton, pp 134–141

Sundheim BM (1995) Overview of results of the MUC-6 evaluation. In: Proceedings of MUC, Columbia, MD, pp 13–31

Tatar S, Çiçekli İ (2011) Automatic rule learning exploiting morphological features for named entity recognition in Turkish. J Inf Sci 37(2):137–151

Tjong Kim Sang EF (2002) Introduction to the CoNLL-2002 shared task: language-independent named entity recognition. In: Proceedings of CONNL, Taipei, pp 1–4

Tjong Kim Sang EF, De Meulder F (2003) Introduction to the CoNLL-2003 Shared Task: language-independent named entity recognition. In: Proceedings of CONLL, Edmonton, pp 142–147

Traboulsi HN (2006) Named entity recognition: a local grammar-based approach. PhD thesis, Surrey University, Guildford

Tür G (2000) A statistical information extraction system for Turkish. PhD thesis, Bilkent University, Ankara

Tür G, Hakkani-Tür DZ, Oflazer K (2003) A statistical information extraction system for Turkish. Nat Lang Eng 9:181–210

Yavuz SR, Küçük D, Yazıcı A (2013) Named entity recognition in Turkish with Bayesian learning and hybrid approaches. In: Proceedings of ISCIS, Paris, pp 129–138

Yeniterzi R (2011) Exploiting morphology in Turkish named entity recognition system. In: Proceedings of ACL-HLT, Portland, OR, pp 105–110

Chapter 7
Dependency Parsing of Turkish

Gülşen Eryiğit, Joakim Nivre, and Kemal Oflazer

Abstract Syntactic parsing is the process of taking an input sentence and producing an appropriate syntactic structure for it. It is a crucial stage in that it provides a way to pass from core NLP tasks to the semantic layer and it has been shown to increase the performance of many high-tier NLP applications such as machine translation, sentiment analysis, question answering, and so on. Statistical dependency parsing with its high coverage and easy-to-use outputs has become very popular in recent years for many languages including Turkish. In this chapter, we describe the issues in developing and evaluating a dependency parser for Turkish, which poses interesting issues and many different challenges due to its agglutinative morphology and freeness of its constituent order. Our approach is an adaptation of a language-independent data-driven statistical parsing system to Turkish.

7.1   Introduction

Parsers for natural languages have to cope with a high degree of ambiguity and nondeterminism, and thus they are typically based on different techniques than the ones used for parsing well-defined formal languages—for example, those used for compilers for programming languages. Hence, the mainstream approach to natural language parsing uses algorithms that efficiently derive a potentially very large set of analyses in parallel, typically making use of dynamic programming,

G. Eryiğit
Istanbul Technical University, Istanbul, Turkey
e-mail: gulsen.cebiroglu@itu.edu.tr

J. Nivre
Uppsala University, Uppsala, Sweden
e-mail: joakim.nivre@lingfil.uu.se

K. Oflazer (✉)
Carnegie Mellon University Qatar, Doha-Education City, Qatar
e-mail: ko@cs.cmu.edu

© Springer International Publishing AG, part of Springer Nature 2018
K. Oflazer, M. Saraçlar (eds.), *Turkish Natural Language Processing*,
Theory and Applications of Natural Language Processing,
https://doi.org/10.1007/978-3-319-90165-7_7

and well-formed substring tables or charts. When disambiguation is required, this approach can be coupled with a statistical model for parse selection that ranks competing analyses with respect to plausibility. Although, for efficiency reasons, it is often necessary to prune the search space prior to the ranking of complete analyses, this type of parser always has to handle multiple analyses. By contrast, parsers for formal languages are usually based on deterministic parsing techniques, which are maximally efficient in that they only derive one analysis. This is possible because the formal language can be defined by an unambiguous formal grammar that assigns a single canonical derivation to each string in the language, a property that cannot be maintained for any realistically-sized natural language grammar. Consequently, these deterministic parsing techniques have been much less popular for natural language parsing, except as a way of modeling human sentence processing, which appears to be at least partly deterministic in nature (Marcus 1980; Shieber 1983).

More recently, however, it has been shown that accurate syntactic disambiguation for natural language can be achieved using a pseudo-deterministic approach, where treebank-induced classifiers are used to predict the optimal next derivation step when faced with a nondeterministic choice between several possible actions. Compared to the more traditional methods for natural language parsing, this can be seen as a severe form of pruning, where parse selection is performed incrementally so that only a single analysis is derived by the parser. This has the advantage of making the parsing process very simple and efficient, but also has the potential disadvantage that overall accuracy suffers because of the early commitment enforced by the greedy search strategy. Somewhat surprisingly though, research has shown that, with the right parsing algorithm and classifier, this type of parser can achieve state-of-the-art accuracy, especially when used with dependency-based syntactic representations.

Transition-based dependency parsing, the term used nowadays for this approach, was pioneered by Kudo and Matsumoto (2002) for unlabeled dependency parsing of Japanese with head-final dependencies only. The algorithm was later generalized to allow both head-final and head-initial dependencies by Yamada and Matsumoto (2003), who reported very good parsing accuracy for English using dependency structures extracted from the Penn Treebank for training and testing. The approach was then extended to labeled dependency parsing of Swedish by Nivre et al. (2004), and of English by Nivre and Scholz (2004), using a different parsing algorithm first presented in Nivre (2003). In the CoNLL-X shared task on multilingual dependency parsing with data from 13 different languages (Buchholz and Marsi 2006), the transition-based parser by Nivre et al. (2006) reached top performance together with the system of McDonald et al. (2006), which is based on a global discriminative model with online learning. These results indicate that, at least for dependency parsing, deterministic parsing is possible without a drastic loss in accuracy.

Turkish, being an agglutinative and free constituent order language, can be seen as the representative of a wider class of languages. Work on dependency parsing of Turkish starts with Oflazer (2003) who proposed an extended finite-state approach for marking dependencies between words in a morphologically analyzed and disambiguated Turkish sentence. Eryiğit and Oflazer (2006) were the first to attempt to apply statistical approaches to dependency parsing of Turkish. They proposed a wide-coverage parser based on three components:

1. a parsing algorithm for building the dependency analyses (Eisner 1996; Sekine et al. 2000),
2. a conditional probability model to score the analyses (Collins 1996), and
3. maximum likelihood estimation to make inferences about the underlying probability models (Collins 1996; Chung and Rim 2004).

Eryiğit and Oflazer (2006) was also the first study to argue that parsing accuracy for Turkish has to be defined based on morphological units, and to show that accuracy can be improved by taking such units rather than word forms as the basic units of syntactic structure. However, the algorithm could only parse head-final dependencies and could not use the lexical information efficiently because of the modest size of the treebank.

This approach of using morphological units was then adopted in a series of studies (Eryiğit et al. 2008, 2006; Nivre et al. 2006), all based on MaltParser (Nivre et al. 2007). The dependency parser presented in the rest of this chapter has been shown to obtain the highest accuracy scores for the dependency parsing of Turkish.

In the remainder of the chapter, we provide an overview of dependency parsing and follow-up with a discussion of the important syntactic and morphological properties of Turkish that need to be considered in designing a suitable syntactic parsing system. We then present an incremental data-driven statistical dependency parsing system and its adaptation for Turkish together with a relevant evaluation methodology, and discuss some recent additional studies and ongoing work.

7.2   Dependency Parsing

Dependency-based syntactic theories are founded on the idea that syntactic structure can be analyzed in terms of binary, asymmetric dependency relations holding between the words of a sentence. This basic conception of syntactic structure underlies a variety of different linguistic theories, such as Structural Syntax (Tesnière 1959), Functional Generative Description (Sgall et al. 1986), Meaning-Text Theory (Mel'čuk 1988), and Word Grammar (Hudson 1990). In recent years, dependency-based syntactic representations have been used primarily in data-driven models that learn to produce dependency structures for sentences solely from an annotated corpus. One potential advantage of such models is that they are easily ported to any domain or language for which annotated resources exist.

In this kind of framework, the syntactic structure of a sentence is modeled by a dependency graph, which represents each word and its syntactic dependents through labeled directed arcs. Figure 7.1 exemplifies this on a Turkish sentence.[1] The arrows

---

[1]Please note that arrows in this representations point from dependents to heads and we do not include punctuation in dependency relations.

Fig. 7.1 The dependency graph of a Turkish sentence for "He regards his pen as his only weapon"

in the figure show the directed dependency relationships between a dependent and a head word, and the labels on top of the arrows denote the types of the dependencies (e.g., subject, modifier, object, etc.) The aim of a dependency parser is to assign the correct dependent-head relationships for each word in a sentence.

Statistical dependency parsing and parsers have made very rapid progress during the last decade. Most of these parsers can be grouped into two broad classes: transition-based parsers and graph-based parsers. Transition-based parsers parameterize the parsing problem by the structure of an abstract state machine, or transition system, and learn to score parsing actions. Early transition-based parsers like those of Yamada and Matsumoto (2003), Nivre et al. (2004), and Attardi (2006) all used locally trained classifiers and deterministic parsing to achieve very efficient parsing. More recent developments have focused on improving predictions using techniques such as beam search (Zhang and Clark 2008) and dynamic oracles (Goldberg and Nivre 2012). Graph-based parsers instead parameterize the parsing task by the structure of the dependency graph, and learn to score entire dependency graphs. Early graph-based parsers like those of McDonald et al. (2005) used only local features to permit exact inference. More recent work has focused on extending the scope of features while maintaining reasonable efficiency, often using approximate inference (McDonald and Pereira 2006; Martins et al. 2009; Koo and Collins 2010; Koo et al. 2010). In this chapter, we will focus on transition-based dependency parsing, which has been the dominant approach for Turkish.

7.3 Morphology and Dependency Relations in Turkish

Turkish displays rather different characteristics compared to the more well-studied languages in the parsing literature. Most of these characteristics are also found in agglutinative languages such as Basque, Estonian, Finnish, Hungarian, Japanese, and Korean.[2] Figure 7.2 (Eryiğit 2007) shows the dependency graphs of the same sentence "He regards his pen as his only weapon." in different languages (Turkish,

---

[2]We however do not necessarily suggest that the morphological sub-lexical representation that we use for Turkish later in this paper is applicable to these languages.

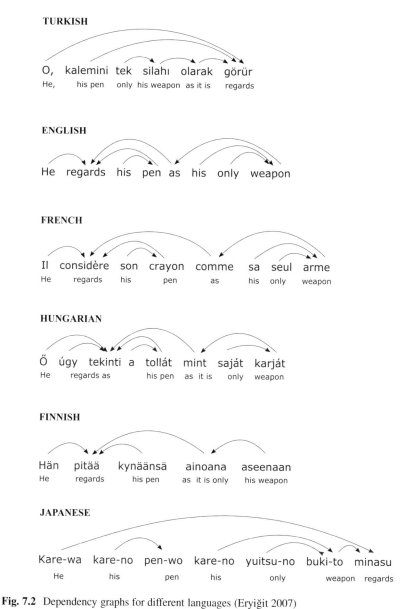

Fig. 7.2 Dependency graphs for different languages (Eryiğit 2007)

English, French, Hungarian, Finnish, and Japanese). One can note in this figure that languages from the same language family (e.g., English and French) have very similar dependency representations whereas Turkish dependency representation could be considered rather close to that of Japanese.

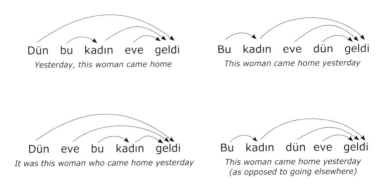

Fig. 7.3 Free constituent ordering in an example Turkish sentence ("Yesterday, this woman came home.") (Eryiğit 2007)

Turkish is a flexible constituent order language. Even though, in written texts, the constituent order predominantly conforms to the Subject-Object-Verb order, constituents may freely be ordered, depending on the discourse context (Erguvanlı 1979; Hoffman 1994). However, from a dependency structure point of view, Turkish is predominantly (but not exclusively) head final. Figure 7.3 gives examples of different orders for a Turkish sentence which means "Yesterday, this woman came home." propositionally. In all of the four combinations, the general meaning of the sentence remains the same but the contextual interpretations emphasized by the order difference are important. For example, in the first ordering, it is the word "home" which is emphasized whereas in the second one, it is the word "yesterday." The dependencies, however, remain to be head-final unless the verb moves to a non-final position.[3] One should notice that there exist also other possible orderings when we write the sentence in an inverted manner, such as "Eve geldi dün bu kadın." or "Eve dün geldi bu kadın."

Turkish has a very rich agglutinative morphological structure (see Chap. 2 for details.) Nouns can give rise to over one hundred inflected forms and verbs to many more. Furthermore, Turkish words may be formed through very productive derivations, increasing substantially the number of possible word forms that can be generated from a root word. It is not uncommon to find up to four or five derivations in a single word. Previous work on Turkish (Hakkani-Tür et al. 2002; Oflazer et al. 2003; Oflazer 2003; Eryiğit and Oflazer 2006) has represented the morphological structure of Turkish words by splitting them into inflectional groups (IGs). The root and derivational elements of a word are represented by different IGs, separated from each other by derivational boundaries (DB). Each IG is then annotated with its own

---

[3]In Turkish, such sentences are called "inverted sentences" and are mostly used in spoken language but rarely in written form.

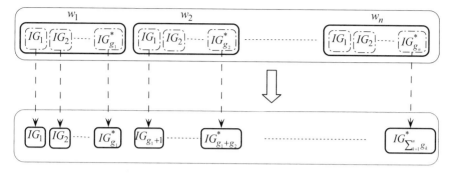

Fig. 7.4 Mapping from word-based to IG-based representation of a sentence

part-of-speech and any inflectional features as illustrated in the following example[4]:

$$arabanızdaydı$$
("it was in your car")
$$arabanızda\; ˆDB\; ydı$$

$$\underbrace{araba + \text{Noun+A3sg+P2pl+Loc}}_{IG_1} \quad DB \quad \underbrace{\text{+Verb+Zero+Past+A3sg}}_{IG_2}$$

"in your car"   "it was"

In this example, the root of the word *arabanızdaydı* is *araba* ("car") and its part-of-speech is noun. From this, a verb is derived in a separate IG. So, the word is composed of two IGs where the first one *arabanızda* ("in your car") is a noun with a locative case marker and in second plural person possessive marker, and the second one is a verbal derivation from this noun, with a past tense marker and third person singular agreement.

Figure 7.4 shows the mapping of the units of word-based sentence model to the IG-based model. However, one still needs to know which IGs are word-final (shown in the figure by asterisks) as they mediate the dependency relations of the dependents.

7.3.1 Dependency Relations in Turkish

Since most syntactic information is mediated by morphology, it is not sufficient for the parser to only find dependency relations between orthographic words—the

---

[4]+A3sg: Third person singular agreement, +P2pl: Second person plural possessive agreement, +Loc: Locative Case.

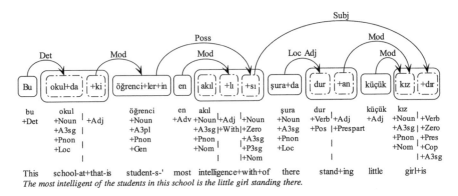

Fig. 7.5 Dependency links in an example Turkish sentence. Plus (+) signs indicate morpheme boundaries. The rounded rectangles show words, while IGs within words that have more than one IG are indicated by the dashed rounded rectangles. The inflectional features of each IG, as produced by the morphological analyzer, are listed below the IG (Oflazer 2014) (reprinted with permission)

correct IGs involved in the relations should also be identified.[5] We may motivate this with the following very simple example: In the phrase *spor arabanızdaydı* ("it was in your sports car"), the adjective *spor* ("sports") should be connected to the first IG of the second word. It is the word *araba* ("car") which is modified by the adjective, not the derived verb form *arabanızdaydı* ("it was in your car"). So a parser should not just say that the first word is a dependent on the second but also state that the syntactic relation is between the last IG of the first word and the first IG of the second word as shown below.

In Fig. 7.5, we see a complete dependency tree for a Turkish sentence laid on top of the words segmented along IG boundaries. The rounded rectangles show the words, while IGs within words are marked with dashed rounded rectangles. The first thing to note in this figure is that the dependency links always emanate from the last IG of a word, since it is the last IG of a word that determines the role of that word as a dependent. The dependency links land on one of the IGs of a (head) word (almost always to the right). The non-final IGs (e.g., the first IG of the word *okuldaki* in Fig. 7.5) may only have incoming dependency links and are assumed to be morphologically linked to the next IG to the right but we do not explicitly show these links.

---

[5]Bozşahin (2002) uses morphemes as sub-lexical constituents in a CCG framework. Since the lexicon was organized in terms of morphemes each with its own CCG functor, the grammar had to account for both the morphotactics and the syntax at the same time.

The noun phrase formed by the three words *öğrencilerin en akıllısı* in this example highlights the importance of the IG-based representation of syntactic relations. Here in the word *akıllısı*, we have three IGs: the first contains the singular noun *akıl* ("intelligence"), the second IG indicates the derivation into the adjective *akıllı* ("intelligence-with" → "intelligent"). The preceding word *en* ("most"), an intensifier adverb, is linked to this IG as a modifier (thus forming "most intelligent"). The third IG indicates another derivation into a noun ("a singular entity that is most intelligent"). This last IG is the head of a dependency link emanating from the word *öğrencilerin* with genitive case-marking ("of the students" or "students") which acts as the possessor of the last noun IG of the third word *akıllısı*. Finally, this word is the subject of the verb IG of the last word, through its last IG.

7.4 An Incremental Data-Driven Statistical Dependency Parsing System

In this section, we introduce a deterministic classifier-based parser using discriminative learning (from now on referred to as the *classifier-based parser*). This parser is based on a parsing strategy that has achieved a high parsing accuracy across a variety of different languages (Nivre et al. 2006, 2007). This strategy consists of the combination of the following three techniques:

1. Deterministic parsing algorithms for building dependency graphs (Kudo and Matsumoto 2002; Yamada and Matsumoto 2003; Nivre 2003),
2. History-based models for predicting the next parser action (Black et al. 1992; Magerman 1995; Ratnaparkhi 1997; Collins 1999),
3. Discriminative classifiers to map histories to parser actions (Kudo and Matsumoto 2002; Yamada and Matsumoto 2003; Nivre et al. 2004).

A system of this kind employs no grammar but relies completely on inductive learning from treebank data for the analysis of new sentences, and on deterministic parsing for disambiguation. This combination of methods guarantees that the parser is robust, never failing to produce an analysis for an input sentence, and efficient, typically deriving this analysis in time that is linear in the length of the sentence.

In the following subsections, we will first present the parsing methodology and then show its adaptation for Turkish using inflectional groups. We will then explore how we can further improve the accuracy by exploiting the advantages of this parser.

7.4.1 Methodology

In this article, we use a variant of the parsing algorithm proposed by Nivre (2003, 2006) that derives a labeled dependency graph in one left-to-right pass over the input, using a stack to store partially processed tokens and a list to store remaining input tokens. However, in contrast to the original arc-eager parsing strategy, we use an arc-standard bottom-up algorithm, as described in Nivre (2004). Like many algorithms used for dependency parsing, this algorithm is restricted to projective dependency graphs.

The parser uses two elementary data structures, a stack of partially analyzed tokens, σ, and an input list of remaining input tokens, τ. The parser is initialized with an empty stack and with all the tokens of a sentence in the input list; it terminates as soon as the input list is empty. In the following, we use subscripted indices, starting from 0, to refer to particular tokens in σ and τ. Thus, σ_0 is the token on top of the stack σ (the *top token*) and τ_0 is the first token in the input list τ (the *next token*); σ_0 and τ_0 are collectively referred to as the *target tokens*, since they are the tokens considered as candidates for a dependency relation by the parsing algorithm.

There are three different parsing actions, or transitions, that can be performed in any non-terminal configuration of the parser:

1. SHIFT: Push the next token onto the stack.
2. $D \xrightarrow{r} H$: Add a dependency arc between the top token (as dependent D) and the next token (as head H), labeled r, then pop the stack.
3. $H \xleftarrow{r} D$: Add a dependency arc between the next token (as dependent D) and the top token (as head H), labeled r, then replace the next token by the top token at the head of the input list.

Figure 7.6 presents the state of the stack and the queue after each action.

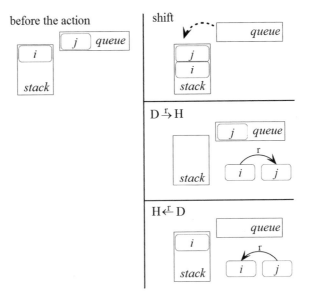

Fig. 7.6 Parsing actions

In order to perform deterministic parsing in linear time, we need to be able to predict the correct parsing action (including the choice of a dependency type r for $D \xrightarrow{r} H$ and $H \xleftarrow{r} D$), at any point during the parsing of a sentence. This is why we use a history-based classifier.

The features of the history-based model can be defined in terms of different linguistic features of tokens, in particular the target tokens. In addition to the target tokens, features can be based on neighboring tokens, both on the stack and in the remaining input, as well as dependents or heads of these tokens in the partially built dependency graph. The linguistic attributes available for a given token are the following:

- Lexical form and root (LEX/LEMMA)
- Main and fine-grain part-of-speech category (CPOS/POS)
- Inflectional features (INF)
- Dependency type to the head if available (DEP)

To predict parser actions from histories, represented as feature vectors, we use support vector machines (SVM), which combine the maximum margin strategy introduced by Vapnik (1995), with the use of kernel functions to map the original feature space to a higher-dimensional space. This type of classifier has been used successfully in deterministic parsing by Kudo and Matsumoto (2002), Yamada and Matsumoto (2003), and Sagae and Lavie (2005), among others. To be more specific, we use the LIBSVM library for SVM learning (Chang and Lin 2011), with a polynomial kernel of degree 2, with binarization of symbolic features, and with the one-versus-one strategy for multi-class classification.[6]

This approach has the following advantages over the previously proposed methods for dependency parsing of Turkish (Oflazer 2003; Eryiğit and Oflazer 2006), in that

- it can process both left-to-right and right-to-left dependencies due to its parsing algorithm,
- it assigns dependency labels simultaneously with dependencies and can use these as features in the history-based model, and
- it does not necessarily require expert knowledge about the choice of linguistically relevant features to use in the representations since SVM training involves implicit feature selection.

---

[6]Experiments have also been performed using memory-based learning (Daelemans and van den Bosch 2005). They were found to give lower parsing accuracy.

7.4.2 Modeling Turkish

Morphological information plays an important role in the syntactic analysis of languages, especially for morphologically rich languages such as Turkish. As explained in previous chapters, Turkish needs a special treatment in this respect. In this section, we are going to investigate the preparation of the training and test data (to be used in the parsing system) and the feature templates that are going to be used for representing the morphological structure of Turkish.

Table 7.1 gives the CoNLL form (Buchholz and Marsi 2006) representation of the sentence "Bu okuldaki öğrencilerin en akıllısı şurada duran küçük kızdır." first introduced in Fig. 7.5. The columns in this figure carry the information listed below.

Column	Represents
1	Token id (ID)
2	Surface form of the token (LEX)
3	Lemma of the token (LEMMA)
4	Coarse part-of-speech category (CPOS)
5	Fine-grained part-of-speech category (POS)
6	Morphological features (INF)
7	Head index (HEAD)
8	Dependency type (DEP)

In Table 7.1, each IG is represented as a separate token. In order to keep the relationships between the IGs in a word, the dependency relations between the IGs

Table 7.1 CoNLL representation of the sentence from Fig. 7.5

ID	LEX	LEMMA	CPOS	POS	INF	HEAD	DEP
1	Bu	bu	Det	Det	_	2	DETERMINER
2	_	okul	Noun	Noun	A3sg|Pnon|Loc	3	DERIV
3	okuldaki	_	Adj	Adj	Rel	4	MODIFIER
4	öğrencilerin	öğrenci	Noun	Noun	A3pl|Pnon|Gen	8	POSSESSOR
5	en	en	Adv	Adv	_	7	MODIFIER
6	_	akıl	Noun	Noun	A3sg|Pnon|Nom	7	DERIV
7	_	_	Adj	Adj	With	8	DERIV
8	akıllısı	_	Noun	Zero	A3sg|P3sg|Nom	14	SUBJECT
9	şurada	şura	Noun	Noun	A3sg|Pnon|Loc	10	LOCATIVE.ADJUNCT
10	_	dur	Verb	Verb	Pos	11	DERIV
11	duran	_	Adj	PresPart	_	13	MODIFIER
12	küçük	küçük	Adj	Adj	_	13	MODIFIER
13	_	kız	Noun	Noun	A3sg|Pnon|Nom	14	DERIV
14	kızdır	_	Verb	Zero	Pres|Cop|A3sg	15	SENTENCE
15	.	.	Punc	Punc	_	0	ROOT

are tagged with a special dependency label "DERIV". One may notice from the table that only the last IG of a word carries the surface form information.[7] A similar situation holds for the lemma field: only the first IG of a word contains the lemma and the others just have empty value for this field. With this representation, the rich derivational structure does not pose any problem anymore and the entire underlying information of a word is kept within the representation without any loss.

Since we now have our parsing units (IGs) between which we will try to discover the dependency relationships, the next stage is to create the machine learning instances from our training data (given in Table 7.1). As explained in previous section, our machine learning component (namely SVM) is used to guess the actions of the parser, e.g., shift, $D \xrightarrow{r} H$ and $H \xleftarrow{r} D$. While creating the training instances for SVM, what we do is to first parse the training sentences according to the deterministic algorithm presented in previous section and keep a record of the required parsing actions in order to obtain the dependency tree in the gold standard treebank structure. For example, let's assume that the target tokens (σ_0, τ_0) are the second IG of the word "okuldaki" and the first (and single) IG of the word "öğrencilerin". In this state, a feature instance that is going to be extracted from the data will be as in Fig. 7.7. Please notice that the DEP features are coming from the partially built dependency graph so far. The history based feature model used in this configuration is shown in Fig. 7.8. This feature model uses five POS features, defined by the POS of the two topmost stack tokens (σ_0, σ_1), the first two tokens of the remaining input (τ_0, τ_1) and the token which comes just after the topmost stack token in the actual sentence. $(\sigma_0 + 1)$.[8] The dependency type features involve the top token on the stack (σ_0), its leftmost and rightmost dependent $(l(\sigma_0), r(\sigma_0))$, and the leftmost dependent of the next input token $(l(\tau_0))$. In addition to these, two INF features and three LEMMA features are used.

After preparing our entire training instances as above, the SVM model is trained to predict parser actions on unseen test data. One should keep in mind that all the features used in the instance representation are categorical features and they should be provided after binarization. This results in a large number of binary features for SVM, represented via a sparse matrix. For example, if there are 14 different CPOS categories (i.e., Adj, Adv, Conj, Det, Dup, Noun, etc.), then we will need 14 binary features to represent only the CPOS feature of a single token.

---

[7] A recent study by Sulubacak and Eryiğit (2013) extends this representation and assigns different lemma and surface form information for each IG.

[8] The token indexes within the actual token sequence are represented by their relative positions to the stack and queue elements. In this representation $\sigma_0 + 1$ refers directly to the right neighbor of the σ_0 within the actual sequence. Similarly, $\sigma_0 - 1$ refers to the left neighbor.

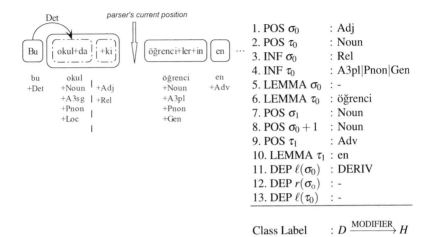

	: Adj
1. POS σ_0	: Adj
2. POS τ_0	: Noun
3. INF σ_0	: Rel
4. INF τ_0	: A3pl|Pnon|Gen
5. LEMMA σ_0	: -
6. LEMMA τ_0	: öğrenci
7. POS σ_1	: Noun
8. POS $\sigma_0 + 1$	: Noun
9. POS τ_1	: Adv
10. LEMMA τ_1	: en
11. DEP $\ell(\sigma_0)$	: DERIV
12. DEP $r(\sigma_0)$	: -
13. DEP $\ell(\tau_0)$	: -

Class Label   : $D \xrightarrow{\text{MODIFIER}} H$

Fig. 7.7 Sample feature instance

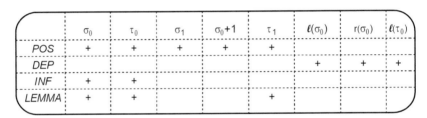

Fig. 7.8 Feature model

7.4.3 Evaluation Metrics

The main evaluation metrics that are used in the evaluation of dependency parsing are the *unlabeled attachment score* (AS_U) and *labeled attachment score* (AS_L), i.e., the proportion of dependents attached to the correct head (with the correct label for AS_L). In Turkish, a correct attachment is one in which the dependent IG (the last IG in the dependent word) is *not only* attached to the correct head word *but also to the correct IG within the head word*. Where relevant, we also report for Turkish the (unlabeled) word-to-word score (WW_U), which only measures whether a dependent word is connected to (some IG in) the correct head word. It should be clear from the previous discussions and from Fig. 7.5 that the IG-to-IG evaluation is the right one to use for Turkish even though it is more stringent than word-to-word evaluation. Dependency links emanating from punctuation are excluded in all evaluation scores. Non-final IGs of a word are assumed to link to the next IG within the word, but these links, referred to as *InnerWord* links, are not considered as dependency relations and are excluded from evaluation scoring.

7.5   Related Work

The first results on the Turkish Treebank come from Eryiğit and Oflazer (2006) who used only a subset of the treebank sentences containing exclusively head-final and projective dependencies. The first results on the entire treebank appear in Nivre et al. (2007), who use memory-based learning to predict parser actions.

The Turkish Treebank has been parsed by seventeen research groups in the CoNLL-X shared task on multilingual dependency parsing (Buchholz and Marsi 2006), where it was seen as the most difficult language by the organizers and most of the groups. Buchholz and Marsi (2006) write:

> The most difficult data set is clearly the Turkish one. It is rather small, and in contrast to Arabic and Slovene, which are equally small or smaller, it covers eight genres, which results in a high percentage of new FORM and LEMMA values in the test set.

The results for Turkish are given in Table 7.2. The parser presented in this chapter obtained the best results for Turkish (with $AS_U = 75.8$ and $AS_L = 65.7$) and also for Japanese which is the only agglutinative and head-final language in the shared task other than Turkish (Nivre et al. 2006). The groups were asked to find the correct IG-to-IG dependency links. When we look at the results, we observe that most of the best performing parsers use one of the parsing algorithms of Eisner (1996), Nivre (2003), or Yamada and Matsumoto (2003) together with a learning method based on the maximum margin strategy. We can also see that a common property of the

Table 7.2 CoNLL-X shared task results on Turkish (from Table 5 in Buchholz and Marsi 2006)

Teams	AS_U	AS_L
Nivre et al. (2006)	75.8	65.7
Johansson and Nugues (2006)	73.6	63.4
McDonald et al. (2006)	74.7	63.2
Corston-Oliver and Aue (2006)	73.1	61.7
Cheng et al. (2006)	74.5	61.2
Chang et al. (2006)	73.2	60.5
Yuret (2006)	71.5	60.3
Riedel et al. (2006)	74.1	58.6
Carreras et al. (2006)	70.1	58.1
Wu et al. (2006)	69.3	55.1
Shimizu (2006)	68.8	54.2
Bick (2006)	65.5	53.9
Canisius et al. (2006)	64.2	51.1
Schiehlen and Spranger (2006)	61.6	49.8
Dreyer et al. (2006)	60.5	46.1
Liu et al. (2006)	56.9	41.7
Attardi (2006)	65.3	37.8

parsers which fall below the average ($AS_L = 55.4$) is that they do not make use of inflectional features, which is crucial for Turkish.[9]

The Turkish dependency parser described in this chapter has been used so far in many downstream applications for Turkish: discriminative language modeling (Arısoy et al. 2012), extracting word sketches (Ambati et al. 2012), question answering (Derici et al. 2014), analyzing interaction patterns (Saygın 2010), constructing parallel treebanks (Megyesi and Dahlqvist 2007; Megyesi et al. 2008), information retrieval (Hadımlı and Yöndem 2011, 2012), text watermarking (Meral et al. 2009), social media monitoring and sentiment analysis (Eryiğit et al. 2013; Yıldırım et al. 2014).

7.6 Conclusions

In recent years, with the usage of data-driven statistical methods, multilingual parsers have been able to yield high performances for many languages. In this chapter, we have presented the issues in dependency parsing of Turkish along with results from its evaluation.

Up to now, the majority of relevant research has been on formal (well-written and edited) texts and gold standard treebanks with manually verified annotations. However, the requirements of parsing real life text brings in additional issues and complications, especially with the rise in the need for processing language used social media. Such data differs significantly from formal texts. It is thus necessary to automate the preprocessing tasks and handle the noise and nonstandard language use in such texts in a very robust manner to improve the performance of parsers. A recent TÜBİTAK project (112E276) "Parsing Web2.0 Sentences" (Eryiğit 2014) focuses on this aspect of parsing for Turkish. This project focuses on research related to all NLP stages required to perform sentence analysis of raw Turkish text as found in social media.

References

Ambati BR, Reddy S, Kilgarriff A (2012) Word sketches for Turkish. In: Proceedings of LREC, Istanbul, pp 2945–2950

Arısoy E, Saraçlar M, Roark B, Shafran I (2012) Discriminative language modeling with linguistic and statistically derived features. IEEE Trans Audio Speech Lang Process 20(2):540–550

Attardi G (2006) Experiments with a multilanguage non-projective dependency parser. In: Proceedings of CONLL, New York, NY, pp 166–170

---

[9]Actually, there are two parsers (Bick and Attardi in Table 7.2) in this group that try to use parts of the inflectional features under special circumstances.

Bick E (2006) Lingpars, a linguistically inspired, language-independent machine learner for dependency treebanks. In: Proceedings of CONLL, New-York, NY, pp 171–175

Black E, Jelinek F, Lafferty JD, Magerman DM, Mercer RL, Roukos S (1992) Towards history-based grammars: using richer models for probabilistic parsing. In: Proceedings of the DARPA speech and natural language workshop, New York, NY, pp 31–37

Bozşahin C (2002) The combinatory morphemic lexicon. Comput Linguist 28(2):145–186

Buchholz S, Marsi E (2006) CoNLL-X shared task on multilingual dependency parsing. In: Proceedings of CONLL, New York, NY, pp 149–164

Canisius S, Bogers T, van den Bosch A, Geertzen J, Sang ETK (2006) Dependency parsing by inference over high-recall dependency predictions. In: Proceedings of CONLL, New York, NY, pp 176–180

Carreras X, Surdeanu M, Marquez L (2006) Projective dependency parsing with perceptron. In: Proceedings of CONLL, New York, NY, pp 181–185

Chang CC, Lin CJ (2011) LIBSVM: a library for support vector machines. ACM Trans Intell Syst Technol 2(3), 1–27

Chang MW, Do Q, Roth D (2006) A pipeline model for bottom-up dependency parsing. In: Proceedings of CONLL, New York, NY, pp 186–190

Cheng Y, Asahara M, Matsumoto Y (2006) Multi-lingual dependency parsing at NAIST. In: Proceedings of CONLL, New York, NY, pp 191–195

Chung H, Rim HC (2004) Unlexicalized dependency parser for variable word order languages based on local contextual pattern. In: Proceedings of CICLING, Seoul, pp 109–120

Collins M (1996) A new statistical parser based on bigram lexical dependencies. In: Proceedings of ACL, Santa Cruz, CA, pp 184–191

Collins M (1999) Head-driven statistical models for natural language parsing. PhD thesis, University of Pennsylvania, Philadelphia, PA

Corston-Oliver S, Aue A (2006) Dependency parsing with reference to Slovene, Spanish and Swedish. In: Proceedings of CONLL, New York, NY, pp 196–200

Daelemans W, van den Bosch A (2005) Memory-based language processing. Cambridge University Press, Cambridge

Derici C, Çelik K, Özgür A, Güngör T, Kutbay E, Aydın Y, Kartal G (2014) Rule-based focus extraction in Turkish question answering systems. In: Proceedings of IEEE signal processing and communications applications conference, Trabzon, pp 1604–1607

Dreyer M, Smith DA, Smith NA (2006) Vine parsing and minimum risk reranking for speed and precision. In: Proceedings of CONLL, New York, NY, pp 201–205

Eisner J (1996) Three new probabilistic models for dependency parsing: an exploration. In: Proceedings of COLING, Copenhagen, pp 340–345

Erguvanlı EE (1979) The function of word order in Turkish grammar. PhD thesis, UCLA, Los Angeles, CA

Eryiğit G (2007) Dependency parsing of Turkish. PhD thesis, Istanbul Technical University, Istanbul

Eryiğit G (2014) ITU Turkish NLP web service. In: Proceedings of EACL, Gothenburg, pp 1–4

Eryiğit G, Oflazer K (2006) Statistical dependency parsing of Turkish. In: Proceedings of EACL, Trento, pp 89–96

Eryiğit G, Nivre J, Oflazer K (2006) The incremental use of morphological information and lexicalization in data-driven dependency parsing. In: Proceedings of the international conference on the computer processing of oriental languages, Singapore, pp 498–507

Eryiğit G, Nivre J, Oflazer K (2008) Dependency parsing of Turkish. Comput Linguist 34(3):357–389

Eryiğit G, Çetin FS, Yanık M, Temel T, Çiçekli İ (2013) Turksent: a sentiment annotation tool for social media. In: Proceedings of the workshop on linguistic annotation workshop and interoperability with discourse, Sofia, pp 131–134

Goldberg Y, Nivre J (2012) A dynamic oracle for arc-eager dependency parsing. In: Proceedings of COLING, Mumbai, pp 959–976

Hadımlı K, Yöndem MT (2011) Information retrieval from Turkish radiology reports without medical knowledge. In: Proceedings of the international conference on flexible query answering systems, Ghent, pp 210–220

Hadımlı K, Yöndem MT (2012) Two alternate methods for information retrieval from Turkish radiology reports. In: Proceedings of ISCIS, London, pp 527–532

Hakkani-Tür DZ, Oflazer K, Tür G (2002) Statistical morphological disambiguation for agglutinative languages. Comput Hum 36(4):381–410

Hoffman B (1994) Generating context appropriate word orders in Turkish. In: Proceedings of the international workshop on natural language generation, Kennebunkport, ME, pp 117–126

Hudson RA (1990) English word grammar, vol 108. Basil Blackwell, Oxford

Johansson R, Nugues P (2006) Investigating multilingual dependency parsing. In: Proceedings of CONLL, New York, NY, pp 206–210

Koo T, Collins M (2010) Efficient third-order dependency parsers. In: Proceedings of ACL, Uppsala, pp 1–11

Koo T, Rush AM, Collins M, Jaakkola T, Sontag D (2010) Dual decomposition for parsing with non-projective head automata. In: Proceedings of EMNLP, Cambridge, MA, pp 1288–1298

Kudo T, Matsumoto Y (2002) Japanese dependency analysis using cascaded chunking. In: Proceedings of CONLL, Taipei, pp 63–69

Liu T, Ma J, Zhu H, Li S (2006) Dependency parsing based on dynamic local optimization. In: Proceedings of CONLL, New York, NY, pp 211–215

Magerman DM (1995) Statistical decision-tree models for parsing. In: Proceedings of ACL, Cambridge, MA, pp 276–283

Marcus M (1980) A theory of syntactic recognition for natural language. MIT Press, Cambridge, MA

Martins AF, Smith NA, Xing EP (2009) Concise integer linear programming formulations for dependency parsing. In: Proceedings of the ACL-IJCNLP, Singapore, pp 342–350

McDonald R, Pereira F (2006) Online learning of approximate dependency parsing algorithms. In: Proceedings of EACL, Trento, pp 81–88

McDonald R, Crammer K, Pereira F (2005) Online large-margin training of dependency parsers. In: Proceedings of ACL, Ann Arbor, MI, pp 91–98

McDonald R, Lerman K, Pereira F (2006) Multilingual dependency analysis with a two-stage discriminative parser. In: Proceedings of CONLL, New York, NY, pp 216–220

Megyesi B, Dahlqvist B (2007) The Swedish-Turkish parallel corpus and tools for its creation. In: Proceedings of the nordic conference on computational linguistics, Tartu, pp 136–143

Megyesi B, Dahlqvist B, Pettersson E, Nivre J (2008) Swedish-Turkish parallel treebank. In: Proceedings of LREC, Marrakesh, pp 470–473

Mel'čuk IA (1988) Dependency syntax: theory and practice. SUNY Press, Albany, NY

Meral HM, Sankur B, Özsoy AS, Güngör T, Sevinç E (2009) Natural language watermarking via morphosyntactic alterations. Comput Speech Lang 23(1):107–125

Nivre J (2003) An efficient algorithm for projective dependency parsing. In: Proceedings of IWPT, Nancy, pp 149–160

Nivre J (2004) Incrementality in deterministic dependency parsing. In: Proceedings of the workshop on incremental parsing: bringing engineering and cognition together, Barcelona, pp 50–57

Nivre J (2006) Inductive dependency parsing. Springer, Dordrecht

Nivre J, Scholz M (2004) Deterministic dependency parsing of English text. In: Proceedings of COLING, Geneva, pp 64–70

Nivre J, Hall J, Nilsson J (2004) Memory-based dependency parsing. In: Proceedings of CONLL, Boston, MA, pp 49–56

Nivre J, Hall J, Nilsson J, Eryiğit G, Marinov S (2006) Labeled pseudo-projective dependency parsing with support vector machines. In: Proceedings of CONLL, New York, NY, pp 221–225

Nivre J, Hall J, Nilsson J, Chanev A, Eryiğit G, Kübler S, Marinov S, Marsi E (2007) MaltParser: a language-independent system for data-driven dependency parsing. Nat Lang Eng 13(2):95–135

Oflazer K (2003) Dependency parsing with an extended finite-state approach. Comput Linguist 29(4):515–544

Oflazer K (2014) Turkish and its challenges for language processing. Lang Resour Eval 48(4):639–653

Oflazer K, Say B, Hakkani-Tür DZ, Tür G (2003) Building a Turkish treebank. In: Treebanks: building and using parsed corpora. Kluwer Academic Publishers, Berlin

Ratnaparkhi A (1997) A linear observed time statistical parser based on maximum entropy models. In: Proceedings of EMNLP, Providence, RI, pp 1–10

Riedel S, Çakıcı R, Meza-Ruiz I (2006) Multi-lingual dependency parsing with incremental integer linear programming. In: Proceedings of CONLL, New York, NY, pp 226–230

Sagae K, Lavie A (2005) A classifier-based parser with linear run-time complexity. In: Proceedings of IWPT, Vancouver, pp 125–132

Saygın AP (2010) A computational analysis of interaction patterns in the acquisition of Turkish. Res Lang Comput 8(4):239–253

Schiehlen M, Spranger K (2006) Language independent probabilistic context-free parsing bolstered by machine learning. In: Proceedings of CONLL, New York, NY, pp 231–235

Sekine S, Uchimoto K, Isahara H (2000) Backward beam search algorithm for dependency analysis of Japanese. In: Proceedings of COLING, Saarbrücken, pp 754–760

Sgall P, Hajicová E, Panevová J (1986) The meaning of the sentence in its semantic and pragmatic aspects. Springer, Dordrecht

Shieber SM (1983) Sentence disambiguation by a shift-reduce parsing technique. In: Proceedings of ACL, Cambridge, MA, pp 113–118

Shimizu N (2006) Maximum spanning tree algorithm for non-projective labeled dependency parsing. In: Proceedings of CONLL, New York, NY, pp 236–240

Sulubacak U, Eryiğit G (2013) Representation of morphosyntactic units and coordination structures in the Turkish dependency treebank. In: Proceedings of the workshop on statistical parsing of morphologically rich languages, Seattle, WA, pp 129–134

Tesnière L (1959) Eléments de syntaxe structurale. Librairie C. Klincksieck, Paris

Vapnik VN (1995) The nature of statistical learning theory. Springer, New York, NY

Wu YC, Lee YS, Yang JC (2006) The exploration of deterministic and efficient dependency parsing. In: Proceedings of CONLL, New York, NY, pp 241–245

Yamada H, Matsumoto Y (2003) Statistical dependency analysis with support vector machines. In: Proceedings of IWPT, Nancy, pp 195–206

Yıldırım E, Çetin FS, Eryiğit G, Temel T (2014) The impact of NLP on Turkish sentiment analysis. In: Proceedings of the international conference on Turkic language processing, Istanbul

Yuret D (2006) Dependency parsing as a classification problem. In: Proceedings of CONLL, New York, NY, pp 246–250

Zhang Y, Clark S (2008) A tale of two parsers: investigating and combining graph-based and transition-based dependency parsing using beam-search. In: Proceedings of EMNLP, Honolulu, HI, pp 562–571

Chapter 8
Wide-Coverage Parsing, Semantics, and Morphology

Ruket Çakıcı, Mark Steedman, and Cem Bozşahin

Abstract Wide-coverage parsing poses three demands: broad coverage over preferably free text, depth in semantic representation for purposes such as inference in question answering, and computational efficiency. We show for Turkish that these goals are not inherently contradictory when we assign categories to sub-lexical elements in the lexicon. The presumed computational burden of processing such lexicons does not arise when we work with automata-constrained formalisms that are trainable on word-meaning correspondences at the level of predicate-argument structures for any string, which is characteristic of radically lexicalizable grammars. This is helpful in morphologically simpler languages too, where word-based parsing has been shown to benefit from sub-lexical training.

8.1 Introduction

Wide-coverage parsing aims to develop broad computational coverage of text. It requires efficient parsers, and also lexicons that are not only large, but are rich in information. Without rich information, it seems difficult to make further use of results that we work so hard to achieve by providing wide coverage. One such information is semantics. To be of subsequent use, such as in making inferences in question answering, rich semantic information must arise from its source,

R. Çakıcı (✉) · C. Bozşahin
Middle East Technical University, Ankara, Turkey
e-mail: ruken@ceng.metu.edu.tr; bozsahin@metu.edu.tr

M. Steedman
University of Edinburgh, Edinburgh, UK
e-mail: steedman@inf.ed.ac.uk

© Springer International Publishing AG, part of Springer Nature 2018
K. Oflazer, M. Saraçlar (eds.), *Turkish Natural Language Processing*,
Theory and Applications of Natural Language Processing,
https://doi.org/10.1007/978-3-319-90165-7_8

which is usually the head of a syntactic construction.[1] The information is not too difficult to capture in languages with simple morphology. In morphologically rich languages, in many cases the head of a construction is not the whole word but a sub-lexical element, for example suffixes in Turkish. This aspect opens a degree of freedom that has been harnessed in various ways in computational models of Turkish, some of which we report here. It also has theoretical ramifications, such as morphological bracketing paradoxes and the nature of the lexicon, with computational repercussions. In this work we report on working computationally with lexicons of words and of sub-lexical elements, for example morphemes.[2]

With these aspects in mind, when we talk about "morphemes in the lexicon" we mean morphological processes that yield discernible substrings to which we can assign categories, to the extent that with these categories they can regulate constituency and predicate-argument structures. Whether that means "morpheme" or a "morpholexical process" in the theory is not going to be a concern for models under discussion in the paper.

To put Turkish morphology in a parsing perspective, let us start with some findings about Turkish word forms. By one estimate, Oflazer (2003), there are on average two to three morphemes per Turkish word, plus the stem. The number of derivational morphemes per word is 1.7 on average, according to the same source. The language has approximately 110 derivational morphemes (Oflazer et al. 1994). The nominal paradigm for inflections engenders 2^3 possible continuations per stem (for number, possession, and case, all of them being optional in morphology but may be required by syntax). The verbal paradigm produces 2^9 such forms per stem, the details of which we skip here. These are paradigmatic/morphemic counts; the actual number of inflections to consider is higher. For example, the number of phonologically realizable nominal inflections is 118. There are 19 nominal inflectional morphs, 41 verbal inflections, 74 derivational morphs onto nominals, and 30 onto verbal forms, giving a total of 164 morph inventory (*ibid.*) Given the fact that almost all of these suffixes have a harmonizing vowel (two possibilities for back/front vowels, and four for high/low, round/nonround), the number of distinct

---

[1]Lewis and Steedman (2014) describe what is at stake if we incorporate distributional semantics of content words but not compositional semantics coming out of such heads.

[2]The notion of morpheme is controversial in linguistics. Matthews (1974), Stump (2001), and Aronoff and Fudeman (2011) provide some discussion. Without delving into morphological theory, we shall adopt the computational view summarized by Roark and Sproat (2007): morphology can be characterized by finite-state mechanisms. Models of morphological processing differ in the way they handle lexical access. For example, two-level morphology is finite-state and linear-time in its morphotactics, but it incurs an exponential cost to do surface form-lexical form pairings during morphological processing (see Koskenniemi 1983; Koskenniemi and Church 1988, and Barton et al. 1987 for extensive discussion). On the other hand, if we have lexical categories for sub-lexical items, then, given a string and its decomposition, we can check efficiently (in polynomial time) whether the category-string correspondences are parsable: the problem is in **NP** (nondeterministic polynomial time) not because of parsing but because of ambiguity. Lexical access could then use the same mechanism for words, morphemes, and morpholexical rules if it wants to.

surface realizations of morphs is approximately 500. This is the number a free text parser is faced with.

The morphologically simplest word is just a bare stem, and the morphologically most complex word in the BOUN corpus of half-billion words (Sak et al. 2011) has eight suffixes on it. In theory, nominal morphology can produce infinitely many word forms because of the potential for iterative nature of the relativization marker *-ki*, which can appear on a nominal stem repeatedly. But, in the entire Turkish fragment of CHILDES (MacWhinney 2000), which is a children's speech database, 30 out of 20,000 nouns bear this suffix, and only once when it appears (Çöltekin and Bozşahin 2007). In the BOUN corpus, which is an adult text corpus, approximately 2.9 million nominal word forms carry it, out of half-billion words, of which 82 have multiple occurrences, but none with three or more instances of *-ki*.

Given these results over large data sets, one is tempted to ask: What would be the benefit of wide-coverage parsing which starts with the morpheme or morpholexical process in the lexicon, rather than (or in addition to) the word form? We might in the long run compile out all word forms, and leave the "outliers" aside, such as three or more *-ki*'s, to do parsing with words.

There are two answers which we find convincing: (a) It has been shown that novel word forms are very likely in a Turkish text even if we calculate our statistics and morphology from a very large data set. This much we know from linguistic theory, but the numbers are striking: Sak et al. (2011) report that approximately 52,000 unique lexical endings, which they define as possible morpheme sequences discounting the stem, have been detected after processing 490 million words.[3] 268 more novel endings have been found in the next ten million words. In a narrower window, 61 novel forms are reported in the 491st millionth data set, i.e., in the next one-million words after having seen 490 million.[4] There can be a lot of unseen word forms. (b) Word forms manifest a good amount of ambiguity. Göksel (2006) shows that pronominal participles are at least four-way ambiguous, and this does not arise from morphology but from syntax. Annotators find these ambiguities very difficult to spot, if not impossible, unless they are trained linguists. That seems difficult to lexicalize at the level of word form. Even for the word forms that have been seen, because of the ambiguity of Turkish morphology there are many occasions in which a word form has been seen, but not with the meaning that is needed for the new instance of the form. Because we cannot compile out all possible forms for a word, we cannot train a parser with the assumption that it has seen all the categories of the word. But then there are problems in defining a model for the unseen words (we need a model for the morphological derivations themselves).

---

[3]For example: for the word *el-ler-im-de-ki* "the ones in my hands" with the morphological breakdown of el-PLU-POSS.1S-LOC-REL, *el* for "hand," the lexical ending is -PLU-POSS.1S-LOC-REL.

[4]Notice that, left unconstrained we face $n \times 2^{164} \approx n \times 10^{49.4}$ word-like forms in Turkish, from 164 morphemes and n lexemes. A much smaller search space is attested because of morphological, semantic, and lexical constraints, but 50,000 and counting is still an enormous search space.

After a brief review of other options regarding meanings of sub-lexical elements in parsing (Sect. 8.2), we show that actively doing morphology can address these problems for the size of datasets we commonly deal with. We cover four aspects in the current paper: (1) radical lexicalization (Sect. 8.3), which is a necessary step in pinning down head information, (2) a mechanism to provide a rich vocabulary for parser training of categorial lexicons with structured representation of meaning (Sect. 8.4), for which we use CCG as the theory, (3) modeling the categorial lexicon (Sect. 8.5), for (4) efficient wide-coverage parsing (Sect. 8.6).

8.2 Morphology and Semantics

For all the points above, we can exploit the degrees of freedom in parsing with respect to the level of detail we expect to get from its delivered semantics. For example, we can take the last inflection on the word to be of relevance to syntax only, as is done by Oflazer (2003) and Eryiğit et al. (2008). The rest of the word must then bear rich lexical knowledge to do semantics, which, as we pointed out earlier, is a difficult task. We can alleviate the ambiguity problem by employing morphological disambiguators, as is done by Sak et al. (2011) and Yuret and Türe (2006). But if the ambiguity is genuinely syntactic, which is argued to be the case for some inflections by Göksel (2006), then there is not much a morphological disambiguator can do about it. Lastly, we can identify the head of a syntactic construction as the whole word rather than part of the word such as the relative marker, as is done by computational work originating from LFG's lexical integrity principle (Sells 1995). However, as the statistics for lexical endings show, there is no end to novel word forms which might bear these heads, therefore they are not strictly lexicalizable in this sense. Moreover, they are ambiguous, for example the same form -*dik* can be the head of a relative clause or a complementizer clause.

The counteracting force for these concerns is computational efficiency. How can we parse efficiently if the number of inputs to the parser multiplies with respect to the number of words in the input string? This is the expected case when we work with uniform lexical access for words and morphemes or morpholexical rules. The complexity class does not change, but practical solutions to wide-coverage parsing may be negatively impacted by this choice. In this chapter we show that such concerns for wide-coverage parsing are overrated. If we can manage to build a lexicon of heads with rich grammatical information, then training the parser to deal with **NP**-hardness of the overall problem may be transparent.[5]

---

[5]Using complexity results in these aspects has been sometimes controversial; see, for example, Berwick and Weinberg (1982), Barton et al. (1987), and Koskenniemi and Church (1988). One view, which we do not follow, is to eliminate alternatives in the model, by insisting on using tractable algorithms, as in Tractable Cognition thesis (van Rooij 2008). The one we follow addresses complexity as a complex mixture of source and data that in the end allows efficient parsing, feasible and transparent training, and scalable performance. For example, Clark and

In the end we might face a choice in wide-coverage parsing, in one case expecting good amount of work and knowledge from morphological analyzers during parsing, at the expense of delivering a rough approximate representation for semantics. In the other case we would expect very little or no analysis from morphological analyzers during parsing, at the expense of doing more morphological training offline, for delivery of a rich proxy for semantics. Or we strike a balance. (We assume that's what adults do.) We show that, before such considerations, the case for going all the way in the lexicon (for the child and machine) to pin down the minimal span of heads for syntactic constructions is viable. Such lexicons can be transparent with respect to morphology, which we think makes model training and development easier without sacrificing delivery of logical forms at a level of detail one would expect from heads. This has been shown for Turkish (Bozşahin 2002; Çakıcı 2008; McConville 2006), Korean (Cha et al. 2002), Hindi (Ambati et al. 2013), and English (Honnibal et al. 2010; Wang et al. 2014).

8.3   Radical Lexicalization and Predicate-Argument Structure of Sub-lexical Elements

We exemplify the fine-grained semantics we obtain for affixes by radically lexicalizing a fragment of a schematized Turkish grammar shown below (fin is "finite," "pn" is person-number).

(1)
		syntax	semantics
S_{fin}	$\rightarrow$	$S_{\text{case=nom,agr=fin,pn}}$	$S'_{1,\text{fin}}=S'_{2,\text{case=nom,agr=fin,pn}}$
S_{agr}	$\rightarrow$	$NP_{\text{case}}\ VP_{\text{agr}}$	$S'=NP'(VP')$
VP_f	$\rightarrow$	$NP_{\text{case}}\ V_f$	$VP'=NP'(V')$
VP_f	$\rightarrow$	V_f	$VP'=V'$
NP_f	$\rightarrow$	$RC\ NP_f$	$NP'_1 = \lambda x . RC'x \wedge NP'_2 x$
RC	$\rightarrow$	$VP_{rel}+an$	$RC'=VP'$

The grammar highlights the syntax, semantics and morphology of the relative clauses identified by the constraints in (2). They are enforced by the head of the construction -*an*, which is a relative marker attached morphologically (by "+") to a certain subparadigm of verbal morphology, which we symbolize above as VP_{rel} for simplicity:

(2)   a. [VP hedef-i vur]-an okçu

  target-ACC hit-REL archer

  "The archer who hit/hits/is hitting the target"

Curran's (2007) CCG parser is cubic time, whereas A* CCG parser is exponential time in the worst case, but with training, it has a superlinear parsing performance on long sentences (Lewis and Steedman 2014). Another example of this view is PAC learnability of Valiant (2013).

b. *[*VP okçu   vur]-an hedef
   archer.NOM hit-REL target

The rules in (1) are written with generalized quantification in mind, for example arguments are functions and verbs are their input, in the right column.

In categorial notation, we can rewrite the second rule on the left in two ways, one from the perspective of *NP*, and the other from that of *VP*. Both will yield a function because of adopting a single perspective of one of the constituents on the right-hand side, and because there is only one result in the left-hand side. We get $NP_{\text{nom,agr}} = S_{\text{agr}}/VP_{\text{agr}}$ from the first perspective, and $VP_{\text{agr}} = S_{\text{agr}}\backslash NP_{\text{nom,agr}}$ from the second. Combining them we get $NP_{\text{nom,agr}} = S_{\text{agr}}/(S_{\text{agr}}\backslash NP_{\text{nom,agr}})$ with the semantics $\lambda P.P\ np'$, where np' is the semantics of the NP. This preserves the generalized quantification interpretation without further assumption. It also captures a property of Turkish grammar that sentences bearing agreement can be subcategorized for in the language, because of S_{agr} category. This is the case for some complement-taking verbs requiring certain types of nominalized clauses. Finite matrix clauses are a special case requiring nominative subjects. (Non-finite agreement is with the genitive case. The first rule on the left eschews these aspects.)

From VP rules we get $V_f = VP_f \backslash NP_{\text{case}}$, where f is a feature bundle and "case" is some case governed by the verb, for example accusative or dative. Substituting for VP, we get $V_f = (S_f \backslash NP_{\text{nom,agr}})\backslash NP_{\text{case}}$ with semantics $\lambda x \lambda y.verb'xy$, and also $V_f=S_f\backslash NP_{\text{nom,agr}}$, meaning $\lambda x.verb'x$.

The reduction of grammar to its lexicon has so far produced explicit semantics for arity of verbs, unlike (1) semantics, where it is implicit. We can add morphology to this process of reduction. From the NP rule, we get $RC = NP_f/NP_f$, and from the relative clause rule, $-an=RC\backslash VP_{\text{rel}}$. We therefore obtain $-an=(NP_f/NP_f)\backslash(S_{\text{rel}}\backslash NP_{\text{nom}})$ with the semantics $\lambda P\lambda Q\lambda x.(Px) \wedge (Qx)$. This is the predicate-argument structure of the head of relativization, and it is easy to check its equivalence with the semantics of relative clauses captured by a conspiracy of several rules in (1). It can capture in the lexicon some semantic constraints we expect to arise for relativized nouns as well. For example, we can constrain Q to be a predicate, say (e, t), and P to be a property, say $((e, t), (e, t))$, because proper names as such (non-properties) cannot be relativized. Notice that we would need to manipulate two rules in (1) to impose this unique property of relativization.

Thus $-an=(NP_f/NP_f)\backslash(S_{\text{rel}}\backslash NP_{\text{nom}})$: $\lambda P\lambda Q\lambda x.(Px) \wedge (Qx)$, the fully lexicalized category of the relative marker *-an*, manifests in a single knowledge source everything that the grammar needs to impose syntactic, semantic, and (soon) morphological constraints on this type of relativization. In this respect we would expect radically lexicalized grammars to be easier to train for wide-coverage, as Hockenmaier and Steedman (2007) have shown for WSJ. If we can capture some training data at the level of affixes, including their dependency structures which are translatable to predicate-argument structures, then it will be feasible for Turkish categorial lexicon too, as we summarize in the current paper, from Bozşahin (2002), Hockenmaier and Steedman (2007), and Çakıcı (2008).

8.4 Combinatory Categorial Grammar: CCG

The preceding discussion might suggest that CCG and radical lexicalization are committed to an Item-and-Arrangement view of morphology.[6] For example, our derivation of a category for the relative marker *-an* makes use of apparently concatenative rewrite rules for NPs and VPs. As Schmerling (1983) and Hoeksema and Janda (1988) noted, Item-and-Process view is compatible with categorial grammar as well, perhaps more so, given an automata-theoretic understanding of IP. Category adjacency is the key to combination, rather than segmental concatenation. Steedman and Bozşahin (2018) discusses these issues in detail. For present purposes, it is best to think of Turkish morphology as a finite-state realizable subset of morpholexical rules that in the end gives rise to categories that can be checked with pieces within a word. We show combination of such categories as "LEX" for convenience.

A model is trained on morphological tags to deal with **NP**-hardness of the problem mentioned in the introduction (rather than dealing with two such problems, one in morphology for matching surface forms, and one in parsing for lexical access). The algorithm explained in Sect. 8.3 can be made to work with morpholexical rules for morphology, rather than with rewrite rules, by conceiving every step of agglutination as a unary lexical rule that operates on the stem. In other words, "beads on a string" morphology of Turkish can be seen as a special case of Item-and-Process, and the idiosyncrasy of the model can be confined to the lexicon.

This is a point of departure for CCG: building a model for a language is the practice of finding lexicalizable combinatory categories subject to invariant combinatorics. Combination is entirely blind to logical form and to ontology of a category-bearing string or process in the lexicon; whether it came from morphology (affixes), syntax (words), or phonology (tonal elements). Rather than working with universal parts-of-speech such as N, V, A, P and their projections, as is done in a superficially similar framework of Lieber (1992), we would want to elucidate differentiable syntactic categories (basic or derived) that make the entire system efficiently parsable. Whether syntax or morphology delivers these categories may be important, but the combinatory machinery does not depend on it. It is a system of reference, and the lexicon is its model.

This system is based on combinators in CCG. Steedman (1996, 2000, 2011), Steedman and Baldridge (2011), and Bozşahin (2012) provide exposure to the theory. It has one primitive: categorially adjacent combination, also called *application*.

---

[6]Item-and-Arrangement (IA) morphology treats word structure as consisting of morphemes which are put one after another, like segments. Item-and-Process morphology (IP) uses lexemes and associated processes on them, which are not necessarily segmental. Another alternative is Word-and-Paradigm, which is similar to IP but with word as the basic unit. The terminology is due to Hockett (1959).

It is a function, and as such it is not a stand-in for IA-style adjacency. It is compatible with IP-adjacency, which we syntacticize as follows:[7]   .

(3) a. Forward Application:

$$X/Y: f \quad Y: a \quad \Rightarrow \quad X: fa \qquad\qquad (>)$$

   b. Backward Application:

$$Y: a \quad X\backslash Y: f \quad \Rightarrow \quad X: fa \qquad\qquad (<)$$

The bearer of main functor (X/Y or $X\backslash Y$) is adjacent to the bearer of argument category Y. The lambda-term for logical form is shown after a colon (:), which in (3) is application of f to a. These rules project applicative structures arising from any category-bearing string, for example (2a). Notice its more specific reading below, because of the chosen category for the verb:

(4)

hedef-i	vur	-an	okçu
target-ACC	hit	-REL	archer

$$\dfrac{\underline{\quad}}{(NP/NP)/((NP/NP)\backslash NP_{acc})}{}^{>\,\mathbf{T}} \quad \dfrac{\quad}{S_{past,3s}\backslash NP_{nom,3s}\backslash NP_{acc}} \quad (NP/NP)\backslash(S\backslash NP_{nom}) \quad NP^{\uparrow}\backslash(NP/NP)$$

$$: \lambda p.p\,t' \qquad : \lambda x\lambda y.hit'xy \qquad : \lambda p\lambda q\lambda x.px \wedge qx \qquad : \lambda p\lambda q\lambda x.p\,a'x \wedge qx$$

$$\dfrac{\quad}{(NP/NP)\backslash NP_{acc}}{}_{<\,\mathbf{B}\,\mathrm{LEX}}$$

$$: \lambda z\lambda q\lambda x.hit'zx \wedge qx$$

$$\dfrac{\quad}{NP/NP: \lambda q\lambda x.hit't'x \wedge qx}{}^{<}$$

$$\dfrac{\quad}{NP^{\uparrow}: \lambda q\lambda x.(hit't'x \wedge a'x) \wedge qx}{}^{<}$$

"The archer who hit the target"

We use "↑" as a shorthand for type-raised categories. Type-raising is morpholexical in CCG, and can be defined as follows:[8]

(5) a. Forward Morpholexical Type Raising:

$A : a \Rightarrow T/(T\backslash A) : \lambda f.fa$ (where $T\backslash A$ is a lexical function-type over arguments of type A into results of type T) $(> \mathbf{T})$

   b. Backward Morpholexical Type Raising:

$A : a \Rightarrow T\backslash(T/A) : \lambda f.fa$ (where T/A is a lexical function-type over arguments of type A into results of type T) $(< \mathbf{T})$

The system shown here, along with (6), is known as **BT**, which suffices for current purposes. For brevity we eschew modalities which further constrain these combinations. For example, the derivations in (7) allow only harmonic composition because of diamond modality; see Steedman and Baldridge (2011).

(6) a. Forward Composition:

$$X/Y: f \;\; Y/Z: g \Rightarrow X/Z: \lambda x.f(gx) \qquad\qquad (> \mathbf{B})$$

---

[7]Nonconcatenative and nonsegmental morphological processes, which are not only characteristic of templatic languages but also abundant in diverse morphological typologies, such as German, Tagalog, and Alabama, are painful reminders that IA cannot be a universal model for all lexicons.

[8]What this means is that, if "archer" in (4) were a quantified phrase, for example *her okçu* "every archer," then the quantifier's lexically value-raised category would lead to *her okçu* $:= NP^{\uparrow}\backslash(NP/NP): \lambda p\lambda q\lambda x.(\forall x)pa'x \to qx$. Value-raising is distribution of type-raising to arguments, as shown in the logical form.

b. Backward Composition:

$$Y\backslash Z : g\ X\backslash Y : f \Rightarrow X\backslash Z : \lambda x.f(gx) \qquad\qquad (<\mathbf{B})$$

c. Forward Crossing Composition:

$$X/Y : f\ Y\backslash Z : g \Rightarrow X\backslash Z : \lambda x.f(gx) \qquad\qquad (>\mathbf{B}_\times)$$

d. Backward Crossing Composition:

$$Y/Z : g\ X\backslash Y : f \Rightarrow X/Z : \lambda x.f(gx) \qquad\qquad (<\mathbf{B}_\times)$$

With this mechanism we can see the transparency of combination with respect to morphology and syntax (assuming $VP = S\backslash NP_{\mathrm{nom}}$ for simplicity):

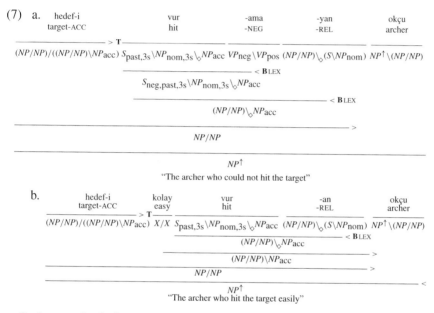

(7) a. "The archer who could not hit the target"

b. "The archer who hit the target easily"

Both examples indicate that lexical integrity can be maintained even if we have categories for things smaller than words.[9] This is why we can deduce categories of word forms from categories of sub-lexical elements.

For example, we could use another independently motivated type-raised category for the embedded object of (4) to show that Turkish relative suffix can in principle attach to the relativized verb *after* the verb takes its adjuncts and residual arguments (8), causing predicate composition, which, with the relative marker *-dik*, leads

---

[9]Here we pass over the mechanism that maintains lexical integrity, which has the effect of doing category combination of bound items before doing it across words. The idea was first stipulated in CCG by Bozşahin (2002) and revised for explanation in Steedman and Bozşahin (2018). In practical parser training the same effect has been achieved in various ways. For example, in a maximum entropy model of Turkish, a category feature for a word is decided based on whether it arises from a suffix of the word (Akkuş 2014). Wang et al. (2014) rely on a morphological analyzer before training, to keep category inference within a word. Ambati et al. (2013) rank intra-chunk (morphological) dependencies higher than inter-chunk (phrasal) dependencies in coming up with CCG categories, which has the same effect.

to unbounded relativization.[10] This is one way CCG avoids bracketing paradoxes, where morphology suggests one combination and syntax another, and the paradox could arise if these two brackets could not be shown to be in the same class of results. That these two derivations have the same result follows from the combinatory base of CCG.

(8)

kolayca	hedef-i	vur	-an	okçu
easily	target-ACC	hit	-REL	archer

$$\frac{(S\backslash NP_{nom})/(S\backslash NP_{nom})}{: \lambda p.easy'\,p} \quad \frac{(S\backslash NP_{nom})/(S\backslash NP_{nom}\backslash NP_{acc})}{: \lambda p.p\,t'} \quad \frac{S_{past,3s}\backslash NP_{nom,3s}\backslash_\diamond NP_{acc}}{: \lambda x \lambda y.hit'xy} \quad \frac{(NP/NP)\backslash_\diamond(S\backslash NP_{nom})}{: \lambda p \lambda q \lambda x.px \wedge qx} \quad \frac{NP\uparrow\backslash(NP/NP)}{: \lambda p \lambda q \lambda x.p\,a'x \wedge qx}$$

$$\frac{\qquad > T}{\dfrac{(S\backslash NP_{nom})/(S\backslash NP_{nom}\backslash NP_{acc})}{: \lambda p.easy'\,(p\,t')}} > B$$

$$\frac{S\backslash NP_{nom}}{: \lambda y.easy'\,(hit'\,t'\,y)} >$$

$$\frac{NP/NP : \lambda q \lambda x.easy'\,(hit'\,t'x) \wedge qx}{} <$$

$$NP\uparrow : \lambda q \lambda x.(easy'\,(hit'\,t'x) \wedge a'x) \wedge qx <$$

"The archer who hit the target easily"

Because of these properties, we shall be able to write the word *vur-an* "that hit" as a nominal modifier later (10). Its category follows from assumptions in (4) and (8). This is crucial for training parsers with rich category inventories, and for models who might opt to train at the level of word forms rather than sub-lexical elements without losing semantic information.[11] The idea relates to supertagging as a proxy for combinatory categories (Clark 2002), and is shared with other radically lexicalized grammars where supertagging originated, epitomized in TAG as "complicate locally, simplify globally';' (Bangalore and Joshi 2010).[12]

---

[10]Notice that the adverb *kolayca* is necessarily a VP modifier, unlike *kolay* of (7b), which is underspecified. We avoid ungrammatical coordinations involving parts of words while allowing suspended affixation, by virtue of radically lexicalizing the conjunction category. For example, [*target-ACC hit and bystander-ACC missed*]-REL *archer* is ungrammatical, and the coordination category $(X_\omega\backslash_\star X_\omega)/_\star X_\omega$ has the constraint (ω), which says that phonological wordhood must be satisfied by all Xs. The left conjunct in this hypothetical example could not project an X_ω because its right periphery—which projects X—would not be a phonological word, as Kabak (2007) showed. It is a forced move in CCG that such constraints on formally available combinations must be derived from information available at the perceptual interfaces.

[11]We note that another wide-coverage parser for Turkish, Eryiğit et al. (2008), which uses dependency parsing, achieves its highest results in terms equivalent to a subset of our sub-lexical training (inflectional groups, in their case). Their comparison includes word-trained lexicons. CCG adds to this perspective a richer inventory of types to train with, and the benefit of naturally extending the coverage to long-range dependencies that are abundant in large corpora, once heads of syntactic constructions bear combinatory categories *in the lexicon*. We say more about these aspects subsequently.

[12]Honnibal and Curran (2009), Honnibal et al. (2010), and Honnibal (2010) have shown that English benefits in parsing performance from sub-lexical training as well, although parsing in their case is word-based. One key ingredient appears to be lexicalizing the unary rules as "hat categories," which indeed makes such CCG categories truly supertags because they can be taken into account in training *before* the parser sees them, whereas the previous usage of supertag in CCG is equivalent to "combinatory lexical category."

8.5   The Turkish Categorial Lexicon

Wide-coverage parsers that deliver meaning face a well-known problem: on one hand we have massive number of word forms many of which are related semantically and morphologically, and on the other we have the logical form, or some kind of dependency encoding, that we expect each word form to have. A training process that avoids information loss as much as it can in connecting these points is needed, summarized above as "supertagging." CCG can take advantage of richness of information in its syntactic types for this process, but this result does not come for free. We mention some work to this end before we move to reporting morphological results for the same task.

Hockenmaier and Steedman (2007) present the first wide-coverage CCG Lexicon from the Penn Treebank. They transform the phrase structure trees in the Penn Treebank into CCG derivation trees. The leaves of the CCG derivation trees then give the words and categories. Tse and Curran (2010) extract a Chinese CCG Bank with a similar method. Extraction of CCG categories from dependency treebanks is different in nature than the original problem of dependency parsing, because dependency trees are in fact graphs that are not trivially mapped to derivation trees. Hockenmaier (2006) extracts a CCG Bank from German dependency corpus TigerBank. With a similar method, Bos et al. (2009) convert Italian dependency trees to constituency trees first to transform them into binary derivation trees in order to extract a CCG Bank for Italian. Çakıcı (2005) suggests a more direct approach by translating the predicate argument structure implied by the dependency graphs into CCG categories without the additional steps of tree transformations. This method does not provide the derivation trees as the others do. Ambati et al. (2013) extract CCG categories for Hindi from a dependency treebank, following Çakıcı (2005), and parse the sentences in the treebank with this lexicon to obtain CCG derivation trees. This method creates a CCG Bank for Hindi automatically.

In the remainder of the paper we will talk about CCG modeling of one data source. The METU-Sabancı Treebank (Chap. 13), hereafter Turkish treebank, is a subcorpus of the METU Turkish Corpus which is a two million word corpus of post-1990 written Turkish. The METU Turkish Corpus includes material taken from three daily newspapers, 87 journal issues, and 201 books (Oflazer et al. 2003; Atalay et al. 2003). The sentences in the treebank are taken from this corpus retaining the proportions of the contributing sources. The Turkish treebank has 5620 sentences and 53,796 tokens (with punctuation). The average sentence length is 9.6 tokens, or 8 words.

Çakıcı (2005) and Çakıcı and Steedman (2009) introduce automatically extracted sets of lexemic (word-based) lexicons and morphemic lexicons for CCG from the treebank. In this paper we use the word "lexemic categories" to mean one category assignment per word form.[13] It was shown that assigning CCG categories

---

[13]In linguistics the term "lexeme" could mean one base lexeme and all its paradigm forms receiving one and same part of speech.

Table 8.1 Ten-fold evaluation of the coverage of lexicons derived from the Turkish treebank

	Word match%	Cat. match%	Pair match%
Mean for morphemic	70.10	94.00	58.50
Std. dev. for morphemic	1.34	1.70	0.92
Mean for lexemic	52.95	90.64	37.06
Std. dev. for lexemic	1.00	1.93	0.63

to morphemes (or morpheme groups) that act as functors over stem categories reduces the sparseness of category types, and it provides correct syntactic analyses– semantic interpretation pairings that could be achieved earlier only manually, by Bozşahin (2002). This is expected for a language like Turkish which has rich yet transparent morphology. Honnibal et al. (2010) show that given the gold-standard morphological analysis for English, an English CCG lexicon with morphemic elements performs better than the word-based version in providing better statistics to overcome sparseness of the category types. For example, the number of verbal category types drops from 159 to 86, including the stem types and the 15 newly introduced inflectional morpheme category types.

Çakıcı (2008) reports that 450 lexemic category types reduce to 311 in the final version of the extracted Turkish CCG lexicon. Given that (1) both lexicons were automatically extracted and that they inherit the residual errors in the dependency treebank, and that (2) the data is too small to have a complete inventory of all the category types, the reduction is significant. The curve of coverage for categories with frequencies greater than 5 starts to converge only after 10K sentences in the Penn Treebank (Hockenmaier 2003).

Table 8.1 shows that the morphemic lexicon derived from the Turkish treebank indeed provides higher coverage than the lexemic lexicon by means of generaliza- tion, despite small size. The table shows the average of a 10-fold cross evaluation of the existence of the category types in the training lexicon, where the training lexicon is 9/10 of the whole treebank and the remaining 1/10 is used as the test partition in each of the ten experiments. The word-pair match shows that the probability of the word and category pair occurring together is 58.5% for the morphemic lexicon, compared to 37.06% for the lexemic one. This means that although the category inventory is not complete, the morphemic lexicon has better coverage than the lexemic one. 94% of the category types are seen in the remaining of the nine parts of the data in the morphemic lexicon, compared to 91% in the lexemic lexicon.

8.5.1 The Lexemic Model

Word-based CCG lexicon extraction (also called "lexemic lexicon extraction" in this chapter) assigns CCG categories to word forms. Most work on automatic or semi- automatic lexicon extraction generally assume word forms as lexical entities; they have lexemic lexicons.

The algorithm for extracting such CCG categories recursively crawls the heads and their arguments, in that order. It assigns the CCG category to the main verb by observing the argument-adjunct distinction of its dependents, then moves on to its arguments and adjuncts recursively. It takes into account information such as part-of-speech tags and other heuristics, when assigning categories. It makes some assumptions for the sake of simplicity and compactness. Some of these assumptions and heuristics are mentioned below. For a detailed discussion, please see Çakıcı (2008).

MODIFIER is the most common label among the dependency labels in the Turkish treebank. A modifier is assigned a category X/X or $X\backslash X$ depending on its direction, where X is the functor (or result) category of what it modifies. An example is given in (9) where *modelle* "with a model" is an adverb modifying another adverb derived from a verb, *yaptığınızda* "when you do (it)" (with the POS tag Adverb_Verb), which in turn modifies an intransitive verb.[14] *Modelle* receives $(S/S)/(S/S)$, and its modifier *iyi* "good" X/X, where X is $(S/S)/(S/S)$, and so on. When uncontrolled these cascaded categories lead to sparseness for the category type set, as different head categories will also lead to different modifier categories, as shown below.

(9) daha|Adv|$((S/S)/(S/S))/((S/S)/(S/S))/((S/S)/(S/S))/((S/S)/(S/S))$ – "more"

iyi|Adj|$((S/S)/(S/S))/((S/S)/(S/S))$ – "good"

bir|Det|NP/NP –"a"

modelle|Noun_Ins|$(S/S)/(S/S)$ –"model+Ins"

yaptığınızda |Adverb_Verb|$(S/S)\backslash NP$ – "do+When"

... *when you do it with a better model* ...

A control mechanism is devised in the algorithm to limit the level of repetition in the category. The combinatory rule of composition will allow the parser to still create a parse while keeping the modifier categories compact, e.g., (S/S). This is similar to what Hockenmaier and Steedman (2007) did for CCGBank. (We also keep in mind that all that optimization depends on a successful parse. In the example above, the determiner category NP/NP would block it.)

Coordination is one area where the need for controlling proliferation of categories is evident. Only like-typed constituents coordinate. A dependency treebank with surface dependency representation might lack information regarding the boundaries of the constituents that coordinate. Çakıcı (2008) discusses these limitations in the Turkish treebank, and provides a solution to the coordination ambiguity by adding extra dependency relations for verbal coordination structures between the coordinating arguments and the verbal heads. Both the lexemic and the morphemic lexicon extraction algorithms make use of this information in order to assign correct categories to the words that involve verb coordination of some

---

[14]The example is from Çakıcı (2008). The convention we follow in display of Turkish treebank data is: word|POS|Category–gloss.

sort. The CCG parser, given the correct CCG categories, recovers dependencies in coordinations that are not originally in the dependency treebank.

The word-based lexicon induction algorithm identifies relative clauses from the relativization morphemes attached to a verb. Çakıcı (2005) provides a solution similar to adding extra dependency links to coordination structures for relative clauses, because of ambiguity in relative clauses. These secondary links provide information about the grammatical role of the extracted argument, and by doing so, it makes possible to assign the correct category to the words involved. Example (10) shows the category assignments to the relative clause in (2); cf. the categories in (4).

(10) hedefi   vuran   okçu   uyudu.
       target-ACC   hit-REL   archer   sleep-PAST.
       NP $(NP_{nom}/NP_{nom})\backslash NP$ NP_{nom} $S\backslash NP_{nom}$.
       "The archer who hit/hits/is hitting the target slept."

The lexemic approach treats the inflected verb that the relative morpheme is attached to as an adjective. This is standard in the word-based lexicon because the word *vuran* behaves like a noun modifier. The relativized verb is assigned the category $(NP_{nom}/NP_{nom})\backslash NP$. Its deduction from semantics of morphemes can be seen in (4) and (8).

This treatment of relativization disregards the predicate argument structure of the verb that is involved in the construction (*vur* in this case), and creates a spurious forest of category types for relativized verbs and/or adjectives in the lexicon. More importantly, it does not provide the correct semantic interpretation. The morphemic lexicon can.

8.5.2 The Morphemic Model

Morphosyntactic interactions mentioned in the previous section speak of a need to have lexical categories for heads of syntactic constructions even if they are sub-lexical elements. The linguistic motivation behind a CCG lexicon with entities smaller than full wordforms is explained in detail in Bozşahin (2002), Çakıcı (2008), and Çakıcı and Steedman (2009). The morphemic lexicon is induced in a similar manner to the lexemic one. The difference is that a word may have more than one morpheme cluster, which is called an inflectional group (IG) in the Turkish treebank. Each such group is assigned a CCG category. Morphological features and other heuristics such as dependency labels are used in assigning the correct category to a lexical entity in a specific context.

There are 27,895 unique word-category pairs for 19,385 distinct tokens in the Turkish lexemic CCG lexicon (Çakıcı 2008). The morphemic lexicon has 13,016 distinct word-category pairs for 6315 distinct word stems and IG stem names. This is a significant improvement because there are more than 69K tokens in morphemic lexicon, compared to about 54K word tokens in the lexemic lexicon.

The average word-category pair frequency goes up from 1.97 to 5.32. These figures not only provide better statistics for training parser models, but also improve the quality of parsing and semantics. Table 8.2 demonstrates the category distribution of one of the most frequent verbs in the treebank—*oku* "read"—with the lexemic approach. Table 8.3 shows the category distribution with the morphemic approach.[15] The morphemic lexicon has four categories for the verb, including the transitive and intransitive forms and their counterparts with pro-drop. Together with the additional morpheme categories, the resulting lexicon is more compact and generalized compared to the lexemic one.

Case-marked modifiers are referred to as adjuncts in the Turkish treebank. They are very common in that subcorpus. Several inflectional morphemes may be in the same inflectional group (IG) with the same CCG category in the morphemic lexicon. Case-marked adjuncts, however, are distinguished by case markers which are part of the inflectional paradigm. Therefore they are a special case, in the sense that the inflectional case morpheme is also treated as a separate lexical entry. The practical aim behind this is to assign correct CCG categories to the words and to their modifiers in order to avoid the category type inflation of adjunct modifiers. For example in (12), küçük "*small*", which is an adjective for *park* in (11), will have the category NP/NP, as opposed to $(S/S)/(S/S)$, as a modifier of the adjunct. (The data line shows the tags matched in the morphemic lexicon. The categories of the zero elements are explained subsequently.)

(11) Parkta geçmiştekiler de var.

In the park, there are the ones from the past, too.

Park	-ta	geçmiş	Adj+Rel	Noun+Zero	de	var	Verb+Zero
park	-LOC	past	the one	-PLU	too	there	is

$$
\frac{\frac{NP \quad (S/S)\backslash NP}{S/S}<\quad \frac{\overline{NP} \quad \frac{NP\backslash NP}{NP} \quad \frac{NP_{nom}\backslash NP}{NP_{nom}\backslash NP}}{} \quad \frac{(NP_{nom}\backslash NP)\backslash (NP_{nom}\backslash NP)}{NP_{nom}\backslash NP}< \quad \overline{NP} \quad \frac{S\backslash NP_{nom}\backslash NP}{S\backslash NP_{nom}}<}{}
$$

(12) a. küçük   park   -ta

$\overline{NP/NP}$ $\overline{NP}$ $\overline{(S/S)\backslash NP}$

"in the small park"

Turkish treebank has some main clauses marked with nominal categories as root. This need presumably arose because Turkish has nominal predication. Rather than defining lexical rules to map some NPs to Ss, the designers appear to have lexicalized the result.[16] This results in S category to be assigned to adjectives and to other nominals since the copular morpheme is empty in present tense for the

---

[15] Figures are from Çakıcı (2008).

[16] In fact, both interpretations are possible, and type-shifting from NP to S would be preferable. For example, "*Arabadaki Mehmet.*" (car-LOC-KI Mehmet) could mean "Mehmet, the one in the car"

Table 8.2 CCG categories (cat) and frequencies (f) of all the derived and inflected forms of verb *oku (read)* in lexemic lexicon

Freq	Word	Cat	Freq	Word	Cat
1	okuyan	(NP/NP)\NP	1	okuyup	(S/S)\NP
1	okuduğu	((S\NP[nom])/ (S\NP[nom]))/((S\NP[nom])/(S\NP[nom]))	1	ok uyup	(NP/NP)/((NP/NP)
1	okuyorsunuz	S\NP	3	okuyor	S\NP
1	okumuşçasına	((S\NP[nom])/(S\NP[nom]))\NP	1	okuyorum	S
1	okuyabilir	S	1	okurum	S\NP
1	okudunuz	S\NP	1	okurken	(S/S)\NP
1	okumadım	S	1	okuttum	NP\NP
1	okuyorum	S\NP[nom]	1	okumayabilir	S/NP
1	okumuştunuz	(S\NP[nom])\NP	1	okudular	S
1	okunacağını	NP/NP[nom]	1	okurduk	S\NP
1	okuyun	S	1	okuyacaklar	S
1	okurdu	S/NP	1	okudu	S\NP
1	okurdu	S\NP	1	okuduk	S
2	okudu	S	1	okudun	S\NP
1	okursa	S/S	1	okumadım	S\NP[nom]
2	okur	S	1	okudum	S\NP
1	okurkenki	NP\NP	1	okumuyorlar	S
1	okuyabilirim	(NP/NP[nom])/NP	1	okunamayan	NP/NP
1	okumalarını	NP/NP	1	okudum	S
1	okuyucunun	NP[nom]	1	okuyor	(S\NP[nom])\NP
1	okumaları	NP\NP	1	okuyayım	S\NP
1	okuyucudan	S/S	1	okudu	(S\NP[nom])\NP
1	okumasını	NP\NP			

Table 8.3 CCG categories (cat) and frequencies (f) of entities of verb *oku (read)* in morphemic lexicon

Freq	Word	Cat
2	oku	S\NP[nom]
4	oku	(S\NP[nom])\NP
20	oku	S
23	oku	S\NP

third person. Categories proliferate. As a solution, *Zero* categories are used in the morphemic lexicon of Çakıcı (2008). Zero categories only occur in the prescribed configurations as functor categories. The nominals they are attached to and their modifiers are thereby assigned the correct category. Example (11) showed how *Verb+Zero* is added to the dependency structure together with its CCG category. These can be compared to unary rules in Hockenmaier and Steedman (2007).

The morphemic approach assigns the relative morpheme *-(y)an* a CCG category of its own. This ensures the correct derivation for the relative clause, shown in (4) and (8), as opposed to its counterpart in the lexemic lexicon as in (10), as discussed earlier.

8.6 Parsing with Automatically Induced CCG Lexicons

CCG categories contain structural information that is crucial to determining the predicate-argument structures. It has been shown by various studies, for example Çakıcı (2008), Çakıcı and Steedman (2009), and Ambati et al. (2014), that CCG categories alone may aid in extracting surface dependencies when used as features for dependency parsers, such as MST parser (McDonald et al. 2005) and Malt Parser (Nivre et al. 2007). Birch et al. (2007) also show that using CCG categories as supertags improves the quality of translations when used as factors in a factored statistical machine translation system for English and Dutch.

The morphemic and the lexemic lexicons for Turkish have been used in different parsing environments. Çakıcı and Steedman (2009) report results for using CCG categories as features to the MST parser (McDonald et al. 2005). The results show that they aid in dependency recovery performance, although the information in the supertags (CCG categories) is used superficially. Table 8.4 shows the dependency accuracy of the MST parser on the CoNLL 2006 shared task data set (Hall and Nilsson 2006), where CCG category is the "fine POS-tag" feature and the POS tag is included as "coarse POS-tag" (Çakıcı 2008). The features in the original

---

or "The one in the car is Mehmet," with the given punctuation. Differences in the interpretations are clear in the following alternative continuations: *Yarın gidiyormuş./Ahmet değil.* "He is leaving tomorrow/Not Ahmet." The first one requires NP reading for the example in the beginning, and the second one S (propositional) reading. Going the other way, i.e., from a lexically specified S for a nominal predicate to an NP, is much more restricted in Turkish. Such type-shifting is in fact headed by verbal inflection.

Table 8.4 MST Parser results on CoNLL 2006 data set

Model	UA	SUA	LA	SLA
CoNLL eval	89.72		84.39	–
CoNLL eval with non-stem IGs included	92.37		88.37	–
MST eval	93.03	54.98	89.63	43.92

Table 8.5 Dependency recovery with C&C parser

Model	Coverage	Cats	UPrec	URec	F
Morphemic	70.10	99.46	72.57	81.18	76.63
Lexemic	65.30	99.43	65.31	72.72	68.82

Table 8.6 Dependency accuracy for object and subject extraction

	Correct	Total	%
Object	74	123	60.2
Subject	173	372	46.5
Total	247	495	49.9

implementation of the MST parser were used apart from the addition of the fine tag feature. UA and SUA are unlabeled accuracy and unlabeled sentence accuracy respectively, and LA and SLA are the labelled counterparts. *MST eval* results are calculated by the evaluation script provided by the MST Parser, and the other results are obtained by the CoNNL shared task evaluation script.

Çakıcı and Steedman (2018) use Clark and Curran's CCG parser with the morphemic and the lexemic lexicons. For the sentences that can be parsed with the assigned categories, a comparable performance in unlabeled dependency recovery is achieved (Table 8.5). Moreover, the parser recovers long-distance dependencies such as object and subject extraction, and some long distance dependencies that arise from coordination that other parsers are not capable of predicting (Table 8.6). For example, from the coverage of *adamın _i okuduğu kitap_i* (man-3S read-REL.3S book) "the book the man read," we also get the coverage of *benim adamın _i okuduğunu bildiğim kitap_i* "the book which I know the man read."

Gold standard CCG categories have been used in these experiments. The parser is trained using the partial training model defined in Clark and Curran (2006). The idea is to extract features from derivation trees that are created by the parser. This facilitates training when derivation trees are not available, which was the case for Turkish.

The Clark and Curran parser outputs deep dependencies that are partially compatible with the treebank dependencies. The results reported in Çakıcı (2008) and Çakıcı and Steedman (2018) are evaluated by comparing the output of the CCG parser with a dependency set that is created by adding the secondary coordination and extraction dependencies to the original surface dependencies in the treebank. Evaluation was done by converting the treebank dependencies into Clark and Curran parser output format. The dependency sets are different, therefore precision, recall, and F-measure are reported for dependency recovery. Coverage of the parser with the morphemic lexicon is around 70%, compared to slightly over 65% of the

lexemic lexicon. *Cats* column in Table 8.5 shows the category accuracy.[17] Parsing
with supertagging experiments on the morphemic lexicon shows that, with 71.55%
accuracy for the supertagger, the parser finds a parse for the 97.3% of the sentences
in the treebank. However, the unlabeled precision and recall for the dependencies in
this case are 55.64% and 63.1%, respectively.

The results in Table 8.5 show dependency recovery of the treebank dependencies,
which are surface dependencies. Table 8.6 reports the long-distance dependency
accuracy for object and subject extraction in Çakıcı and Steedman (2018). The table
lists the total number of each of the dependency link added to the treebank (Çakıcı
2005), for sentences that receive at least one parse, and the number and percentage
accuracy of predicting these dependency links by the Clark and Curran's parser
using the morphemic lexicon. Gold-standard supertags are used in this experiment
since the aim here is to analyze the performance of the CCG parser in retrieving the
long-distance dependencies in *correct* parses, and to evaluate them across the added
gold-standard long-distance dependencies. The cause of long-distance dependency
recovery rate for subject extraction being lower than the one for object extraction is
attributed to subject pro-drop being more common than the object drop in Turkish.
Subject extraction is incorrectly analyzed as pro-drop in some of the examples.

8.7   Conclusion

Modeling the lexicon using CCG categories, which are, by definition, paired with a
logical form, or predicate-argument structure, appears to be helpful in recovery of
logical forms in wide-coverage parsing. The granularity of parser semantics we can
expect to elicit from such lexicons depends on coverage of heads of constructions
along with their logical forms, independent of whether they are words, affixes, tones,
or multi-word elements. Whether parsing itself is word-based or morpheme-based,
the training stage benefits from such knowledge. The results for Turkish in particular
suggest that proliferation of categories due to lexical rules and type-shifting (e.g.,
nominal predicates and adjunction) can be kept under control.

Categories of sub-lexical elements may turn out to be domains of locality
which may require reaching into arguments and adjuncts of the word the element
is attached to. Transparent projection of these properties by composition is the
expedient to natural extension of morphemic lexicons to capture long-range dependencies, which no other Turkish parser we are aware of is capable of doing.
In that respect, parsing performances must also be evaluated over the kinds of
constructions they can deal with, rather than the number of constructions that
can be handled, or reported coverage in a corpus. This aspect directly relates to
richness of head information in the lexicon, be it word, sub-lexical element, tone,

---

[17]Although the gold-standard CCG categories (supertags) are used, this number is slightly less
than 100%. This is possibly caused by an implementation discrepancy.

or a morpholexical process. Model training and wide-coverage parsing with such elements are beneficial and feasible.

References

Akkuş BK (2014) Supertagging with combinatory categorial grammar for dependency parsing. Master's thesis, Middle East Technical University, Ankara

Ambati BR, Deoskar T, Steedman M (2013) Using CCG categories to improve Hindi dependency parsing. In: Proceedings of ACL, Sofia, pp 604–609

Ambati BR, Deoskar T, Steedman M (2014) Improving dependency parsers using combinatory categorial grammar. In: Proceedings of EACL, Gothenburg, pp 159–163

Aronoff M, Fudeman K (2011) What is morphology?, 2nd edn. Wiley-Blackwell, Chichester

Atalay NB, Oflazer K, Say B (2003) The annotation process in the Turkish treebank. In: Proceedings of the workshop on linguistically interpreted corpora, Budapest, pp 33 – 38

Bangalore S, Joshi AK (eds) (2010) Supertagging. MIT Press, Cambridge, MA

Barton G, Berwick R, Ristad E (1987) Computational complexity and natural language. MIT Press, Cambridge, MA

Berwick R, Weinberg A (1982) Parsing efficiency, computational complexity, and the evaluation of grammatical theories. Linguist Inquiry 13:165–192

Birch A, Osborne M, Koehn P (2007) CCG supertags in factored statistical machine translation. In: Proceedings of WMT, pp 9–16

Bos J, Bosco C, Mazzei A (2009) Converting a dependency treebank to a categorial grammar treebank for Italian. In: Proceedings of the international workshop on treebanks and linguistic theories, Milan, pp 27–38

Bozşahin C (2002) The combinatory morphemic lexicon. Comput Linguist 28(2):145–186

Bozşahin C (2012) Combinatory linguistics. Mouton De Gruyter, Berlin

Çakıcı R (2005) Automatic induction of a CCG grammar for Turkish. In: Proceedings of the ACL student research workshop, Ann Arbor, MI, pp 73–78

Çakıcı R (2008) Wide-coverage parsing for Turkish. PhD thesis, University of Edinburgh, Edinburgh

Çakıcı R, Steedman M (2009) A wide-coverage morphemic CCG lexicon for Turkish. In: Proceedings of ESSLLI workshop on parsing with categorial grammars, Bordeaux, pp 11–15

Çakıcı R, Steedman M (2018) Wide coverage CCG parsing for Turkish, in preparation

Cha J, Lee G, Lee J (2002) Korean combinatory categorial grammar and statistical parsing. Comput Hum 36(4):431–453

Clark S (2002) A supertagger for combinatory categorial grammar. In: Proceedings of the TAG+ workshop, Venice, pp 19–24

Clark S, Curran JR (2006) Partial training for a lexicalized grammar parser. In: Proceedings of NAACL-HLT, New York, NY, pp 144–151

Clark S, Curran JR (2007) Wide-coverage efficient statistical parsing with CCG and log-linear models. Comput Linguist 33:493–552

Çöltekin Ç, Bozşahin C (2007) Syllable-based and morpheme-based models of Bayesian word grammar learning from CHILDES database. In: Proceedings of the annual meeting of the cognitive science society, Nashville, TN, pp 880 – 886

Eryiğit G, Nivre J, Oflazer K (2008) Dependency parsing of Turkish. Comput Linguist 34(3): 357 – 389

Göksel A (2006) Pronominal participles in Turkish and lexical integrity. Ling. Linguaggio 5(1):105–125

Hall J, Nilsson J (2006) CoNLL-X shared task: multi-lingual dependency parsing. MSI Report 06060, School of Mathematics and Systems Engineering, Växjö University, Växjö

Hockenmaier J (2003) Data models for statistical parsing with combinatory categorial grammar. PhD thesis, University of Edinburgh, Edinburgh

Hockenmaier J (2006) Creating a CCGbank and a wide-coverage CCG lexicon for German. In: Proceedings of COLING-ACL, Sydney, pp 505–512

Hockenmaier J, Steedman M (2007) CCGbank: a corpus of CCG derivations and dependency structures extracted from the Penn Treebank. Comput Linguist 33(3):356–396

Hockett CF (1959) Two models of grammatical description. Bob-Merrill, Indianapolis, IN

Hoeksema J, Janda RD (1988) Implications of process-morphology for categorial grammar. In: Oehrle RT, Bach E, Wheeler D (eds) Categorial grammars and natural language structures. D. Reidel, Dordrecht

Honnibal M (2010) Hat categories: representing form and function simultaneously in combinatory categorial grammar. PhD thesis, University of Sydney, Sydney

Honnibal M, Curran JR (2009) Fully lexicalising CCGbank with hat categories. In: Proceedings of EMNLP, Singapore, pp 1212–1221

Honnibal M, Kummerfeld JK, Curran JR (2010) Morphological analysis can improve a CCG parser for English. In: Proceedings of COLING, Beijing, pp 445–453

Kabak B (2007) Turkish suspended affixation. Linguistics 45:311–347

Koskenniemi K (1983) Two-level morphology: a general computational model for word-form recognition and production. PhD thesis, University of Helsinki, Helsinki

Koskenniemi K, Church KW (1988) Complexity, two-level morphology and Finnish. In: Proceedings of COLING, Budapest, pp 335–339

Lewis M, Steedman M (2014) A* CCG parsing with a supertag-factored model. In: Proceedings of EMNLP, Doha, pp 990–1000

Lieber R (1992) Deconstructing morphology: word formation in syntactic theory. The University of Chicago Press, Chicago, IL

MacWhinney B (2000) The CHILDES project: tools for analyzing talk, 3rd edn. Lawrence Erlbaum Associates, Mahwah, NJ

Matthews P (1974) Morphology: an introduction to the theory of word-structure. Cambridge University Press, Cambridge

McConville M (2006) An inheritance-based theory of the lexicon in combinatory categorial grammar. PhD thesis, University of Edinburgh, Edinburgh

McDonald R, Crammer K, Pereira F (2005) Online large-margin training of dependency parsers. In: Proceedings of ACL, Ann Arbor, MI, pp 91–98

Nivre J, Hall J, Nilsson J, Chanev A, Eryiğit G, Kübler S, Marinov S, Marsi E (2007) MaltParser: a language-independent system for data-driven dependency parsing. Nat Lang Eng 13(2):95–135

Oflazer K (2003) Dependency parsing with an extended finite-state approach. Comput Linguist 29(4):515–544

Oflazer K, Göçmen E, Bozşahin C (1994) An outline of Turkish morphology. www.academia.edu/7331476/An_Outline_of_Turkish_Morphology (7 May 2018)

Oflazer K, Say B, Hakkani-Tür DZ, Tür G (2003) Building a Turkish treebank. In: Treebanks: building and using parsed corpora. Kluwer Academic Publishers, Berlin

Roark B, Sproat RW (2007) Computational approaches to morphology and syntax. Oxford University Press, Oxford

Sak H, Güngör T, Saraçlar M (2011) Resources for Turkish morphological processing. Lang Resour Eval 45(2):249–261

Schmerling S (1983) Two theories of syntactic categories. Linguist Philos 6(3):393–421

Sells P (1995) Korean and Japanese morphology from a lexical perspective. Linguist Inquiry 26(2):277–325

Steedman M (1996) Surface structure and interpretation. MIT Press, Cambridge, MA

Steedman M (2000) The syntactic process. MIT Press, Cambridge, MA

Steedman M (2011) Taking scope. MIT Press, Cambridge, MA

Steedman M, Baldridge J (2011) Combinatory categorial grammar. In: Boyer R, Börjars K (eds) Non-transformational syntax: formal and explicit models of grammar: a guide to current models, Wiley-Blackwell, West Sussex

Steedman, M. and C. Bozşahin (2018) Projecting from the Lexicon. MIT Press, (submitted)

Stump GT (2001) Inflectional morphology: a theory of paradigm structure. Cambridge University Press, Cambridge

Tse D, Curran JR (2010) Chinese CCGbank: extracting CCG derivations from the Penn Chinese treebank. In: Proceedings of COLING, Beijing, pp 1083–1091

Valiant L (2013) Probably approximately correct: nature's algorithms for learning and prospering in a complex world. Basic Books, New York, NY

van Rooij I (2008) The tractable cognition thesis. Cogn Sci 32(6):939–984

Wang A, Kwiatkowski T, Zettlemoyer L (2014) Morpho-syntactic lexical generalization for CCG semantic parsing. In: Proceedings of EMNLP, Doha, pp 1284–1295

Yuret D, Türe F (2006) Learning morphological disambiguation rules for Turkish. In: Proceedings of NAACL-HLT, New York, NY, pp 328–334

Chapter 9
Deep Parsing of Turkish with Lexical-Functional Grammar

Özlem Çetinoğlu and Kemal Oflazer

Abstract In this chapter we present a large scale, deep grammar for Turkish based on the Lexical-Functional Grammar formalism. In dealing with the rich derivational morphology of Turkish, we follow an approach based on morphological units that are larger than a morpheme but smaller than a word, in encoding rules of the grammar in order to capture the linguistic phenomena in a more formal and accurate way. Our work covers phrases that are building blocks of a large scale grammar, and also focuses on linguistically—and implementation-wise—more interesting cases such as long distance dependencies and complex predicates.

9.1 Introduction

In this chapter we present a large scale, deep grammar for Turkish based on the Lexical-Functional Grammar (LFG) formalism. Our goal in building this grammar was not only to develop a primary NLP resource but also to understand, define, and represent the linguistic phenomena of Turkish in a more formal way. One of the distinguishing aspects of this work is the implementation of the grammar by employing parsing units smaller than words but larger than morphemes, namely *inflectional groups* (IGs hereafter). We have already seen the use of IGs in the context of morphological disambiguation in Chap. 3 and in parsing in Chap. 7. Modeling with inflection groups allows us to incorporate the complex morphology and the syntactic relations mediated by morphological units in a manageable way and handle lexical representations of very productive derivations.

Ö. Çetinoğlu
University of Stuttgart, Stuttgart, Germany
e-mail: ozlem@ims.uni-stuttgart.de

K. Oflazer (✉)
Carnegie Mellon University Qatar, Doha-Education City, Qatar
e-mail: ko@cs.cmu.edu

© Springer International Publishing AG, part of Springer Nature 2018
K. Oflazer, M. Saraçlar (eds.), *Turkish Natural Language Processing*,
Theory and Applications of Natural Language Processing,
https://doi.org/10.1007/978-3-319-90165-7_9

The LFG formalism (Kaplan and Bresnan 1982; Dalrymple 2001) is a well-established unification-based theory. It is widely used with many contributors working on various languages from diverse language families. The experience of these contributors is shared through the ParGram (Parallel Grammars) project (Butt et al. 1999, 2002). The resulting grammars have been used in several projects such as machine learning, modeling syntax/semantics interface, and machine translation.[1]

The Turkish LFG grammar is also part of the ParGram project which has also developed large scale grammars for a range of languages including Arabic, Chinese, English, French, Georgian, German, Hungarian, Japanese, Malagasy, Norwegian, Urdu, and Welsh. Despite the differences among the languages involved, the aim has been to produce parallel syntactic analyses with the assumption that although word order, surface representation, or constituent hierarchy may differ, the function of constituents are the same for equivalent sentences among languages. This assumption enables the sharing of the linguistic know-how on some well studied topics when a new grammar is developed within ParGram.

While developing the Turkish LFG grammar, we benefited from the accumulated general grammar engineering know-how, as well as know-how on linguistic phenomena like coordination, free word order, and long distance dependencies. In turn, we brought attention to topics such as morphology-syntax interaction and implementation of morphological causatives and passives, that were relatively less studied in ParGram due to the linguistic nature of earlier grammars.

9.2 Lexical-Functional Grammar and Xerox Linguistic Environment

LFG (Kaplan and Bresnan 1982; Dalrymple 2001) is a linguistic theory representing syntax at two parallel levels: Constituent structures (c-structures) capture the syntactic structure of a sentence while feature structures (f-structures) define the functional structure and relations. C-structures are represented using context-free phrase structure trees and f-structures are represented by sets of pairs of attributes and values. Attributes may be features, such as tense and gender, or functions, such as subject and object. Values may be atomic such as FUTURE for tense, or a subsidiary f-structure such as a pronoun f-structure for subject. In general, c-structures are more language specific whereas f-structures of the same phrase for different languages are expected to be similar to each other.[2]

The Turkish LFG grammar is implemented using the Xerox Linguistic Environment (XLE) (Kaplan and Maxwell 1996), a grammar development platform

---

[1] ParGram/ParSem. An international collaboration on LFG-based grammar and semantics development: pargram.b.uib.no (Accessed Sept. 14, 2017).

[2] We assume that the reader is familiar with the basic concepts of LFG. Otherwise, please refer to Kaplan and Bresnan (1982) and Dalrymple (2001).

that facilitates the integration of various modules, such as tokenizers, finite-state morphological analyzers, and lexicons. We have integrated into XLE, a series of finite state transducers for morphological analysis and for multi-word processing for handling lexicalized, semi-lexicalized collocations and a limited form of non-lexicalized collocations. We use the Turkish morphological analyzer (Oflazer 1994), (see Chap. 2), as the main analyzer with additional transducers for multi-word processing, also discussed in Chap. 2. The finite state analyzers provide the relevant ambiguous morphological interpretations for words and their split into IGs, but do not provide syntactically relevant semantic and subcategorization information for lemmas. Such information is encoded in a lexicon of lemmas on the grammar side.

9.3   Inflectional Groups as First-Class Syntactic Citizens

As discussed earlier, the morphological analysis of a Turkish word can be represented as a sequence of tags corresponding to overt or covert morphemes, with derivational boundaries delineating inflectional groups. A given word may have multiple such representations depending on any morphological ambiguity brought about by alternative segmentations of the word, and by ambiguous interpretations of morphemes.

We repeat here an example phrase in Fig. 9.1 from earlier chapters (but with more explicit labeling of IGs relevant to topic here) to help clarify how IGs are involved in syntactic relations.

Here, the superlative adverb *en* "most" modifies the adjective *akıllı* "intelligent" (literally "with intelligence"); not *akıl* "intelligence" or the whole word *akıllısı* "the intelligent one of". The genitive noun *öğrencilerin* "students'" specifies the derived noun phrase (NP) *en akıllısı* "the most intelligent one of".

The f-structure representation of the NP in Fig. 9.1 is given in (1).[3] The semantics of the derivational suffix *-li* is represented as 'li⟨↑ OBJ⟩' (the subsidiary f-structure

Fig. 9.1 The dependencies between IGs of the noun phrase *öğrencilerin en akıllısı*. Grey boxes show word boundaries. Morphemes are separated with hyphens and IGs are marked with horizontal brackets. IG$_{ij}$ denotes the *j*th IG of word *i*

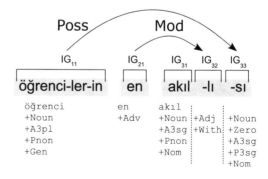

---

[3]All functional structures are simplified in this chapter to emphasize the relevant features only.

corresponding to the ADJUNCT of the main f-structure). First, the f-structure of noun *akıl* "intelligence" is placed as the OBJ of the derivational suffix. Supporting the dependency representation in Fig. 9.1, the f-structure of the adverb *en* is placed as the adjunct of *li⟨akıl⟩*, that is, the adjective *akıllı*. Zero derivation of an adjective to a noun, as exemplified in the given phrase, indicates that there is a generic person modified by the adjective in question. In terms of f-structure representation, this corresponds to a new PRED 'null-pro' with the adjective as the ADJUNCT of the new structure which is shown as the outermost matrix in (1). The derived noun behaves essentially like a lexical noun and can be specified by another noun, here by *öğrencilerin* "of the students".

(1)
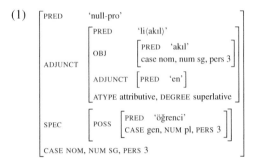

The effect of using IGs as the representative units can be explicitly seen in the c-structure, where each IG corresponds to a separate node as in (2).

(2)
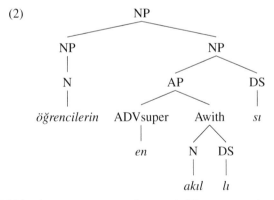

Within the tree representation, each IG corresponds to a separate node. Thus, the LFG grammar rules constructing the c-structures are coded using IGs as units of parsing. If an IG contains the lemma of a word, then the node corresponding to that IG is named as one of the syntactic category symbols. The rest of the IGs are given the node name DS (to indicate derivational suffix), no matter what the content of the IG is.

The representation of derivational suffixes in Turkish has been the most discussed subject since the beginning of the grammar development within the ParGram project. Basically, the IG approach goes against the Lexical Integrity Principle of LFG Theory (Bresnan and Mugane 2006), who state:

> Every lexical head is a morphologically complete word formed out of different elements and by different principles from syntactic phrases.

In our approach, lexical heads might not be morphologically complete words but could be derivational suffixes, causing the words to be separated into several nodes in a c-structure. As shown in (2), the noun *akıllısı* is represented with three different nodes in the c-structure, although it is a single word.

There are five lexical integrity tests employed by Bresnan and Mchombo (1995) to decide whether words constructed by derivational suffixes are lexicalized or not. When these tests are applied to Turkish derived words, one can observe that there are certain suffixes which do not obey the standard definition of suffixes although they are attached to words orthographically. The most distinctive results came from the tests on phrasal recursivity.[4] Bresnan and Mchombo (1995) state that

> Word-internal constituents generally differ from word-external phrases in disallowing the arbitrarily deep embedding of syntactic phrasal modifiers.

But in Turkish the noun-to-adjective derivational suffix *-li* has the phrasal recursivity as given in (3), where we indicate the phrase boundaries by [...].

(3)   a.   elbise-li
          dress-With
          "with a dress"

      b.   [mavi elbise]-li
          [blue dress]-With
          "with the blue dress"

      c.   [[açık mavi] elbise]-li
          [[light blue] dress]-With
          "with the light blue dress"

We can observe that the scope of the derivational suffix is at the phrasal level rather than the word level. If we had attached the suffix *-li* to the stem *elbise* without considering the phrasal scope, the adjective *mavi* would seem to modify the derived adjective *elbiseli* in (3b). Similarly, c-structure in (4) would be the representation of the phrase in Fig. 9.1, if the IG representation had not been used.

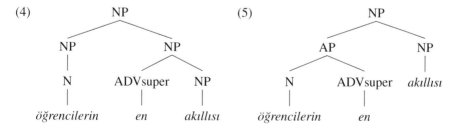

---

[4]The rest of the tests are discussed in Sect. 3.3.2 of Çetinoğlu (2009).

Another proposed alternative was implementing the approach in Bresnan and Mugane (2006). The tree in (5) gives the c-structure of Fig. 9.1 according to this approach. In these alternatives, the lexical integrity is preserved but the c-structure does not reflect the actual syntactic relations between the relevant units, leading to both information loss and a misconception about the grammar rules of the language. For instance, in (4) the adverb *en* seem to modify the derived noun *akıllısı* although adverbs cannot modify noun phrases.

9.4 Previous Work

There have been a number previous studies in developing computational formal grammars for Turkish.[5]

Güngördü and Oflazer (1995) is the earliest such work describing a rather extensive grammar for Turkish using the LFG formalism. Although this grammar had a good coverage and handled phenomena such as free-constituent order, the underlying implementation was based on pseudo-unification. But most crucially, it employed a rather standard approach to represent the lexical units: words with multiple nested derivations were represented with complex nested feature structures where linguistically relevant information could be embedded at unpredictable depths which made access to them in rules extremely complex and unwieldy.

Şehitoğlu (1996) developed a formal grammar of Turkish using a Head-driven Phrase Structure Grammar, another unification-based grammar formalism.

Bozşahin (2002) employed morphemes overtly as lexical units in a CCG framework to account for a variety of linguistic phenomena in a prototype implementation. The drawback was that morphotactics was explicitly raised to the level of the sentence grammar, hence the categorial lexicon accounted for both constituent order and the morpheme order with no distinction (see also Chap. 8).

9.5 LFG Analyses of Various Linguistic Phenomena

In this section, we summarize the rule coverage of our grammar by giving overviews of the basic phrasal components. We start with noun phrases and continue with adjective, adverbial, and postpositional phrases. We then cover deverbal constructions of adjectives and adverbs, and end with temporal phrases.

---

[5]We do not cover previous general LFG work here explicitly and instead cite them when the relevant phenomena are discussed throughout the following sections.

9.5.1 Noun Phrases

Göksel and Kerslake (2005) define a noun phrase as any sequence of words that can function as a subject, or as some kind of a complement, such as an object, a subject complement, or the complement of a postposition. The case and referentiality play an important role in determining the argumenthood of noun phrases. Our grammar covers a wide range of different types of noun phrases, including indefinite and definite noun compounds, possessives, pronouns, proper nouns, derived noun phrases, NPs modified by adjectives, determiners, numbers, measure phrases, postpositions, and combinations of these. In indefinite noun compounds, an NP in nominative case modifies the head NP and the modifying NP functions as MODifier in the LFG representation. In definite noun compounds, an NP in genitive case modifies the head NP, and this time the modifying NP functions as a possessive specifier, namely SPEC POSS. (6) and (7) give the c-structure and the f-structure for the simple definite noun compound *kitabın kapağı* "the book's cover".

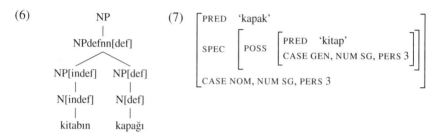

The definiteness feature of nouns is stored in the c-structure by using complex categories, i.e., categories that can take arguments, so that its value can be modified during unification. For noun phrases, the value of the argument is either DEF or INDEF. An example which makes use of this property is given in (8). The head of the NP is *kitap* "book" which is indefinite by itself, but the whole phrase *evdeki kitap* "the book in the house" is definite. During parsing, the f-structure of the head unifies with the f-structure of the whole phrase. Having a feature value pair [DEF -] in the f-structure of *kitap* "book" would result in an unwanted [DEF -] in the final f-structure. Instead, we do not carry the argument INDEF of the NP up the tree but assign the correct value DEF to the argument of the complex category NPadj.

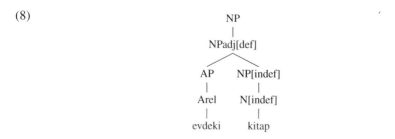

The actual noun f-structures also carry semantic information about nouns (e.g., COMMON, PROPER, COUNT, MASS, MEASURE). This information is crucial for parsing some phrases. The morphological analyzer outputs some semantic information such as PROPER, but most of the semantic details are manually encoded in the lexicon. For instance, measure nouns have a semantic marker in the lexicon and measure phrases have a separate rule in the grammar. (9) and (10) show the c-structure and f-structure of the phrase *iki kilo elma* "two kilos of apple".

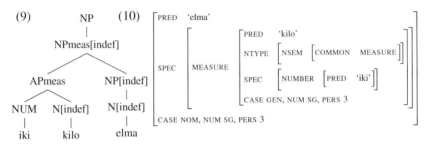

The grammar also employs separate rules for NPs modified by determiners, numbers, adjectives, postpositions, definite and indefinite noun compounds, overt and covert possessives, pronouns, proper names, proper location names, sentential complements and infinitives, and NPs derived from adjectives or numbers.

9.5.2 Adjective Phrases

The adjective phrase grammar includes rules for basic, comparative and superlative adjectival phrases such as *mutlu* 'happy', *daha mutlu* "happier", *en mutlu* "the happiest". The degree of the adjective is also represented in the f-structure, with values POSITIVE, COMPARATIVE, and SUPERLATIVE respectively. (11) and (12) give the c-structure and f-structure for the AP *daha mutlu kedi* "happier cat".

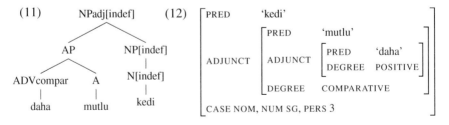

Derived adjectives are handled by encoding two types of rules. If the derivational suffix has phrasal scope it has a separate rule. If the adjective suffix is attached to simple words, for instance *-ci* "-ist" in e.g., *merkez-ci* "centralist", *barış-çı* "pacifist", then a generic rule is used.

9.5.3   Adverbial Phrases

The part of the grammar that handles adverbial phrases consists of rules for parsing simple, comparative, and superlative adverbs, adverbs modifying other adverbs, e.g., *az* "less", *çok* "more", derived adverbs, e.g., *sakin-ce* "calmly", and adverbs formed by duplicating adjectives, e.g., *sakin sakin* "calmly, lit. calm calm" (See Chap 2). There is also a special constituent focus rule for adverbs like *bile* "even", *de/da* "too", *falan/filan* "etc."[6] These attach after every possible phrase. For the basic sentence in (13a), sentence (13b) is one possibility. The phrases *sabah bile* "even in the morning", *yumurtasını bile* "even her egg", *yedi bile* "even ate" are the other options.

(13)   a.   Zeynep   sabah   yumurta-sı-nı ye-di.
            Zeynep.Nom morning.Nom egg-Poss-Acc eat-Past.3sg
            "Zeynep ate her egg in the morning."

       b.   Zeynep   bile sabah   yumurta-sı-nı ye-di.
            Zeynep.Nom even morning.Nom egg-Poss-Acc eat-Past.3sg
            "Even Zeynep ate her egg in the morning."

The c-structure and f-structure of the NP *Zeynep bile* of (13b) is given in (14) and (15) respectively.

(14)   NP[def]   (15)   ⎡ PRED   'Zeynep'   ⎤
                    ╱   ╲   ⎢   ⎥
            NP[def]   ADVfoc   ⎢ ADJUNCT   [PRED 'bile']⎥
               │   │   ⎢   ⎥
            PROP   bile   ⎣ CASE   NOM   ⎦
               │
            Zeynep

9.5.4   Postpositional Phrases

The postposition analysis is rather straightforward. The only crucial information, that is, the case marker of the NP that the postposition subcategorizes for, comes conveniently from the morphological analyzer.[7] For instance, the analysis for *ait* "belonging to" is `ait+Postp+PCDat`. The feature `+PCDat` indicates that the preposition subcategorizes for a dative object, hence the dative marked *Ali'-ye* "to Ali" can function as the OBJect of *ait*. The f-structure of the postpositional phrase

---

[6]This rule is very similar to the one used in the ParGram English grammar.

[7]This feature was originally included in the morphological analyzer as it would help to disambiguate the analysis of the previous token in the text in the early rule based morphological disambiguation tools.

Ali'ye ait "belonging to Ali" is illustrated in (16). Whether the resulting postposition phrase (POSTPP) modifies an NP, e.g., *Ali'ye ait kitap* "the book belonging to Ali", or serves as an adverbial phrase, e.g., *yemekten sonra* "after the dinner", is determined by semantic markers.

$$(16) \quad \begin{bmatrix} \text{PRED} & \text{'ait}\langle\text{Ali}\rangle\text{'} \\ \text{OBJ} & \begin{bmatrix} \text{PRED} & \text{'Ali'} \end{bmatrix} \end{bmatrix}$$

There are also a handful of words that behave as postpositions although they are nouns. They cannot be taken as simple lexicalized postpositions neither by the morphology nor by the syntax due to agreement in person during the phrase construction. *yüzünden* "because of", as one of the members of the set, has the alternations in (17a) and (17b) for 1st and 3rd person singular. The lemma (here, *yüz*) and the case (here, ablative) of the noun acting as postposition are hand-coded in the lexicon. Agreement is handled in grammar rules. (18) gives the f-structure for (17a).

(17) a. ben-im yüz-üm-den
          I-Gen   because.of-P1sg-Abl
          "because of me"

      b. o-nun   yüz-ün-den
          s/he/it-Gen because.of-P2sg-Abl
          "because of him/her/it"

$$(18) \quad \begin{bmatrix} \text{PRED} & \text{'yüz}\langle\text{ben}\rangle\text{'} \\ \text{OBJ} & \begin{bmatrix} \text{PRED} & \text{'ben'} \\ \text{CASE GEN, NUM SG, PERS 1} \end{bmatrix} \\ \text{SPEC} & \begin{bmatrix} \text{POSS} & \begin{bmatrix} \text{PRED} & \text{'null\_pro'} \\ \text{NUM SG, PERS 1} \end{bmatrix} \end{bmatrix} \\ \text{CASE ABL, NUM SG, PERS 3} \end{bmatrix}$$

9.5.5 Temporal Phrases

Our grammar has a specific temporal phrase subgrammar that covers point-in-time expressions, particularly clock-time expressions (*saat 2'de* "at 2 o'clock", *gecenin üçünde* "at three (o'clock) at night"), days of the week (*Salıları* "on Tuesdays", *Cuma günü* "on Friday"), calendar dates (*9 Mart 2007* "9th March 2007", *Ekim 19'da* "on October 19th"), seasons (*yazın* "in the summer", *kış mevsiminde* "in the winter"), and some general phrases (*şimdi* "now", *dün sabah* "yesterday morning").

This specific subgrammar was developed as part of a M.Sc. thesis by Gümüş (2007). The core of this grammar relies on our NP rules, hence the implementation is parallel to ours. Also the features and templates are based on our version for the sake of consistency. Gümüş added new rules to parse temporal phrases that are not covered by the basic NP rules (e.g., a nominative N modifying an N for *dün sabah* 'yesterday morning'). She also semantically marked certain types of words as being temporal with more specific information such as date, clock-time, day, or season.

We then integrated this date-time grammar into our system. The integration process brought about some additional ambiguity issues which were solved by introducing OT-Marks (Frank et al. 2001) that help us rank the parser outputs.

9.6   Sentential Derivations, Sentences and Free Constituent Order

Turkish handles sentential complements, sentential adjuncts, and relative clauses all by morphological derivations. In this section, we go into detail with these derivations by using examples and presenting their LFG analyses. We also explain how different sentence types are handled and discuss the problems we encountered in implementing free constituent order.

9.6.1   Sentential Derivations

Turkish sentential complements and adjuncts are marked by productive verbal derivations into nominals (infinitives, participles) or adverbials. Relative clauses with subject, object, or adjunct gaps are formed by participles which function as a modifier of the head noun (which happens to be the filler of the gap). (19) shows a simple sentence that will be used throughout the following examples. Its c- and f-structures are given in (20a) and (20b), respectively.

(19)   Kız   adam-ı   ara-dı.
       girl.Nom man-Acc call-Past.3sg
       "The girl called the man."

(20)   a.

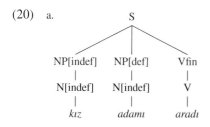

b.

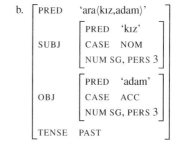

9.6.1.1   Sentential Complements

Example (21) depicts a sentence with a sentential complement that functions as the object for the verb *söyledi* "said". The complement is derived from (19) with a past participle suffix.

(21)   Manav   kız-ın   adam-ı   ara-dığı-nı   söyle-di.
       grocer.Nom girl-Gen man-Acc call-PastPart-Acc say-Past.3sg
       "The grocer said that the girl called the man."

The derivation takes place on the verb of the sentential complement, and the nominal features, e.g., CASE, come from the participle IG. The arguments of the base verb (here, SUBJ and OBJ) are parsed as in a normal sentence. (22) gives the c-structure of the sentence in (21). Note that the participle IG including the derivational morpheme is attached to the base verb in the node Vnom, which is a separate node in the tree. This is necessitated by the free constituent order: the NP *adamı kızın aradığını* is also valid, as well as the NPs with other permutations of the constituents within the participle phrase.

(22)

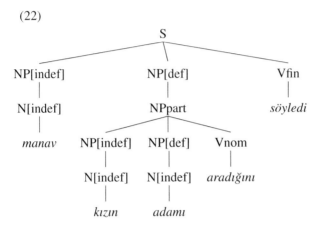

The corresponding f-structure is shown in (23). Since the participle IG has the complete set of syntactic features of a noun, no new rules are needed to incorporate the derived f-structure to the rest of the grammar, that is, the derived phrase can be used as if it is a simple NP.

(23)
$$
\begin{bmatrix}
\text{PRED} & \text{'söyle}\langle\text{manav, ara}\rangle\text{'} \\[2pt]
\text{SUBJ} & \begin{bmatrix} \text{PRED} & \text{'manav'} \\ \text{CASE NOM, NUM SG, PERS 3} \end{bmatrix} \\[10pt]
\text{OBJ} & \begin{bmatrix}
\text{PRED} & \text{'ara}\langle\text{kız, adam}\rangle\text{'} \\[2pt]
\text{SUBJ} & \begin{bmatrix} \text{PRED} & \text{'kız'} \\ \text{CASE GEN, NUM SG, PERS 3} \end{bmatrix} \\[6pt]
\text{OBJ} & \begin{bmatrix} \text{PRED} & \text{'adam'} \\ \text{CASE ACC, NUM SG, PERS 3} \end{bmatrix} \\[6pt]
\text{CHECK} & \begin{bmatrix} \text{PART} & \text{PASTPART} \end{bmatrix} \\[4pt]
\text{CASE ACC, NUM SG, PERS 3, CLAUSE-TYPE NOM}
\end{bmatrix} \\[10pt]
\text{TENSE} & \text{PAST}
\end{bmatrix}
$$

The f-structure and c-structure similarities of sentences in (19) and (21) can easily be observed. In both cases, the structures of (21), in a way, encapsulate the structures of (19). The structures of the basic sentence and the derived sentential complement have many features in common. This is also reflected in the grammar rules. Basically the rules differ in the construction of the verb and some minor constraints, e.g., the case of the subject. To understand whether the parsed sentence is a complete sentence or not, the finite verb requirement is checked.

9.6.1.2   Sentential Adjuncts

Another verbal derivation that follows the same mechanism is the construction of sentential adjuncts. A sentential adjunct example which derives (19) into an adverb is given in (24).

(24)   Kız   adam-ı   ara-r-ken   polis   gel-di
      girl.Nom man-Acc call-Aor-While police.Nom come-Past.3sg
      "The police came while the girl called the man."

The c-structure construction of the adverbial clause in (25) is similar to the sentential complement c-structure in (22). Again, Vadv of the adverbial clause is constructed through sub-lexical rules.

(25)

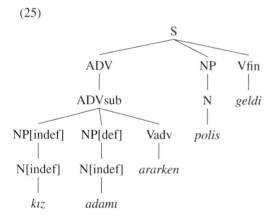

The f-structure for this sentence is shown in (26). Similar to the nominalized clause, which functions as an OBJ in (23), the derived ADJUNCT contains the verb's SUBJect and OBJect as well as the features of the adverb such as ADJUNCT-TYPE. The CHECK feature is important for controlling the SUBJect of the adverbial clause.

(26)

$$
\begin{bmatrix}
\text{PRED} & \text{`gel}\langle\text{polis}\rangle\text{'} \\[2pt]
\text{SUBJ} & \begin{bmatrix} \text{PRED} & \text{`polis'} \\ \text{CASE NOM, NUM SG, PERS 3} \end{bmatrix} \\[12pt]
\text{ADJUNCT} & \begin{bmatrix}
\text{PRED} & \text{`ara}\langle\text{kız, adam}\rangle\text{'} \\[2pt]
\text{SUBJ} & \begin{bmatrix} \text{PRED} & \text{`kız'} \\ \text{CASE NOM, NUM SG, PERS 3} \end{bmatrix} \\[10pt]
\text{OBJ} & \begin{bmatrix} \text{PRED} & \text{`adam'} \\ \text{CASE ACC, NUM SG, PERS 3} \end{bmatrix} \\[10pt]
\text{CHECK} & \begin{bmatrix} \text{SUB} & \text{WHILE} \end{bmatrix} \\[6pt]
\text{ADJUNCT-TYPE SUB}
\end{bmatrix} \\[12pt]
\text{TENSE} & \text{PAST}
\end{bmatrix}
$$

Derived deverbal adverbs can be divided into two groups according to subject control[8]: One group, namely *-yAlH* "having *verb*ed", *-yHncA* "when (s/he) *verb*s", *-ken* "while (s/he is) *verb*ing", *-mAdAn* "without having *verb*ed", *-DHkçA* "as long as (s/he) *verb*s", allows different subjects for the adverbial clause and the main sentence. In the other group, namely *-yHp* "after having *verb*ed" and *-yArAk* "by *verb*ing", the subject of the matrix verb is also the subject of the inner clause. The suffix *-cAsHnA* "as if (someone is) *verb*ing" belongs to both of the groups depending on the tense of the verb. If the verb is in aorist tense, then the subjects of the matrix verb and the inner clause should match, but if the verb is in narrative tense, then the subjects might differ.

9.6.1.3 Relative Clauses

Relative clauses in Turkish are gapped sentences which function as modifiers of nominal heads. Previous studies on Turkish relative clauses (Güngördü and Engdahl 1998; Barker et al. 1990) suggest that these pose interesting issues for linguistic and computational modeling. Here we address only their realization within our LFG grammar.

In LFG, relative clauses, as other types of long distance dependencies, are handled by functional uncertainty (Kaplan and Zaenen 1989). We also follow this approach. Once we derive the participle phrase, we unify the head modified by the participle with the appropriate argument of the verb, using rules based on functional uncertainty. (27) shows a relative clause where a participle form is used as a modifier of the noun *adam* "man".

---

[8] See Chap. 2 for Turkish verbal morphotactics on how these forms are derived.

(27)   manav-ın   kız-ın   [ ]ᵢ   ara-dığı-nı   söyle-diğ-i
       grocer-Gen girl-Gen obj.gap call-PastPart-Acc say-PastPart-P3sg
       adamᵢ
       man.Nom
       "the man the grocer said the girl called"

The c-structure of the sentence in (27) is given in (28). The sentential NP denoted as NPpart in the tree is treated like any regular NP by the rule that parses the participle AP. NPpart has an implicit gap but empty nodes are not allowed in LFG c-structures. The verb *ara* "call" of NPpart subcategorizes for a subject and an object, and the f-structure of NPpart, hence all the f-structures encapsulating it, would be incomplete with a missing object.

(28)

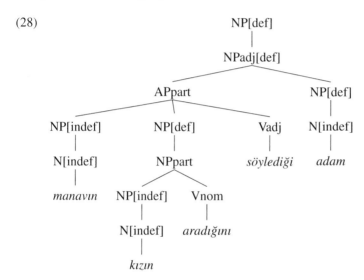

The resulting f-structure can be examined more easily in (29). At the innermost level, the NP *kızın aradığını* "that the girl called" is parsed with a gap object. It then functions as the OBJect of the outer adjectival phrase *manavın kızın aradığını söylediği* "that the grocer said the girl called". The participle then modifies the head NP *adam* "man", hence functions as the ADJUNCT of the topmost level f-structure. The gap in the derived form, the object here, is then unified with the head word *adam* as marked with co-indexation in (29). As a result, *adam* unifies with its ADJUNCT's OBJect's OBJect.

$$
\text{(29)} \quad
\begin{bmatrix}
\text{PRED} & \text{'adam'} \; \boxed{1} \\[2ex]
\text{ADJUNCT} &
\begin{bmatrix}
\text{PRED} & \text{'söyle}\langle\text{manav, ara}\rangle\text{'} \\[2ex]
\text{SUBJ} &
\begin{bmatrix}
\text{PRED} & \text{'manav'} \\
\text{CASE GEN, NUM sg, PERS } 3
\end{bmatrix} \\[3ex]
\text{OBJ} &
\begin{bmatrix}
\text{PRED} & \text{'ara}\langle\text{kız, adam}\rangle\text{'} \\[2ex]
\text{SUBJ} &
\begin{bmatrix}
\text{PRED} & \text{'kız'} \\
\text{CASE GEN, NUM SG, PERS } 3
\end{bmatrix} \\[2ex]
\text{OBJ} &
\begin{bmatrix}
\text{PRED} & \text{'adam'} \; \boxed{1}
\end{bmatrix} \\[2ex]
\text{CHECK} &
\begin{bmatrix}
\text{PART} & \text{PASTPART}
\end{bmatrix} \\[1ex]
\text{CASE ACC, NUM SG, PERS } 3, \text{ CLAUSE-TYPE NOM}
\end{bmatrix} \\[3ex]
\text{CHECK} &
\begin{bmatrix}
\text{PART} & \text{PASTPART}
\end{bmatrix} \\[1ex]
\text{ADJUNCT-TYPE RELATIVE}
\end{bmatrix}
\end{bmatrix}
$$

The example sentence (27) includes (21) as a relative clause with the object extracted, hence the similarity in the f-structures can be observed easily. The ADJUNCT in (29) is almost the same as the whole f-structure of (23), differing only in TENSE and ADJUNCT-TYPE features.

9.6.2 Sentences

The Turkish LFG grammar employs a comprehensive sentence rule that covers constituents such as noun phrases, adverbial phrases, postpositional phrases, temporal phrases, NPs functioning as adverbs, and a finite verb. The finite verb can be a simple or a derived verb, a noun-verb compound, or can have one of valency alternating suffixes. There is a meta sentence rule which checks if the verb is finite, controls whether subcategorization frames are filled and assigns PRO. No matter how complicated the verb formation is, all sentences are parsed with the same rule.

Copular sentences, on the other hand, have a special rule. When the copular suffix -DHr is attached to an NP, AP, or POSTPP (postpositional phrase), the morphological output is parallel to a regular verb, hence sentences containing such copular verbs are parsed with the standard sentence rule. However it is also possible to construct copular sentences by using NP, AP, or POSTPPs as the predicate, without any explicit derivation. (30a) and (30b) give two copular sentences with and without the copular suffix, respectively. The special copular sentence rule covers cases like (30b) to assure that f-structures are identical. Moreover, the representation of the past tense of copular verbs is parallel to that of regular verbs, but the future tense is a construction with the light verb *ol-* "be". (30c) and (30d) give two copular sentences in the past and future tense, respectively.

(30)
   a. Kedi   mutlu-dur.   c. Kedi   mutlu-ydu.
       cat.Nom happy-Cop.3sg   cat.Nom happy-Past.3sg

       "The cat is happy."   "The cat was happy."

   b. Kedi   mutlu.   d. Kedi   mutlu ol-acak.
       cat.Nom happy   cat.Nom happy be-Fut.3sg

       "The cat is happy."   "The cat will be happy."

In the implementation, we pay attention to the parallelism of the structures of different sentence types represented in (30). (31)–(33) illustrate the f-structures of (30b)–(30d). The differences in the f-structures are their TENSE values. Also note that the value of VTYPE in (33) is MAIN instead of COPULAR.

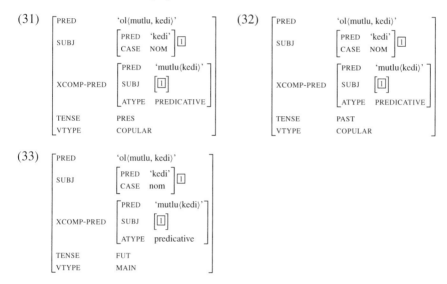

(31)
$$\begin{bmatrix} \text{PRED} & \text{'ol}\langle\text{mutlu, kedi}\rangle' \\ \text{SUBJ} & \begin{bmatrix} \text{PRED} & \text{'kedi'} \\ \text{CASE} & \text{NOM} \end{bmatrix}\boxed{1} \\ \text{XCOMP-PRED} & \begin{bmatrix} \text{PRED} & \text{'mutlu}\langle\text{kedi}\rangle' \\ \text{SUBJ} & \boxed{1} \\ \text{ATYPE} & \text{PREDICATIVE} \end{bmatrix} \\ \text{TENSE} & \text{PRES} \\ \text{VTYPE} & \text{COPULAR} \end{bmatrix}$$

(32)
$$\begin{bmatrix} \text{PRED} & \text{'ol}\langle\text{mutlu, kedi}\rangle' \\ \text{SUBJ} & \begin{bmatrix} \text{PRED} & \text{'kedi'} \\ \text{CASE} & \text{NOM} \end{bmatrix}\boxed{1} \\ \text{XCOMP-PRED} & \begin{bmatrix} \text{PRED} & \text{'mutlu}\langle\text{kedi}\rangle' \\ \text{SUBJ} & \boxed{1} \\ \text{ATYPE} & \text{PREDICATIVE} \end{bmatrix} \\ \text{TENSE} & \text{PAST} \\ \text{VTYPE} & \text{COPULAR} \end{bmatrix}$$

(33)
$$\begin{bmatrix} \text{PRED} & \text{'ol}\langle\text{mutlu, kedi}\rangle' \\ \text{SUBJ} & \begin{bmatrix} \text{PRED} & \text{'kedi'} \\ \text{CASE} & \text{nom} \end{bmatrix}\boxed{1} \\ \text{XCOMP-PRED} & \begin{bmatrix} \text{PRED} & \text{'mutlu}\langle\text{kedi}\rangle' \\ \text{SUBJ} & \boxed{1} \\ \text{ATYPE} & \text{predicative} \end{bmatrix} \\ \text{TENSE} & \text{FUT} \\ \text{VTYPE} & \text{MAIN} \end{bmatrix}$$

9.6.3   Handling Constituent Order Variations

Although Turkish is known to be a free constituent order language, there are still some restrictions on the word order, especially in the constituent order of subordinate clauses. The nominative object is restricted to immediate preverbal position, but accusative objects can move freely.[9] Still, the usage of some adverbs restrict the position of direct objects. (34) exemplifies the different placement of the adverb *hızlı* "fast" in sentences with direct or indirect objects. (34d) is not

---

[9]There are also some exceptions to this rule. In the sentence *yapayım sana yemek* "Let me cook for you", the nominative object *yemek* comes after the verb *yapayım* (Sarah Kennely (p.c.)). Aslı Göksel (p.c.) gives another example: *ekmek ben hiç yemem* "I never eat bread." Here, the nominative object precedes the nominative subject.

grammatical if we want the adverb to modify the verb. This restriction comes from the semantics of the adverb, as *hızlı* can be interpreted both as an adjective and as an adverb, and in (34d) it modifies *kitabı* "book" instead of the verb *read*. If the adverb has no adjective interpretation, it can be placed in a prenominal position and it still modifies the verb as given in (34e).

(34) a. Ben   kitab-ı   hız-lı   oku-r-um.
            I.Nom book-Acc speed-With read-Aor-1sg
            "I read the book fast."

      b. *Ben   kitap   hız-lı   oku-r-um.
            I.Nom book.Nom speed-With read-Aor-1sg
            "I read books fast."

      c. Ben   hız-lı   kitap   oku-r-um.
            I.Nom speed-With book.Nom read-Aor-1sg
            "I read books fast."

      d. *Ben   hız-lı   kitab-ı   oku-r-um.
            I.Nom speed-With book-Acc read-Aor-1sg
            "I read the book fast. (intended)"

      e. Ben   sabahleyin   kitab-ı   oku-r-um.
            I.Nom in.the.morning book-Acc read-Aor-1sg
            "I read the book in the morning."

Our implementation allows the constituents of sentential complements move freely within the participle. But there is also a possibility that the constituents of the sentential complement interfere with the constituents of the main sentence, as in (35a). As can be observed from the subtree NPpart in (22), the whole participle phrase is parsed at once and then used in the main sentence level. Hence, it is not possible to parse non-contiguous chunks of the participle in our approach. Note that the other non-contiguous cases, such as (35b) and (35c) are not grammatical.

(35) a. manav   **adam-ı ara-dığ-ı-nı**   söyle-di   **kız-ın**
            grocer.Nom man-Acc call-PastPart-P3sg-Acc say-Past.3sg girl-Gen
            "The grocer said that the girl called the man."

      b. ***kız-ın** manav   **adam-ı ara-dığ-ı-nı**   söyle-di
            girl-Gen grocer.Nom man-Acc call-PastPart-P3sg-Acc say-Past.3sg
            "The grocer said that the girl called the man."

      c. ***adam-ı ara-dığ-ı-nı** manav   **kız-ın**   söyle-di
            grocer.Nom man-Acc   call-PastPart-P3sg-Acc say-Past.3sg girl-Gen
            "The grocer said that the girl called the man."

In general, wh-question sentences are constructed by simply omitting the target of the question and inserting the question word into its place, as exemplified in (36a) and (36c). But there is an exception for this generalization; although (36e) is grammatical, (36f) is not.

(36) a. kitab-ı   ben   oku-du-m
          book-Acc I.Nom read-Past-1sg
          "I read the book."

      b. kitab-ı   kim   oku-du
          book-Acc who.Nom read-Past.3sg
          "Who read the book?"

      c. ben   kitap   oku-du-m
          I.Nom book.Nom read-Past-1sg
          "I read books."

      d. kim   kitap   oku-du
          who.Nom book.Nom read-Past.3sg
          "Who read books?"

      e. kitap/kitabı   oku-du-m   ben
          book.Nom/Acc read-Past-1sg I.Nom
          "I read books/the book."

      f. *kitap/kitabı   oku-du   kim
          book.Nom/Acc read-Past.3sg who.Nom
          "Who read books/the book?"

Question sentences like (36a) and (36c) are parsed with the standard sentence rule. The major difference is the value of the feature CLAUSE-TYPE. It is DECL for declarative sentences but INT for questions.

The grammar also contains rules to parse yes/no questions. Yes/no questions in Turkish are built by attaching a question clitic to the main predicate to have the whole sentence as its scope, or to a constituent to set the focus on that constituent in the question sentence. (37a) and (37b) give two such examples.

(37) a. Kitab-ı   oku-du-n   mu?
          book-Acc read-Past-2sg Ques
          "Did you read the book?"

      b. Kitab-ı   ben   mi   oku-du-m?
          book-Acc I.Nom Ques read-Past-1sg
          "Was it me who read the book?"

9.7 Coordination

Coordination is an important issue to be solved especially in a computational approach, as the number of possible interpretations of the coordination increases with the number of constituents involved in the coordination, leading to many ambiguous cases. The efforts of ParGram members brought up a common set of rules which facilitate the implementation of coordinated structures in XLE. In simple coordination, coordination is a set consisting the f-structure of each conjunct (Kaplan and Maxwell 1988).

(38) gives the f-structure of the phrase *adam ve kadın* "the man and the woman". Some of the attributes are nondistributive across the members of the set, instead they have their own attribute value pairs in the set itself. For instance, PERS is a nondistributive attribute, so that two singular nouns can form a coordinate structure which is plural. The outermost f-structure does not have a PRED, but the coordinator is represented in COORD-FORM. <S inside the f-structure of *kadın* indicates that *adam* precedes *kadın* in the coordination structure.

$$(38) \quad \left[\begin{array}{l} \left\{ \begin{array}{l} \left[\begin{array}{l} \text{PRED} \quad \text{'adam'} \\ \text{CASE NOM, NUM SG, PERS 3} \end{array} \right] \\ \left[\begin{array}{l} \text{PRED} \quad \text{'kadın'} \\ \langle \text{S} \quad \left[\quad \text{'adam'} \right] \\ \text{CASE NOM, NUM SG, PERS 3} \end{array} \right] \end{array} \right\} \\ \text{CASE NOM, COORD +, COORD-FORM VE, NUM PL, PERS 3} \end{array} \right]$$

In addition to standard coordination, Turkish has other interesting coordination structures using what is called *suspended affixation* (Kabak 2007), in which the inflectional features of the last element in a coordination have phrasal scope, that is, all other coordinated constituents have certain default features which are then *overridden* by the features of the last element in the coordination. A very simple case of such suspended affixation is exemplified in (39a) and (39b). Note that, it is not a derivational but an inflectional morpheme that has phrasal scope in this case.

(39) a. kız   adam   ve kadın-ı   ara-dı
          girl.Nom man.Nom and woman-Acc call-Past.3sg
          "The girl called the man and the woman."

      b. kız   [adam   ve kadın]-ı   ara-dı
          girl.Nom [man.Nom and woman]-Acc call-Past.3sg
          "The girl called the man and the woman."

The f-structure of *adam ve kadını* in (39b) is given in (40). For Turkish, CASE is also one of the nondistributive attributes. The standard coordination rule is modified so that the case of the coordination is the case of the last conjunct if the previous conjuncts are in nominative case. In (40), the CASE of the coordination is ACC although *adam* has CASE NOM.

(40)

$$
\left[
\begin{array}{l}
\left\{
\begin{array}{l}
\left[
\begin{array}{l}
\text{PRED} \quad \text{'adam'} \\
\text{CASE NOM, NUM SG, PERS 3}
\end{array}
\right] \\
\left[
\begin{array}{l}
\text{PRED} \quad \text{'kadın'} \\
\langle \text{S} \quad \left[\quad \text{'adam'} \quad \right] \\
\text{CASE ACC, NUM SG, PERS 3}
\end{array}
\right]
\end{array}
\right\} \\
\text{CASE ACC, COORD +, COORD-FORM VE, NUM PL, PERS 3}
\end{array}
\right]
$$

Although it is possible to parse basic coordinated phrases with or without suspended affixation in the current implementation, the grammar lacks a wide coverage of coordinated structures especially for verbal coordination where one or more arguments are shared by the coordinated verbs.

9.8 Valency Alternations

In this section, we analyze how causatives and passives are constructed and how they interact by looking at double causatives, impersonal passives, and passivization of causatives. We also discuss how these structures should be represented in the LFG theory and give our implementation with example f-structures.

9.8.1 Causatives

Causatives in Turkish are formed as verbal derivations. It is possible for a verb to have multiple causative markers.[10] In (41), we see two sentences with the intransitive verb *uyudu* "he/she/it slept" and its causative form *uyuttu* "he/she/it made (someone else) sleep". The suffix *-t* is attached to the root *uyu* and tense and person markers follow as in standard verb conjugation.

(41)   a.   kedi   uyu-du
            cat.Nom sleep-Past.3sg
            "The cat slept."

       b.   çocuk   kedi-yi uyu-t-tu
            child.Nom cat-Acc sleep-Caus-Past.3sg
            "The child made the cat sleep."

---

[10]Double causatives are quite common, triple causatives are also observed.

In the underlying representation, the causative morpheme introduces a new IG. The morphological analysis for *uyudu* is

```
uyu+Verb+Pos+Past+A3sg
```

and for its causative *uyuttu*, the analysis becomes:

```
uyu+Verb^DB+Verb+Caus+Pos+Past+A3sg
```

We applied several language-dependent tests in order to decide the representation of causatives in Turkish (Çetinoğlu and Butt 2008), and concluded that they should be represented with monoclausal structures. When defining the arguments and the structure of the causative predicate in the implementation, we follow the approach used for Urdu complex predicates (Butt and King 2006). The end result of processing an IG which has a verb with a causative form is to create a flat f-structure whose PRED feature has a value composed of the information coming from both the arguments of causativized verb and the IG containing the causative morpheme.

F-structures (42) and (43) show the initial representation of the base sentence and the resulting structure after causativization. The former subject *kedi* "cat" in nominative case is the object in accusative case when causativized. The subject of the new sentence is *çocuk* "child".

$$
(42) \quad
\begin{bmatrix}
\text{PRED} & \text{'uyu}\langle\text{kedi}\rangle\text{'} \\
\text{SUBJ} & \begin{bmatrix} \text{PRED} & \text{'kedi'} \\ \text{CASE} & \text{NOM} \end{bmatrix} \\
\text{TENSE} & \text{PAST}
\end{bmatrix}
\qquad
(43) \quad
\begin{bmatrix}
\text{PRED} & \text{'caus}\langle\text{çocuk, uyu}\langle\text{kedi}\rangle\rangle\text{'} \\
\text{SUBJ} & \begin{bmatrix} \text{PRED} & \text{'çocuk'} \\ \text{CASE} & \text{NOM} \end{bmatrix} \\
\text{OBJ} & \begin{bmatrix} \text{PRED} & \text{'kedi'} \\ \text{CASE} & \text{ACC} \end{bmatrix} \\
\text{TENSE} & \text{PAST}
\end{bmatrix}
$$

The c-structures of transitive verbs have no representational difference from intransitive ones. Their f-structures also follow the same representation approach. (44) gives a transitive verb and its causativized version.

(44) a. Köpek   kedi-yi kovala-dı.
          dog.Nom cat-Acc chase-Past.3sg
          "The dog chased the cat."

      b. Çocuk   köpeğ-e kedi-yi kovala-t-tı.
          child.Nom dog-Dat cat-Acc chase-Caus-Past.3sg
          "The child made the dog chase the cat."

(45) and (46) give the f-structures of (44a) and (44b) respectively. PRED value of the base verb is the second argument of the causativized verb's PRED. The first argument is the SUBJect and comes from the causativized verb. The OBJect of the base verb is still the OBJect when the verb is causativized and the SUBJect of the base verb becomes the thematically restricted object OBJ-TH. The c-structure of causatives is a flat tree due to free word order.

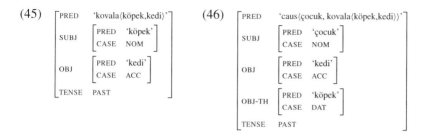

Double causativization of verbs is also frequent in Turkish, especially if a verb is intransitive. (47) demonstrates the double causativization of the intransitive in (41a). Once an intransitive verb is causativized, the resulting predicate `caus⟨SUBJ, pred⟨OBJ⟩⟩`' bears the grammatical functions of a canonical transitive. Therefore it will be parsed without any need for modifications in the grammar rules.

(47) Anne   çocuğ-a kedi-yi uyu-t-tur-du.
       mother.Nom child-Dat cat-Acc sleep-Caus-Caus-Past.3sg
       "The mother made the child make the cat sleep."

Double causativization of transitives, however, is controversial. A single causativization example along with two double causativization examples are given in (48). As exemplified in (48b), it is not considered to be grammatical to overtly state both of the intermediaries between the agent and the theme of the event. Unlike (48b), the sentence in (48c) is grammatical when one of the intermediaries is covert. But then, the ranking is ambiguous although it is certain that somebody else is involved in the causation hierarchy. We give both possible interpretations in (48c).

(48) a. Çocuk   köpeğ-e kedi-yi kovala-t-tı.
          child.Nom dog-Dat cat-Acc chase-Caus-Past.3sg
          "The child made the dog chase the cat."

      b. *Çocuğ-a köpeğ-e kedi-yi kovala-t-tır-dı.
          child-Dat dog-Dat cat-Acc chase-Caus-Caus-Past.3sg
          "S/he made the child make the dog chase the cat."

      c. Çocuk   köpeğ-e kedi-yi kovala-t-tır-dı.
          child.Nom dog-Dat cat-Acc chase-Caus-Caus-Past.3sg
          "The child made someone make the dog chase the cat."
          "The child made the dog make someone chase the cat."

The simplified f-structure of (48c) is depicted in (49). *kedi* "cat" is chased by *köpek* "dog", and *çocuk* "child" is the agent that starts the causation. The intermediary person between the child and dog is not explicit in the sentence, hence is represented as NULL in the f-structure.

(49)
$$
\begin{bmatrix}
\text{PRED} & \text{`caus⟨çocuk, caus⟨NULL, kovala⟨köpek,kedi⟩⟩⟩'} \\
\text{SUBJ} & \begin{bmatrix} \text{PRED} & \text{`çocuk'} \\ \text{CASE} & \text{NOM} \end{bmatrix} \\
\text{OBJ} & \begin{bmatrix} \text{PRED} & \text{`kedi'} \\ \text{CASE} & \text{ACC} \end{bmatrix} \\
\text{OBJ-TH} & \begin{bmatrix} \text{PRED} & \text{`köpek'} \\ \text{CASE} & \text{DAT} \end{bmatrix} \\
\text{TENSE} & \text{PAST}
\end{bmatrix}
$$

9.8.2 Passives

The passive construction is also a morphological process in Turkish. (50) gives a basic example on passivization of a transitive verb. The direct object in the accusative case becomes the subject in the nominative case after passivization. The verb agrees with the subject.

(50)   a.   Köpek   ben-i kovala-dı.
            dog.Nom I-Acc chase-Past.3sg
            "The dog chased me."

       b.   Ben   (köpek   tarafından) kovala-n-dı-m.
            I.Nom (dog.Nom by)   chase-Pass-Past-1sg
            "I was chased (by the dog)."

In Turkish, it is possible to passivize intransitives with constituents other than direct object, as in (51). In these cases, passivization is impersonal, that is, the constituent preserves its function (and also its case marking) and there is no subject in the passivized sentence. Kornfilt (1997) shows such passives are impersonal by stating the two properties that do not obey the subjecthood rules: the constituent is not in nominative case and it does not agree with the verb in person and number (51b). Still, we can derive a participle from the passivized sentence and extract the constituent in the same way as subject, as in (51c).

(51)   a.   Ali   okul-a/biz-e   git-ti.
            Ali.Nom school-Dat/we-Dat go-Past.3sg
            "Ali went to the school/to our place."

       b.   Okul-a/Biz-e   gid-il-di.
            school-Dat/we-Dat go-Pass-Past.3sg
            "The school/Our place was gone to."

       c.   gid-il-en   okul
            go-Pass-Prespart school.Nom
            'the school that was gone to'

To prevent a confusion that might arise, (51a) is only given to show the passivization of an intransitive verb in terms of syntactic and morphological modifications; it does not necessarily mean that (51b) is the passive form of (51a). In all cases of impersonal passivization, the agent is uncertain, yet can be identified as a group of people, not a single person.

When the 'group' meaning is intended in the sentence, transitive verbs can also be impersonally passivized by using double passivization. (52) gives two double passivized sentences, both having the meaning that the actions are taken together with a group. It may also contain the generic meaning, as exemplified in (53).[11]

(52)　a.　Film　　　izle-n-il-di.
　　　　　movie.Nom watch-Pass-Pass-Past.3sg
　　　　　"The movie was watched."

　　　b.　Tatlı-lar　　　ye-n-il-di.
　　　　　Dessert-Pl.Nom eat-Pass-Pass-Past.3sg
　　　　　"Desserts were eaten."

(53)　Harp-te　vur-ul-un-ur.
　　　war-Loc shoot-Pass-Pass-Aor.3sg

　　　"One is shot (by someone) in war." (Özkaragöz 1986)

The simplified f-structures of (50a) and its passivized form (50b) are given in (54) and (55), respectively. In the passive f-structure, the predicate representation changes, the OBJect becomes SUBJect and the feature PASSIVE is added in the top level.

(54)
$$
\begin{bmatrix}
\text{PRED} & \text{'kovala}\langle\text{köpek, ben}\rangle\text{'} \\
\text{SUBJ} & \begin{bmatrix} \text{PRED} & \text{'köpek'} \\ \text{CASE} & \text{NOM} \end{bmatrix} \\
\text{OBJ} & \begin{bmatrix} \text{PRED} & \text{'ben'} \\ \text{CASE} & \text{ACC} \end{bmatrix} \\
\text{TENSE} & \text{PAST}
\end{bmatrix}
$$

(55)
$$
\begin{bmatrix}
\text{PRED} & \text{'kovala}\langle\text{NULL, ben}\rangle\text{'} \\
\text{SUBJ} & \begin{bmatrix} \text{PRED} & \text{'ben'} \\ \text{CASE} & \text{NOM} \end{bmatrix} \\
\text{TENSE} & \text{PAST} \\
\text{PASSIVE} & +
\end{bmatrix}
$$

The rule for the implementation of impersonal passivization is similar to the canonical rule. (56) gives the f-structure for the impersonally passivized sentence in (51b).

(56)
$$
\begin{bmatrix}
\text{PRED} & \text{'git}\langle\text{NULL}\rangle\text{'} \\
\text{ADJUNCT} & \begin{bmatrix} \text{PRED} & \text{'okul'} \\ \text{CASE} & \text{DAT} \end{bmatrix} \\
\text{TENSE} & \text{PAST} \\
\text{PASSIVE} & +
\end{bmatrix}
$$

---

[11] The single passivization of the sentences in (52) can be assumed to have the same interpretation with the double passivization, but (53) does not have such a parallelism.

Passivization of causatives is straightforward from a theoretical point of view but poses interesting issues in terms of implementation. Causativization increases the valency of the verb by one. If the verb is intransitive, the result is a transitive verb. Therefore, one would expect the passivization of causatives to be like the passivization of transitive verbs. We repeat the causative sentence (41b) as (57a) here and give its passive form in (57b).

(57)   a.   Çocuk   kedi-yi uyu-t-tu.
            child.Nom cat-Acc sleep-Caus-Past.3sg
            "The child made the cat sleep."

       b.   Kedi   uyu-t-ul-du.
            cat.Nom sleep-Caus-Pass-Past.3sg
            "The cat was made to sleep."

Similarly we repeat the f-structure corresponding to (57a) in (58) and give the passivized causative sentence in (59). The f-structures are parallel to those of transitive examples in (54) and (55) as expected.

$$
(58) \quad
\begin{bmatrix}
\text{PRED} & \text{'caus}\langle\text{çocuk, uyu}\langle\text{kedi}\rangle\rangle\text{'} \\[4pt]
\text{SUBJ} & \begin{bmatrix} \text{PRED} & \text{'çocuk'} \\ \text{CASE} & \text{NOM} \end{bmatrix} \\[10pt]
\text{OBJ} & \begin{bmatrix} \text{PRED} & \text{'kedi'} \\ \text{CASE} & \text{ACC} \end{bmatrix} \\[10pt]
\text{TENSE} & \text{PAST}
\end{bmatrix}
\qquad
(59) \quad
\begin{bmatrix}
\text{PRED} & \text{'caus}\langle\text{NULL, uyu}\langle\text{kedi}\rangle\rangle\text{'} \\[4pt]
\text{SUBJ} & \begin{bmatrix} \text{PRED} & \text{'kedi'} \\ \text{CASE} & \text{NOM} \end{bmatrix} \\[10pt]
\text{TENSE} & \text{PAST} \\[4pt]
\text{PASSIVE} & +
\end{bmatrix}
$$

Hovewer, the implementation does not go in parallel with the linguistic theory. This is because XLE handles passivization as a lexical rewrite rule placed in the suffix lexicon and causativization is handled in the actual grammar rules. We refer the interested reader to Sect. 4.3.3 of Çetinoğlu (2009) for implementation details.

9.9   Non-canonical Objects

Turkish has a well-known case alternation on objects that correlates with the semantics of specificity (Enç 1991). A nonspecific direct object generally bears nominative case and a specific direct object is marked with the accusative. (60a,b) exemplify this well-known contrast. In addition to this alternation, an ablative object indicates partitivity when the object is consumable (Dede 1981; Kornfilt 1990), as in (60c).

(60)   a.   **Su**   içtim.
            water.Nom drink.Past.1sg
            "I drank water."

    b. **Su-yu**   içtim.
        water-Acc drink.Past.1sg
        "I drank the water."

    c. **Su-dan**   içtim.
        water-Abl drink.Past.1sg
        "I drank some of/from the water."

In addition to signaling partitivity, case in Turkish also appears to make distinctions between the degree of *affectedness* of an object, which sometimes result in different object cases for a verb depending on its sense. The examples in (61) illustrate this type of case alternation (Dede 1981).

(61)   a. Ali   **çocuğ-u** vur-du.
           Ali.Nom child-Acc shot-Past.3sg
           "Ali shot the child."

       b. Ali   **çocuğ-a** vur-du.
           Ali.Nom child-Dat hit-Past.3sg
           "Ali hit the child."

Another type of non-canonical case marking on objects found with a large subset of psych verbs. Although all the verbs given in (62) are similar in meaning, only (62a) bears the canonical accusative case. (62b) and a group of verbs such as *nefret et* "hate", *kork* "fear", *şüphelen* "suspect", *iğren* "be disgusted with" have ablative objects and (62c), and another subset of pysch verbs such as *yalvar* "beg", *kız* "be angry", *inan* "believe" have dative objects.

(62)   a. Ali   **Ayşe'-yi** sev-iyor.
          Ali.Nom Ayşe-Acc love.Prog.3sg
          "Ali loves Ayşe."

      b. Ali   **Ayşe'-den** hoşlan-ıyor.
          Ali.Nom Ayşe-Abl like.Prog.3sg
          "Ali likes Ayşe."

      c. Ali   **Ayşe'-ye** tap-ıyor.
          Ali.Nom Ayşe-Dat adore.Prog.3sg
          "Ali adores Ayşe."

There is also another set of verbs which simply take non-canonical objects. These verbs do not have a common semantic property and can have either ablative or dative objects. *bin* "ride" in (63) and *yardım et* "help" are from this class.

(63)   Hasan   at-a   bin-di.
      Hasan.Nom horse-Dat ride.Past.3sg
      "Hasan rode the horse."

We examined how these non-canonical objects should be analyzed within the LFG framework and observed the behavior of different subsets under passivization, causativization and raising as our tests.

(64) demonstrates the behavior of the verb *bin* "ride" in (63) when it undergoes causativization and passivization in (64a) and (64b) respectively. In contrast with a canonical case, this time, the subject of the base verb has the accusative case when the sentence is causativized. And again, unlike canonical cases, the case-marking of the object is preserved when the sentence is passivized.

(64) a. Babası   Hasan'-ı   at-a   bin-dir-di.
          father.P3sg Hasan.Acc horse.Dat ride-Caus-Past.3sg
          "His father made Hasan ride the horse."

      b. At-a   bin-il-di.
          horse-Dat ride-Pass-Past.3sg
          "The horse was ridden."

Based on our examination of the data, we concluded that ablative partitives and affectedness alternation involve OBJects, however, psych verbs and other non-canonical case marking verbs subcategorize for thematic objects OBJ-TH.

The corresponding f-structures of (63) and (64) are given in (65)–(67). In the f-structures, SUBJ of (65) becomes OBJ of (66) and OBJ-TH remains the same. In (67), however, it is not the accusative OBJECT but the dative OBJ-TH takes the SUBJect role. Moreover, unlike canonical passive subjects, it preserves its case.

$$
(65)\ \begin{bmatrix} \text{PRED} & \text{'bin}\langle\text{Hasan, at}\rangle\text{'} \\ \text{SUBJ} & \begin{bmatrix} \text{PRED} & \text{'Hasan'} \\ \text{CASE} & \text{NOM} \end{bmatrix} \\ \text{OBJ-TH} & \begin{bmatrix} \text{PRED} & \text{'at'} \\ \text{CASE} & \text{DAT} \end{bmatrix} \\ \text{TENSE} & \text{PAST} \end{bmatrix}
\qquad
(66)\ \begin{bmatrix} \text{PRED} & \text{'caus}\langle\text{baba, bin}\langle\text{Hasan, at}\rangle\rangle\text{'} \\ \text{SUBJ} & \begin{bmatrix} \text{PRED} & \text{'baba'} \\ \text{CASE} & \text{NOM} \end{bmatrix} \\ \text{OBJ} & \begin{bmatrix} \text{PRED} & \text{'Hasan'} \\ \text{CASE} & \text{ACC} \end{bmatrix} \\ \text{OBJ-TH} & \begin{bmatrix} \text{PRED} & \text{'at'} \\ \text{CASE} & \text{DAT} \end{bmatrix} \\ \text{TENSE} & \text{PAST} \end{bmatrix}
$$

$$
(67)\ \begin{bmatrix} \text{PRED} & \text{'bin}\langle\text{NULL, at}\rangle\text{'} \\ \text{SUBJ} & \begin{bmatrix} \text{PRED} & \text{'at'} \\ \text{CASE} & \text{DAT} \end{bmatrix} \\ \text{TENSE} & \text{PAST} \\ \text{PASSIVE} & + \end{bmatrix}
$$

More in-depth investigation of non-canonical objects as well as implementation details are discussed in Çetinoğlu and Butt (2008).

9.10 Evaluation

Testing is one of the crucial steps of developing an accurate large-scale grammar. We tested our grammar both with manual test sets tailored to monitor the development of the grammar and its parallelism to other ParGram grammars, and with test suites that measure its coverage against real world text.

9.10.1 Manual Test Sets

The first group of manual test sets was constructed incrementally as new sets of rules were added to the grammar. A total of 384 sentences/phrases cover noun phrases, basic and complex sentence structures, participles, copular sentences, and date-time phrases. The second group of manual test sets comprises the ParGram sentences. In a course of 3 years, 118 sentences were shared among participants to discuss a wide range of linguistic phenomena and parallelism among their analyses. 110 of these sentences have a counterpart in Turkish in terms of parallel linguistic structure, and 99 of those were parsed successfully.

9.10.2 Sentence Test Suite

The first test set on real world data includes complex sentences. We randomly picked file 00007121.txt from METU Corpus (Say et al. 2004) which contains an excerpt from the fiction book *Öykümü Kim Anlatacak* "Who will tell my story" (İşigüzel 1994). We took the first four paragraphs of the text and prepared an XLE test file by removing punctuation marks and placing one sentence per line. Table 9.1 shows the basic statistics concerning the test file.

The shortest sentence contains a single word and the longest sentence contains 27 words. The average sentence length is 7 words. In terms of IGs, the shortest sentence has only one IG and the longest sentence has 35 IGs. The average number of IGs per sentence is 8.83. The number of morphemes in Table 9.1 and the number of IGs per sentence indicate that the sentences are more complex than the word counts indicate.

Table 9.1 Basic statistics about the sentence test suite

Type	Count
Sentences	43
Words	301
Unique words	245
Morphological analyses	636
Unique morphological analyses	482

Table 9.2 Types of phrases
used in the noun phrase test

Type	Count	Parsed
Simple noun phrases	194	182
Relative clauses	48	37
Sentential complements	36	30
Coordination	19	5
Total	297	254

Of the 43 sentences, 33 were parsed successfully. (68) is one of the parsed sentences. The complete set of sentences is given in Appendix C of Çetinoğlu (2009).

(68)   Yol-um-un   üzeri-nde-ki dev   alışveriş merkez-i-ne
       way-P1sg-Gen on-Loc-Rel huge shopping center-P3sg-Dat
       gir-ip   vitrin-ler-e   bak-ıyor-um.
       enter-AfterDoingSo shopwindow-Pl-Dat look.at-Prog-1sg
       "I look at the shop windows after entering the huge shopping center on my way."

9.10.3   Noun Phrase Test Suite

The second test suite on real world data is devoted to noun phrases. We randomly picked file `00033224.txt` and file `00129176.txt` from the literature section of METU Corpus (Say et al. 2004). Then the noun phrases in these files were manually extracted and divided into four groups. Table 9.2 gives the number of phrases in each subset of the test NPs. The complete list of phrases is given in Appendix D of Çetinoğlu (2009).

The set of simple noun phrases is composed of simple nouns, derived nouns, indefinite and definite noun compounds, adjective-modified NPs, pronouns and alike. Since these simple noun phrases are the base constituents of more complex noun phrases, the success rate is high in this set (93.8%). The groups of relative clauses and sentential complements are important in that the rules parsing these phrases are parallel to the rules parsing sentences. Hence this subtest also gives us some idea about the coverage of the sentences. Coordination has the lowest success rate among all kinds of noun phrases as the coordination rules do not cover different types of coordinated noun phrases. The overall accuracy is 85.5%.

9.11   Conclusions

This chapter has described the highlights of our work on developing a LFG grammar for Turkish employing sub-lexical constituents, that we have called inflectional groups. Such a sub-lexical constituent choice has enabled us to handle the very productive derivational morphology in Turkish in a rather principled way and has made the grammar more or less oblivious to morphological complexity.

Our wide coverage grammar contains rules parsing an extensive set of NPs and other basic phrases, sentential complements, adjuncts, relative clauses, basic coordinated phrases, and sentences composed of these constituents. We thoroughly examined some of the linguistic phenomena, such as causativization, passivization, and non-canonical objects, and proposed solutions on how they can be represented structurally and how we can implement them within the LFG architecture. We tested our grammar coverage on sentences and noun phrases with real world data.

The results of the tests conducted also address a major drawback: highly ambiguous output. In LFG, one widely used solution for the problem is applying Optimality Theory (OT) (Prince and Smolensky 2004) by using OT-marks (Frank et al. 2001). With the help of the OT-marks, it is possible to mark the rules that cause a phrase to have different parses and to rank those rules in a user defined order. The use of OT-marks is limited to very few cases and can be enriched with the help of linguistic heuristics and statistical information. OT-Marks are also a key to robustness by allowing parses with common mistakes in written data or daily speech although they are not strictly grammatical (Frank et al. 2001). In addition, XLE facilitates integrating statistical methods into the system to output the most probable one among correct parses (Kaplan et al. 2004). Additional details on preparing the statistical input to train the system can be found in previous work (Riezler et al. 2002; Riezler and Vasserman 2004).

The Turkish LFG Grammar is no longer an active member of the ParGram project since 2009, yet participates in the collaborations that brings several grammars together. A recent example is the ongoing development of ParGramBank (Sulger et al. 2013), a parallel treebank that contains c- and f-structures of a diverse set of linguistic phenomena, aligned in ten languages.

References

Barker C, Hankamer J, Moore J (1990) Wa and Ga in Turkish. In: Dziwirek K, Farrell P, Meijas-Bikandi E (eds) Grammatical relations. CSLI Publications, Stanford, CA

Bozşahin C (2002) The combinatory morphemic lexicon. Comput Linguist 28(2):145–186

Bresnan J, Mchombo SA (1995) The lexical integrity principle: evidence from Bantu. Nat Lang Linguist Theory 13(2):181–254

Bresnan J, Mugane J (2006) Agentive nominalizations in Gikuyu and the theory of mixed categories. In: Butt M, Dalrympe M, King TH (eds) Intelligent linguistic architectures: variations on themes by Ronald M. Kaplan. CSLI Publications, Stanford, CA

Butt M, Niño M, Segond F (1999) A grammar writer's cookbook. CSLI Publications, Stanford, CA

Butt M, Dyvik H, King TH, Masuichi H, Rohrer C (2002) The parallel grammar project. In: Proceedings of the workshop on grammar engineering and evaluation, Taipei, pp 1–7

Butt M, King TH (2006) Restriction for morphological valency alternations: the Urdu causative. In: Butt M, Dalrympe M, King TH (eds) Intelligent linguistic architectures: variations on themes by Ronald M. Kaplan. CSLI Publications, Stanford, pp 235–258

Çetinoğlu Ö (2009) A large scale LFG grammar for Turkish. PhD thesis, Sabancı University, Istanbul

Çetinoğlu Ö, Butt M (2008) Turkish non-canonical objects. In: Proceedings of the LFG'08 conference, Sydney, pp 214–234

Dalrymple M (2001) Lexical-functional grammar. Syntax and semantics. Academic, New York, NY

Dede M (1981) Grammatical relations and surface cases in Turkish. In: Proceedings of Berkeley Linguistic Society, Berkeley, CA, vol 7

Enç M (1991) The semantics of specificity. Linguist Inq 22(1):1–25

Frank A, King TH, Kuhn J, Maxwell JT (2001) Optimality Theory style constraint ranking in large-scale LFG grammars. In: Sells P (ed) Formal and empirical issues in optimality theoretic syntax. CSLI Publications, Stanford, CA

Göksel A, Kerslake C (2005) Turkish: a comprehensive grammar. Routledge, London

Gümüş T (2007) LFG for Turkish point-in-time expressions. Master's thesis, Istanbul Technical University, Istanbul

Güngördü Z, Engdahl E (1998) A relational approach to relativization in Turkish. In: Proceedings of joint conference on formal grammar, HPSG and categorial grammar, Saarbrücken

Güngördü Z, Oflazer K (1995) Parsing Turkish using the lexical-functional grammar formalism. Mach. Transl 10(4):515–544

İşigüzel Ş (1994) Öykümü Kim Anlatacak. Can Yayınları, Istanbul

Kabak B (2007) Turkish suspended affixation. Linguistics 45:311–347

Kaplan RM, Bresnan J (1982) Lexical-functional grammar: a formal system for grammatical representation. In: Bresnan J (ed) The mental representation of grammatical relations. The MIT Press, Cambridge, MA, pp 173–281

Kaplan RM, Maxwell JT (1988) Constituent coordination in lexical-functional grammar. In: Proceedings of COLING, Budapest, pp 303–305

Kaplan RM, Maxwell JT (1996) LFG grammar writer's workbench. Tech. rep., Xerox PARC, Palo Alto, CA

Kaplan RM, Zaenen A (1989) Long-distance dependencies, constituent structure, and functional uncertainty. In: Baltin M, Kroch A (eds) Alternative conceptions of phrase structure. Chicago University Press, Chicago, IL, pp 17–42

Kaplan R, Riezler S, King T, Maxwell J, Vasserman A, Crouch R (2004) Speed and accuracy in shallow and deep stochastic parsing. In: Proceedings of NAACL-HLT, Boston, MA, pp 97–104

Kornfilt J (1990) Remarks on headless partitives and case in Turkish. In: Mascaro J, Nespor M (eds) Grammar in progress — GLOW essays for Henk van Riemsdijk, vol 30. Foris Publications, Providence, RI, pp 285–303

Kornfilt J (1997) Turkish. Routledge, London

Oflazer K (1994) Two-level description of Turkish morphology. Lit Linguist Comput 9(2):137–148

Özkaragöz İ (1986) Monoclausal double passives in Turkish. In: Slobin DI, Zimmer K (eds) Studies in Turkish linguistics. John Benjamins, Amsterdam

Prince A, Smolensky P (2004) Optimality theory: constraint interaction in generative grammar. Blackwell, Oxford

Riezler S, Vasserman A (2004) Incremental feature selection and L1 regularization for maximum-entropy modeling. In: Proceedings of EMNLP, Barcelona, pp 174–181

Riezler S, King TH, Kaplan RM, Crouch R, Maxwell JT, Johnson M (2002) Parsing the wall street journal using a lexical-functional grammar and discriminative estimation techniques. In: Proceedings of ACL, Philadelphia, PA, pp 271–278

Say B, Zeyrek D, Oflazer K, Özge U (2004) Development of a corpus and a treebank for present-day written Turkish. In: Proceedings of the international conference on Turkish linguistics, Magosa, pp 183–192

Şehitoğlu O (1996) A sign-based phrase structure grammar for Turkish. Master's thesis, METU, Ankara

Sulger S, Butt M, King TH, Meurer P, Laczkó T, Rákosi G, Dione CB, Dyvik H, Rosén V, De Smedt K, Patejuk A, Çetinoglu Ö, Arka IW, Mistica M (2013) ParGramBank: the ParGram parallel treebank. In: Proceedings of ACL, Sofia, pp 550–560

Chapter 10
Statistical Machine Translation and Turkish

Kemal Oflazer, Reyyan Yeniterzi, and İlknur Durgar-El Kahlout

Abstract Machine translation is one of the most important applications of natural language processing. The last 25 years have seen tremendous progress in machine translation, enabled by the development of statistical techniques and availability of large-scale parallel sentence corpora from which statistical models of translation can be learned. Turkish poses quite many challenges for statistical machine translation as alluded to in Chap. 1, owing mainly to its complex morphology. This chapter discusses in more detail the challenges of Turkish in the context of statistical machine translation and describes two widely different approaches that have been employed in the last several years to English to Turkish machine translation.

10.1 Introduction

Statistical machine translation from English to Turkish poses a number of challenges. Typologically English and Turkish are rather distant languages: while English has very limited morphology and rather fixed SVO constituent order, Turkish is an agglutinative language with a very rich and productive derivational and inflectional morphology, and a very flexible (but SOV dominant) constituent order. One implication of complex morphology is that, in parallel texts, Turkish words usually align to multiple words on the English side. When done at the word level, alignment is very noisy and masks the more (statistically) meaningful alignments at

K. Oflazer (✉)
Carnegie Mellon University Qatar, Doha-Education City, Qatar
e-mail: ko@cs.cmu.edu

R. Yeniterzi
Özyeğin University, Istanbul, Turkey
e-mail: reyyan.yeniterzi@ozyegin.edu.tr

İ. Durgar-El Kahlout
TÜBİTAK-BİLGEM, Gebze, Kocaeli, Turkey
e-mail: ilknur.durgar@tubitak.gov.tr

© Springer International Publishing AG, part of Springer Nature 2018
K. Oflazer, M. Saraçlar (eds.), *Turkish Natural Language Processing*,
Theory and Applications of Natural Language Processing,
https://doi.org/10.1007/978-3-319-90165-7_10

207

Table 10.1 Inflected forms of the word *ev*

Count	Word	Surface segmentation	Gloss
271	ev	ev	house
258	eve	ev+e	to the house
136	evde	ev+de	at the house
72	evi	ev+i	the house
60	evin	ev+in	your/of house
46	evden	ev+den	from the house
21	evdeki	ev+de+ki	(that is) in the house
21	evime	ev+im+e	to my house
15	evim	ev+im	my house
13	evimde	ev+im+de	at my house
13	evimiz	ev+imiz	our house
5	evimden	ev+im+den	from my house
4	evlerdeki	ev+ler+de+ki	(that are) in the houses
3	evidir	ev+i+dir	it is his/her house
3	evimdeki	ev+im+de+ki	(that is) in my house
2	evlerden	ev+ler+den	from the houses
2	evindeydi	ev+in+de+ydi	it was at your house
2	evdeyim	ev+de+yim	I am at the house
1	evdekilerle	ev+de+ki+ler+le	with the ones(that are) in the house
1	evdekinden	ev+de+ki+nden	from the ones (that are) in the house
1	evdesiniz	ev+de+siniz	you are at the house
1	evdeydim	ev+de+ydi+m	I was at the house
1	evdeyken	ev+de+yken	when he/she is in the house
1	evdeysen	ev+de+yse+n	if you are at the house
1	evinizle	ev+iniz+le	with your house

the sub-lexical level. Another issue of practical significance is the lack of large-scale parallel text—the critical resource used in building machine translation systems, although this has recently been improved in the last couple of years through the availability of miscellaneous parallel texts. In 2016, for the first time, Turkish–English translation has been selected as one of the new competitive tasks in the annual Conference on Machine Translation (WMT16).[1]

To point out the implications of a very large vocabulary enabled by the productive morphology, we present the data in Table 10.1 extracted from the IWSLT'13 evaluation campaign Turkish–English parallel corpus.[2] The root word *ev* (house) occurs in this corpus with 77 different inflected forms with 50 of these forms occurring less than 10 times and 20 being singleton forms. Table 10.1 shows some

---

[1]www.statmt.org/wmt16/ (Accessed Sept. 14, 2017).

[2]International Workshop on Spoken Language Translation: workshop2013.iwslt.org/ (Accessed Sept. 14, 2017).

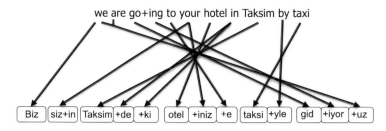

Fig. 10.1 Morpheme alignments for a pair of hand-aligned English–Turkish sentences

of these forms and their segmentations to surface morphemes and the corresponding English glosses.

Clearly the alignments need to figure out that the word *ev* corresponds to *house*. However, in the absence of any explicit indication of the morphological structure, we can only get rather weak alignments of word forms to English phrases at best. Further, we would be unable to deal with any other inflected forms that are unknown or not previously seen in the training corpus.

Figure 10.1 shows a pair of hand-aligned Turkish and English sentences with the internal morphological structure of the words made explicit wherever possible, so that the alignments of English content words and function words to Turkish content words and morphemes can be observed.

10.2   Handling Morphology in Statistical Machine Translation

Incorporating morphology when working with morphologically rich languages has been addressed by several researchers for many years. For German, Niessen and Ney (2004) have used morphological decomposition with base forms and part-of-speech tags to introduce a hierarchical lexicon model for improving word alignment quality. Corston-Oliver and Gamon (2004) normalized inflectional morphology by stemming all of the words in German and English texts. Yang and Kirchhoff (2006) morphologically decomposed unknown source words at the test time and translated words that are unknown to the decoder by using phrase-based back-off models. For Arabic, Lee (2004), Zollmann et al. (2006), and Sadat and Habash (2006) exploited morphology by using morphologically-analyzed and/or tagged resources. Popovic and Ney (2004) presented different ways of improving translation quality from inflected languages Serbian, Catalan, and Spanish by using stems, suffixes, and part-of-speech information. Goldwater and McClosky (2005) replaced Czech words with lemmas and pseudo-words to obtain improvements in Czech-to-English statistical machine translation. Talbot and Osborne (2006) reduced source and target vocabulary by clustering related words to translate from Czech, French, and Welsh. Minkov et al. (2007) used morphological post-processing on the target side by using structural information and information from the source side in order to improve translation quality for Russian and Arabic. Carpuat (2009) replaced words from specific morphological classes with their lemmas in French-English SMT. Luong et al. (2010) proposed a hybrid morpheme-word representation in the translation models of morphologically-rich languages. Naradowsky and Toutanova (2011), Nguyen et al. (2010), Mermer and Akın (2010), and Chung and Gildea (2009) have used automatically induced morphological segmentations. Recently, Eyigöz et al. (2013a,b) explored a multi-level alignment scheme specifically for morphologically complex languages where both word and morphemic representations of parallel texts were used for more accurate alignment.

In the last decade, Turkish and English statistical machine translation has been addressed by several researchers. Early efforts (Durgar-El Kahlout and Oflazer 2006, 2010; Oflazer and Durgar-El Kahlout 2007; Durgar-El Kahlout 2009) used morphological analysis to separate some Turkish inflectional morphemes that have counterparts on the English side in English-to-Turkish statistical machine translation. Later Durgar-El Kahlout et al. (2012) explored morphological segmentation experiments on the tourism domain (BTEC data) with Bayesian word alignment. Bisazza and Federico (2009) explored a series of segmentation schemes to explore the optimal segmentation for statistical machine translation of Turkish. Yılmaz and Durgar-El Kahlout (2014) have also explored the use of recurrent neural network language models for Turkish machine translation.

On a very different direction Yeniterzi (2009) and Yeniterzi and Oflazer (2010) applied syntactic transformations such as joining function words on the English side to the related content words to make English side more like Turkish.

In the rest of this chapter, we present an overview of two statistical machine translation approaches for English to Turkish statistical machine translation. Although we have experimented a bit with Turkish to English translation, we do not address this direction here, as we believe that translating into English has been quite well studied and translating from Turkish does not necessarily present additional fundamental complications.

In the first approach, we deal explicitly with morphology—we make Turkish more like English and segment Turkish words in their overt morphemes, perform some segmentation on the English side separating suffixes like the plural or verb suffixes. We call this the *morpheme segmentation approach*. In the second one, we do substantially more preprocessing, including syntactic parsing, on the English side so that we make English more like Turkish, based on the observation that many kinds of specific (possibly discontinuous) phrases in English actually correspond solely to morphology on the Turkish side and identifying these. We call this the *syntax-to-morphology mapping approach*.

10.3   The Morpheme Segmentation Approach

Our initial experiments with statistical machine translation into Turkish (Durgar-El Kahlout and Oflazer 2006) showed that when English–Turkish parallel data were aligned at the word level, a Turkish word would typically have to align with a complete phrase on the English side, and that sometimes these phrases on the English side could be discontinuous, and suggested that exploiting sub-lexical structure would be a fruitful avenue to pursue. For instance, the Turkish word *tatlandırabileceksek* could be translated as (and hence would have to align to) something equivalent to) "if we were going to be able to make [something] acquire flavor." This word could be aligned as follows (shown with co-indexation of Turkish surface morphemes and English words)[3]:

$$(\text{tat})_1(\text{lan})_2(\text{dır})_3(\text{abil})_4(\text{ecek})_5(\text{se})_6(\text{k})_7$$

$$(\text{if})_6(\text{we are})_7(\text{going to})_5(\text{be able})_4(\text{to make})_3[\text{something}](\text{acquire})_2(\text{flavor})_1$$

As mentioned numerous times, the productive morphology of Turkish implies potentially a very large vocabulary size, as noun roots have about 100 inflected form and verbs have much more. These numbers are much higher when derivations are considered: one can generate thousands of words from a single root when, say, only at most two derivations are allowed.[4] Thus, sparseness is an important issue given that we have very modest parallel resources available. However, Turkish employs about 30,000 root words and about 150 distinct suffixes, so when morphemes are used as the units in the parallel texts, the sparseness problem can be alleviated to some extent.

Our approach in this section represents Turkish words with their morphological segmentation. We use lexical morphemes instead of surface morphemes, as most surface distinctions are manifestations of word-internal phonological phenomena such as vowel harmony, and morphotactics which are not relevant for translation. With lexical morpheme representation, we can abstract away such word-internal details *and* conflate statistics for seemingly different suffixes, as at this level of representation words that look very different on the surface look very similar. For instance, although the words *evinde* "in his house" and *masasında* "on his table" look quite different, the lexical morphemes except for the root are the same: ev+sH+ndA vs. masa+sH+ndA (see Oflazer and Durgar-El Kahlout (2007) for details.)

We should however note that although employing a morpheme-based represen-tations dramatically reduces the vocabulary size on the Turkish side, it also runs the risk of overloading the decoder mechanisms to account for *both* word-internal morpheme sequencing and sentence level word ordering.

---

[3]Note that on the English side, the filler for [something] would come in the middle of this phrase.

[4]See Chap. 1 for details.

Our parallel data consists mainly of documents in international relations and legal documents from sources such as the Turkish Ministry of Foreign Affairs, EU, etc. We process these as follows:

1. We segment the words in our Turkish corpus into lexical morphemes whereby differences in the surface representations of morphemes due to word-internal phenomena are abstracted out to improve statistics during alignment. Note that as with many similar languages, the segmentation of a surface word is generally ambiguous, we first generate a representation using our morphological analyzer (Oflazer 1994) that contains both the lexical segments and the morphological features encoded for all possible segmentations and interpretations of the word and perform morphological disambiguation using morphological features (Yuret and Türe 2006). Once the contextually salient morphological interpretation is selected, we discard the features leaving behind the lexical morphemes making up a word, though we could have used as well the corresponding feature names.[5]

2. We tag the English side using TreeTagger (Schmid 1994), which provides a *lemma* and a *part-of-speech* for each word. We then remove any tags which do not imply an explicit morpheme or an exceptional form. So for instance, if the word *book* gets tagged with +*NN*, we keep *book* in the text, but remove +*NN*. For *books* tagged with +*NNS* or *booking* tagged with +*VVG*, we keep *book* and +*NNS*, and *book* and +*VVG*. A word like *went* is replaced by *go* +*VVD*.[6]

3. From these morphologically segmented corpora, we also extract for each sentence, the sequence of roots for open class content words (nouns, adjectives, adverbs, and verbs). For Turkish, this corresponds to removing *all* morphemes and any roots for closed classes. For English, this corresponds to removing all words tagged as closed class words along with the tags such as +*VVG* above that signal a morpheme on an open class content word. We use this to augment the training corpus and bias content word alignments, with the hope that such roots may get a better chance to align without any additional "noise" from morphemes and other function words.

Table 10.2 presents various statistical information about this parallel corpus. One can note that Turkish has many more distinct word forms (about twice as many as English), but has much less number of distinct content words than English.[7] For language models in decoding and *n*-best list rescoring, we use, in addition

---

[5]This disambiguator has about 94% accuracy.

[6]Ideally, it would have been very desirable to actually do derivational morphological analysis on the English side, so that one could, for example, analyze *accession* into *access* plus a marker indicating nominalization.

[7]The training set in the first row of Table 10.2 was limited to sentences on the Turkish side which had at most 90 tokens (roots and bound morphemes) in total in order to comply with the limitations of the GIZA++ alignment tool. However when only the content words are included, we have more sentences to include since much less number of sentences violate the length restriction when morphemes/function words are removed.

Table 10.2 Statistics on Turkish and English training and test data, and Turkish morphological structure

	Sent.	Words (UNK)	Unique words	Morph.	Unique morph.	Morph./ word	Unique roots	Unique suffixes
Turkish								
Train	45,709	557,530	52,897	1,005,045	15,081	1.80	14,976	105
Content	56,609	436,762	13,767					
Tune	200	3258	1442	6240	859	1.92	810	49
Test	649	10,334 (545)	4355	18,713	2297	1.81	2220	77
English								
Train	45,709	723,399	26,747					
Content	56,609	403,162	19,791					
Test	649	13,484 (231)	3220					

to the training data, a monolingual Turkish text of about 100,000 sentences (in a segmented and disambiguated form).

A typical sentence pair in our (fully-segmented) data looks like the following, where we have highlighted the content root words with bold font, co-indexed them to show their alignments and bracketed the "words" that BLEU evaluation on test would consider.

T: [**kat**$_1$ +hl +ma] [**ortaklık**$_2$ +sH +nHn] [**uygula**$_3$ +Hn +mA +sH] [,]
[**ortaklık**$_4$] [**anlaşma**$_5$ +sH] [**çerçeve**$_6$ +sH +ndA]
[**izle**$_7$ +Hn +yAcAk +DHr] [.]
E: the **implementation**$_3$ of the **accession**$_1$ **partnership**$_2$ will be
monitor$_7$ +vvn in the **framework**$_6$ of the **association**$_4$ **agreement**$_5$.

Note that when the morphemes/tags (tokens starting with a +) are concatenated, we get the "word-based" version of the corpus, since surface words are directly recoverable from the concatenated representation. We use this word-based representation also for word-based language models used for rescoring.

10.3.1 *Experiments and Results*

We employed the phrase-based statistical machine translation framework (Koehn et al. 2003), and used the Moses toolkit (Koehn et al. 2007), and the SRILM language modelling toolkit (Stolcke 2002), and evaluated our decoded translations using the BLEU measure (Papineni et al. 2002), using a *single* reference translation.

We performed four sets of experiments employing different morphological representations on the Turkish side and adjusting the English representation accordingly wherever needed.

1. **Baseline:** English and Turkish sentences are represented with full words. For example, `kitap+sH+nHn` (representing *kitabının* (of his book) would be used on the Turkish side and `book+NNS` (representing *books*) on the English side.
2. **Full Morphological Segmentation:** English and Turkish sentences are represented with tokens that are root words and bound morphemes/tags. For instance, for the example in the above paragraph, the three tokens `kitap +sH +nHn` would be used on the Turkish side and the two tokens `book +NSS` would be used on the English side.
3. **Root+Morphemes Segmentation:** Turkish sentences are represented with roots and combined morphemes. For English sentences, we used the same representation in (2). For example, for the Turkish word above, only two tokens `kitap +sH+nHn` would be used.
4. **Selective Morphological Segmentation:** A systematic analysis of the alignment files produced by GIZA++ for a small subset of the training sentences showed that certain morphemes on the Turkish side were almost consistently never aligned with anything on the English side: e.g., the compound noun marker morpheme in Turkish (+sH) does not have a corresponding unit on the English side, as English noun–noun compounds do not carry any overt markers. Such markers were never aligned to anything or were aligned almost randomly to tokens on the English side. Further, since we perform derivational morphological analysis on the Turkish side but not on the English side, we also noted that most verbal nominalizations on the English side were just aligned to the verb roots on the Turkish side and the additional markers on the Turkish side indicating the nominalization, and various agreement markers, etc., were mostly unaligned.

   For just these cases, we selectively attached such morphemes (and in the case of verbs, the intervening morphemes) to the root, but otherwise kept other morphemes, especially any case morphemes, still by themselves, as they almost often align with prepositions on the English side quite accurately.[8]

   In this case, the Turkish word above would be represented by the two tokens `kitap+sH +nhn`. English words are still represented as in case 2 above.

For each of the four representational schemes we went through the following process:

1. The training corpus was augmented with the content word parallel data.[9]
2. A 5-gram morpheme-based language model was constructed for Turkish (to be used by the decoder) using the Turkish side of the training data along with an

---

[8] It should be noted that what to selectively attach to the root should be considered on a per-language basis; if Turkish were to be aligned with a language with similar morphological markers, this perhaps would not have been needed.

[9] Using the content word data improved performance for all representations *except* the baseline.

additional monolingual Turkish text of about 100K sentences represented in the same scheme as the Turkish side of the training data.

3. Training was performed and the phrase table was extracted using a maximum phrase size of 7. Minimum error rate training with the tune set did not provide any tangible improvements.[10]

4. The test corpus was decoded using the Moses decoder with modified parameters *-dl -1* to allow for long distance movement and *-weight-d 0.1* to avoid penalizing long distance movement.[11] The decoder also produced 1000-best candidate translations.

5. For representation schemes 2–4, the 1000-best candidates were then converted into word-based representation (by just attaching any morpheme/tag tokens to the stem to the left) and rescored using weighted combination of the 4-gram *word-based* language model score and the translation score produced by the decoder. The combination weights were optimized on the tune corpus.

6. The top rescored candidate translations were selected and compared with the (single) reference translation using the BLEU measure.

The results of these experiments are presented in Table 10.3.

The best BLEU results are obtained with selective morphological segmentation (24.61) and represent a relative improvement of 23%, compared to the respective baseline of 19.77. One should also note that the default decoding parameters used by the Moses decoder produce much worse results especially for the fully segmented model.

Our further experiments are only executed on top of the results of the best performing representation—selective morphological segmentation.

Table 10.3 BLEU results for the four representational schemes

Experiment/decoder parameters	BLEU
Word-based baseline/default parms	16.13
Word-based baseline/modified parms	19.77
Full morphological segmentation/default parms	13.55
Full morphological segmentation/modified parms	22.18
Root+morphemes segmentation/modified parms	20.12
Selective morphological segmentation/modified parms	**24.61**

---

[10]We ran MERT on the baseline model and the morphologically segmented models forcing *-weight-d* to range a very small around 0.1, but letting the other parameters range in their suggested ranges. Even though the procedure came back claiming that it achieved a better BLEU score on the tune set, running the new model on the test set did not show any improvement at all. This may have been due to the fact that the initial choice of *-weight-d* along with *-dl* set to -1 provides such a drastic improvement that perturbations in the other parameters do not have much impact.

[11]We arrived at this combination by experimenting with the decoder to avoid the almost monotonic translation we were getting with the default parameters. These parameters boosted the BLEU scores substantially compared to default parameters used by the decoder.

10.3.1.1 Augmenting the Training Data

In order to overcome the disadvantages of the small size of our parallel data, we experimented with ways of using portions of the phrase table that is generated by the training process, as additional training data.

The phrase extraction process performs English–Turkish and Turkish–English alignments using the GIZA++ tool and then combines these alignments with some additional post-processing and extracts "phrases," sequences of source and target tokens that align to tokens in the other sequence. Such phrases do not necessarily correspond to linguistic phrases.

The following is a very small portion of the phrase table generated by the Moses training process for the selective morphological segmentation representation:

```
good word ||| müjde ||| 1 0.25 0.5 0.0037281 2.718
good ||| düzgün ||| 0.0714286 0.0322581 0.00487805 0.0018382 2.718
good ||| en iyi ||| 0.388889 0.25589 0.0341463 0.00605536 2.718
good ||| en ||| 0.00833333 0.0194715 0.00487805 0.0257353 2.718
good ||| eşya ||| 0.030303 0.32967 0.00487805 0.0551471 2.718
good ||| güzel ||| 0.2 0.0645161 0.0097561 0.0036765 2.718
good ||| iyi bir ||| 0.2 0.492308 0.0146341 0.0126794 2.718
good ||| iyi ||| 0.85 0.492308 0.497561 0.235294 2.718
good +NNS ||| mal +lar ||| 0.540741 0.605839 0.356098 0.152574 2.718
```

The first and second parts of any entry in the phrase table are the English (e) and Turkish (t) parts of a pair of aligned phrases. Among the sequences of the numbers that follow, the first is $p(e|t)$, the conditional probability that the English phrase is e given that the Turkish phrase is t; the third number is $p(t|e)$ and captures the probability of the symmetric situation.

Among these phrase table entries, those with $p(e|t) \approx p(t|e)$ and $p(t|e) + p(e|t)$ larger than some threshold can be considered as reliable mutual translations in that they mostly translate to each other and not much to others. So we extracted those phrases with $0.9 \leq p(e|t)/p(t|e) \leq 1.1$ and $p(t|e) + p(e|t) \geq 1.5$ and added them to further bias the alignment process.

The six steps listed earlier were repeated for this augmented selectively morphologically segmented training corpus. The BLEU result that was obtained was 26.16, showing a 32.3% relative improvement over the 19.77 baseline, and 6.3% relative improvement over the previous result.

10.3.2 Word Repair

The detailed BLEU results of 26.16 showing the 1-, 2-, 3-, and 4-gram match scores, [53.0/29.9/20.3/14.6] for our best performing model, indicated that only 53% of the words in the candidate translations are determined correctly. However, when *all words* in both the candidate and reference translations are reduced to roots and BLEU is computed again we get the *root* BLEU results of 30.62 with corresponding matches [64.6/35.7/23.4/16.3]. This shows that we are getting 64.6% of the roots in the translations correct but only 53% of the words forms are correct, indicating that for many cases, the roots are correct but the full word forms are either incorrect or correct but do not match the existing word form in the reference translation. Such words can be classified into three groups:

1. Morphologically malformed words—words with the correct root word but with morphemes that are either categorically incorrect (e.g., case morpheme on a verb) or morphotactically incorrect (e.g., morphemes in the wrong order).
2. Morphologically well-formed words which are out-of-vocabulary (OOV) relative to the training corpus and the language model corpus.
3. Morphologically well-formed words which are *not* out-of-vocabulary relative to the training corpus and the language model corpus, but do not match the reference.

Words in groups 1 and 2 can be identified easily: Words in group 1 would be rejected using our morphological analyzer, while words for group 2 would be accepted by the morphological analyzer but would not be in the vocabulary of the training and language model corpora. However, there is no way knowing whether a word falls in group 3 without looking at the reference.

The approach we have taken to deal with the words for case 1 is as follows:

1. Using a finite state model of lexical morpheme structure of possible Turkish words, with morphemes being as the symbols (except for the letters in roots), we use error-tolerant finite state recognition (Oflazer 1996) to generate morphologically correct word forms with the same root, but with morpheme structures up to 2 unit morpheme edit operations (add, delete, substitute, transpose morphemes) away. We do this for every morphologically malformed word in a candidate translation sentence. For instance, the word form (in lexical morpheme representation) $gel+dA+ydH$ is malformed and possible corrections at distance 1 are $\{gel\underline{+yAcAk}+ydH, gel\underline{+mHs}+ydH, gel\underline{+dH}+ydH, gel\underline{+sA}+ydH, gel\underline{+yA}+ydH\}$. We convert the sentence to a lattice representation replacing each malformed with the correct alternatives.
2. The resulting lattice is then rescored with the language model to pick the best alternative for each malformed word. In this step, the morpheme-based language model performed better than the word-based-language model.

When words that are one morpheme operation away were considered as possible alternatives, the BLEU score improved to 26.46. The BLEU score improved go 26.49 when words that are two morpheme operations away were included.

We took a similar approach for handling words for case 2. We generated alternatives for these morphologically correct but OOV words that were 1 and 2 morpheme operation distance away, but this time we restricted the alternatives to the vocabulary of the training and language model corpora. With both distances 1 and 2, performance of the system improved further to 26.87 BLEU points. All in all, word repair provides an additional improvement of 2.7% relative (compared to 26.16) and the final BLEU score represents a relative improvement of 35.9% over the baseline score.

10.3.3 Sample Translations

When we consider input English sentences that are between 5 and 15 words, the translation quality of our system is considerably better than the quality for the complete test set. Below we present translations of three sentences from the test data along with the literal paraphrases of the translation and the reference versions. The first two are quite accurate and acceptable translations while the third clearly has missing, incorrect but also interesting parts: we see that the English *key* is translated to the Turkish *kilit* (lock) which is the correct collocational translation.[12]

Input: 1. everyone's right to life shall be protected by law.
Translation: 1. herkesin yaşama hakkı kanunla korunur.
Literally: 1. everyone's living right is protected with law.
Reference: 1. herkesin yaşam hakkı yasanın koruması altındadır.
Literally: 1. everyone's life right is under the protection of the law.

Input: promote protection of children's rights in line with EU and international standards.
Translation: çocuk haklarının korunmasının ab ve uluslararası standartlara uygun şekilde geliştirilmesi.
Literally: develop protection of children's rights in accordance with EU and international standards.
Reference: ab ve uluslararası standartlar doğrultusunda çocuk haklarının korunmasının teşvik edilmesi.
Literally: in line with EU and international standards promote/motivate protection of children's rights.

Input: as a key feature of such a strategy, an accession partnership will be drawn up on the basis of previous European council conclusions.
Translation: bu stratejinin kilit unsuru bir katılım ortaklığı belgesi hazırlanacak kadarın temelinde, bir önceki avrupa konseyi sonuçlarıdır.
Literally: a lock feature of this strategy accession partnership document will be prepared based on the previous European council resolutions.
Reference: bu stratejinin kilit unsuru olarak, daha önceki ab zirve sonuçlarına dayanılarak bir katılım ortaklığı oluşturulacaktır.
Literally: as a lock feature of this strategy an accession partnership based on earlier EU summit resolutions will be formed.

---

[12]We should also note that all sentences were lowercased so that we would not have to deal with exact capitalization issue at that stage.

10.3.4   Observations on the Morpheme Segmentation Approach

For English-to-Turkish statistical machine translation, employing a language-pair specific morphological representation somewhere in between using full word-forms and fully morphologically segmented representations along with augmenting the limited training data with content words and highly reliable phrases provides the most leverage. We observed that given the typical complexity of Turkish words, there was a substantial percentage of words whose morphological structure was incorrect: either the morphemes were not applicable for the part-of-speech category of the root word selected, or they were in the wrong order. The main reason for these problems was most likely that the same statistical translation, reordering and language modeling mechanisms were being employed to *both* determine the morphological structure of the words *and*, at the same time, get the global order of the words correct. Repairing morphologically malformed words and OOV words provided some minor additional improvement but nothing significant that would make a dent in the overall performance.

Translation into Turkish seems to involve processes that are somewhat more complex than standard statistical translation models. For example, we observed cases where the morphological structure of a single word on the Turkish side was synthesized from the translations of two or more phrases, and errors in any translated morpheme or its morphotactic position rendered the synthesized word incorrect, even though the rest of the word was quite fine. This indirectly implies that BLEU is particularly harsh for Turkish and the morpheme-based approach, because of the all-or-none nature of token comparison when computing the BLEU score. Furthermore, there are also cases where words with different morphemes have very close morphosemantics, convey the relevant meaning, and are almost interchangeable:

- *gel+Hyor* (*geliyor*—he is coming) vs. *gel+mAktA* (*gelmekte*—he is (in a state of) coming) are essentially the same. On a scale of 0–1, one could rate these at about 0.95 in similarity.
- *gel+yAcAk* (*gelecek*—he will come) vs. *gel+yAcAk+dhr* (*gelecektir*—he *will* come) in a sentence final position. Such pairs could be rated perhaps at 0.90 in similarity.
- *gel+dH* (*geldi*—he came (evidential past tense)) vs. *gel+mHs* (*gelmiş*—he came (hearsay past tense)). These essentially mark past tense but differ in how the speaker relates to the event and could be rated at perhaps 0.70 similarity.

We have also developed a tool, BLEU+ (Tantuğ et al. 2008) that implements *a slightly different formulation of token similarity* in BLEU computation considering (1) root word similarity, by considering synonyms (e.g., as in Meteor) and hypernyms, using a WordNet, and (2) morphosemantic similarity considering (almost) synonymous morphemes. BLUE+ can also compute METEOR scores, oracle BLEU scores assuming all morphologically malformed words are perfectly corrected, and also root BLEU scores providing for a better understanding of the quality and the limits of the output translation. This tool is discussed in some more detail in Chap. 11.

10.4   The Syntax-to-Morphology Mapping Approach

Work described in the previous section has used an approach which relied on identifying the contextually correct parts-of-speech, roots, and any morphemes on the English side, and the complete sequence of roots and overt derivational and inflectional morphemes for each word on the Turkish side. Once these were identified as separate tokens, they were then used as "words" in a standard phrase-based (PB) framework (Koehn et al. 2003).

Motivated by the observation that many local and some non-local syntactic structures in English essentially map to morphologically complex words in Turkish, this section presents a radically different approach which does not segment Turkish words into morphemes, but uses a representation equivalent to the full word form. On the English side however, it relies on a full syntactic analysis using a dependency parser. This analysis then lets us abstract and encode many local and some non-local English syntactic structures as complex tags on words which for the purpose of translation look like additional morphemes on those words. Thus we can bring the representation of English syntax closer to the Turkish morphosyntax—English inflectional morphology now looks like Turkish inflectional morphology.

Such an approach enables the following:

- Driven by the pattern of morphological structures of full word forms on the Turkish side represented as root words and complex tags, we can identify and reorganize phrases on the English side, to "align" English syntax to Turkish morphology wherever possible.
- Continuous and discontinuous variants of certain (syntactic) phrases can be conflated during the SMT phrase extraction process.
- The length of the English sentences can be dramatically reduced, as most function words encoding syntax are now abstracted into complex tags on their respective headwords.
- The representation of both the source and the target sides of the parallel corpus can now be mostly normalized. *This facilitates the use of factored phrase-based (FPB) translation that was not previously applicable due to the morphological complexity on the target side and mismatch between source and target morphologies.*

We find that with the full set of syntax-to-morphology transformations and some additional techniques we can get about 39% relative improvement in BLEU scores over a word-based baseline and about 28% improvement of a factored baseline, all experiments being done over ten training and test sets in a 10-fold way. We also find that constituent reordering taking advantage of the syntactic analysis of the source

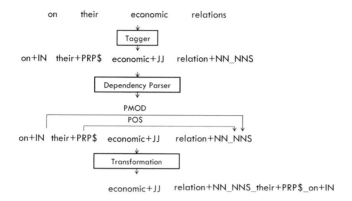

Fig. 10.2 Transformation of an English prepositional phrase

side does not necessarily provide tangible improvements when averaged over the ten data sets.

10.4.1 Mapping Source-Side Syntax to Target-Side Morphology

In this section, we describe how we map between certain source language syntactic structures and target words with complex morphological structures. At the top of Fig. 10.2, we see a pair of (syntactic) phrases, where we have (positionally) aligned the words that should be translated to each other. We can note that the function words *on* and *their* are not really aligned to any of the Turkish words as they really correspond to two of the morphemes of the last Turkish word.

The basic idea in this approach is to take various function words on the English side, whose syntactic relationships are identified by the parser, and then package them as complex tags on the related content headwords. When we tag and syntactically analyze the English side into dependency relations, we get the representation in the bottom of Fig. 10.2.[13] Figure 10.3 shows how the transformed representation maps to the parallel Turkish phrase which has been morphologically analyzed and disambiguated.

---

[13]The meanings of various tags are as follows: Dependency Labels: **PMOD**—Preposition Modifier; **POS**—Possessive. Part-of-Speech Tags for the English words: **+IN**—Preposition; **+PRP$**—Possessive Pronoun; **+JJ**—Adjective; **+NN**—Noun; **+NNS**—Plural Noun. Morphological Feature Tags in the Turkish Sentence: **+A3pl**—3rd person plural; **+P3sg**—3rd person singular possessive; **+Loc**—Locative case. Note that we mark an English plural noun as **+NN_NNS** to indicate that the root is a noun and there is a plural morpheme on it. Note also that *economic* is also related to *relations* but we are not interested in such content words and their relations.

Fig. 10.3 How English syntax maps to Turkish morphology

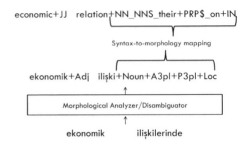

In this example, if we move the first two function words from the English side and attach them as *syntactic tags* to the word they are in dependency relation with, we get the aligned representation at the bottom of Fig. 10.3.[14, 15] Here we can note that all root words and tags that correspond to each other are nicely structured and are in the same relative order. In fact, we can treat each token as being composed of two factors: the roots and the accompanying tags. The tags on the Turkish side encode morphosyntactic information encoded in the morphology of the words, *while the (complex) tags on the English side encode local (and sometimes, non-local) syntactic information.* Furthermore, we can see that before the transformations, the English side has four words, while afterwards it has only two words. We find (and elaborate later) that this reduction in the English side of the training corpus, in general, is about 30%, and is correlated with improved BLEU scores. We believe the removal of many function words and their folding into complex tags (which do not get involved in GIZA++ alignment—we only align the root words) seems to improve alignment as there are less number of "words" to worry about during that process.

Another interesting side effect of this representation is the following. As the complex syntactic tags on the English side are based on syntactic relations and not necessarily positional proximity, the tag for *relations* in a phrase like *in their cultural, historical, and economic relations* would be exactly the same as above. Thus phrase extraction algorithms can conflate all constructs like *in their ... relations* as one phrase, regardless of the intervening modifiers, assuming that parser does its job properly. Not all cases can be captured as cleanly as this example, but most transformations capture local and non-local syntax involving many function words and then encode syntax with complex tags resembling full morphological tags on the Turkish side. *These transformations, however, are not meant to perform sentence level constituent reordering on the English side.* We explore these later.

We developed a set of about 20 linguistically-motivated syntax-to-morphology transformations which had variants parameterized depending on what, for instance, the preposition or the adverbial was, and how they map to morphological structure

---

[14]We use _ to prefix such syntactic tags on the English side.

[15]The order is important in that we would like to attach the same sequence of function words in the same order so that the resulting tags on the English side are the same.

on the Turkish side. For instance, one general rule handles cases like *while ... verb* and *if ... verb*, etc., mapping these to appropriate complex tags. It is also possible that multiple transformations can apply to generate a single English complex tag: a portion of the tag can come from a verb complex transformation, and another from an adverbial phrase transformation involving a marked such as *while*. Our transformations handle the following cases:

- Prepositions attach to the head-word of their complement noun phrase as a component in its complex tag.
- Possessive pronouns attach to the head-word they specify.
- The possessive markers following a noun (separated by the tokenizer) attach to the noun.
- Auxiliary verbs and negation markers attach to the lexical verb that they form a verb complex with.
- Modals attach to the lexical verb they modify.
- Forms of *be* used as predicates with adjectival or nominal dependents attach to the dependent.
- Forms of *be* or *have* used to form passive voice with past participle verbs, and forms of *be* used with *-ing* verbs to form present continuous verbs, attach to the verb.
- Various adverbial clauses formed with *if*, *while*, *when*, etc., are reorganized so that these markers attach to the head verb of the clause.

As stated earlier, these rules are linguistically motivated and are based on the morphological structure of the *target* language words. Hence for different target languages these rules will be different. The rules recognize various local and non-local syntactic structures in the source side parse tree that correspond to complex morphological of target words and then remove source function words folding them into complex tags. For instance, the transformations in Fig. 10.2 are handled by scripts that process MaltParser's (Nivre et al. 2007) dependency structure output and essentially implement the following sequence of rules expressed as pseudo code:

```
1) if  (<Y>+PRP$ POS <Z>+NN<TAG>)
   then {
        APPEND <Y>+PRP$ TO <Z>+NN<TAG>
        REMOVE <Y>+PRP$
     }

2) if (<X>+IN   PMOD <Z>+NN<TAG>)
   then {
        APPEND <X>+IN TO <Z>+NN<TAG>
        REMOVE <X>+IN
     }
```

Here <X>, <Y> and <Z> can be considered as Prolog-like variables that bind to patterns (mostly root words), and the conditions check for specified dependency relations (e.g., PMOD) between the left and the right sides. When the condition is satisfied, then the part matching the function word is removed and its syntactic information is appended to form the complex tag on the noun (<TAG> would either match null string or any previously appended function word markers).[16]

There are several other rules that handle more mundane cases of date and time constructions (for which the part of the date construct which the parser attaches a preposition is usually different than the part on the Turkish side that gets inflected with case markers, and these have to be reconciled by overriding the parser output).

10.4.2 Experimental Setup and Results

This section presents an example of a sentence with multiple transformations applied, after discussing the preprocessing steps.

Let's assume we have the following pair of parallel sentences:

```
E: if a request is made orally the authority must make a record
   of it
T: istek sözlü olarak yapılmışsa yetkili makam bunu kaydetmelidir
```

On the English side of the data, we use the Stanford Log-Linear Tagger (Toutanova et al. 2003), to tag the text with Penn Treebank Tagset. On the Turkish side, we perform a full morphological analysis (Oflazer 1994) and morphological disambiguation (Yuret and Türe 2006) to select the contextually salient interpretation of words. We then remove any morphological features that are not explicitly marked by an *overt morpheme*.[17] So for both sides we get

```
E: if+IN a+DT request+NN is+VBZ made+VBN orally+RB the+DT
   authority+NN must+MD make+VB a+DT record+NN of+IN it+PRP
T: istek+Noun sözlü+Adj olarak+Verb+ByDoingSo
   yap+Verb+Pass+Narr+Cond yetkili+Adj makam+Noun bu+Pron+Acc
   kaydet+Verb+Neces+Cop
```

Finally we parse the English sentences using MaltParser (Nivre et al. 2007), which gives us labeled dependency parses. On the output of the parser, we make one more transformation. We replace each word with its root, and possibly add an additional tag for any inflectional information conveyed by overt morphemes or exceptional forms. This is done by running the TreeTagger (Schmid 1994) on the English side which provides the roots in addition to the tags, and then carrying over this information to the parser output. For example, is is now tagged as be+VB_VBZ, made is now tagged as make+VB_VBN, and a word like books is tagged as book+NN_NNS (and not as books+NNS). On the Turkish side, each marker with a preceding + is a morphological feature. The first marker is the part-of-speech tag of the root and the remainder are the overt inflectional and derivational

---

[16]We outline two additional rules later when we see a more complex example in Fig. 10.4.

[17]For example, the morphological analyzer outputs +A3sg to mark a singular noun, if there is no explicit plural morpheme. Such markers are removed.

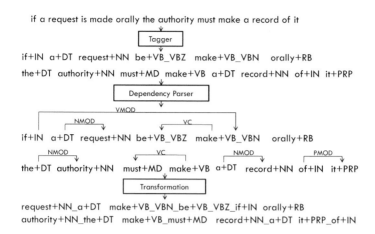

Fig. 10.4 An English sentence with multiple transformations applied

markers of the word. For example, the analysis `kitap+Noun+A3pl+P2pl+Gen` for a word like `kitap+lar+ınız+ın` (*of your books*) represents the root `kitap` (*book*), a `Noun`, with third person plural agreement `A3pl`, second person plural possessive agreement, `P2pl` and genitive case `Gen`.

Figure 10.4 shows how multiple transformations are applied to an English sentence. For example, two rules process the if-clause in Fig. 10.4 and these rules are applied sequentially: The first rule recognizes the passive construction mediated by `be+VB<AGR>` forming a verb complex (`VC`) with `<Y>+VB_VBN` and appends the former to the complex tag on the latter and then deletes the former token. The second rule then recognizes `<X>+IN` relating to `<Y>+VB<TAGS>` with `VMOD` and appends the former to the complex tag on the latter and then deletes the former token. After all the rules are applied, basically all function words are bundled as a complex tag attached to the relevant English content word. Figure 10.5 shows the corresponding Turkish sentence morphologically processed. Here co-indexation on the root words indicates which root words on one side should align to the root words on the other side. Ultimately we would want the alignment process to uncover the *root word alignments* indicated here. *We can also note that the initial form of the English sentence has 14 words and the final form after transformations has 7 words (with complex tags).*

We worked on the same parallel corpus that had been used in the work described in the earlier section. The data set consists of 52,712 parallel sentences. In order to have more confidence in the impact of our transformations, we randomly generated 10 training, test and tune set combinations. For each combination, the latter two

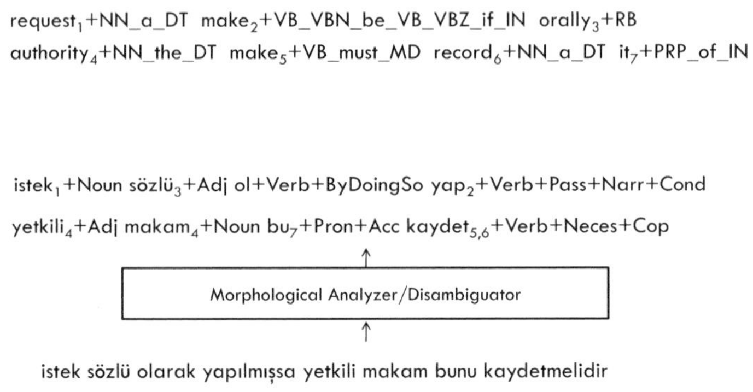

request₁+NN_a_DT make₂+VB_VBN_be_VB_VBZ_if_IN orally₃+RB
authority₄+NN_the_DT make₅+VB_must_MD record₆+NN_a_DT it₇+PRP_of_IN

istek₁+Noun sözlü₃+Adj ol+Verb+ByDoingSo yap₂+Verb+Pass+Narr+Cond
yetkili₄+Adj makam₄+Noun bu₇+Pron+Acc kaydet₅,₆+Verb+Neces+Cop

↑

┌───┐
│   Morphological Analyzer/Disambiguator   │
└───┘

↑

istek sözlü olarak yapılmışsa yetkili makam bunu kaydetmelidir

Fig. 10.5 How syntax in the English sentence in Fig. 10.4 maps to Turkish morphology

were 1000 sentences each and the remaining 50,712 sentences were used as training sets.[18,19]

We performed our experiments with the Moses toolkit (Koehn et al. 2007). In order to encourage long distance reordering in the decoder, we used a distortion limit of −1 and a distortion weight of 0.1, as before.[20] We did not use MERT to further optimize our model.

For evaluation, we used the BLEU metric. Each experiment was repeated over the ten data sets. Wherever meaningful, we reported the average BLEU scores over ten data sets along with the maximum and minimum values and the standard deviation.

10.4.2.1 The Baseline Systems

As a baseline system, we built a standard phrase-based system, using the surface forms of the words without any transformations, and with a 3-gram language model in the decoder. We also built a second baseline system with a factored model. Instead of using just the surface form of the word, we included the root, part-of-speech, and morphological tag information into the corpus as additional factors alongside the surface form.[21] Thus, a token is represented with three factors as Surface|Root|Tags where Tags are complex tags on the English side, and morphological tags on the Turkish side.[22] Table 10.4 shows these factors for two

---

[18]The tune set was not used in this work but reserved for future work so that meaningful comparisons could be made.

[19]It is possible that the ten test sets are not mutually exclusive.

[20]These allow and do not penalize unlimited distortions, but increase decoding time.

[21]In Moses, factors are separated by a '|' symbol.

[22]Concatenating Root and Tags gives the Surface form, in that the surface is unique given this concatenation.

Table 10.4 Factored token representations

Representation	English	Turkish
Surface	make+VB_VBN_be+VB_VBZ_if+IN	yap+Verb+Pass+Narr+Cond
Surface|	make+VB_VBN_be+VB_VBZ_if+IN|	yap+Verb+Pass+Narr+Cond|
Root|	make|	yap|
Tags	+VB_VBN_be+VB_VBZ_if+IN	+Verb+Pass+Narr+Cond

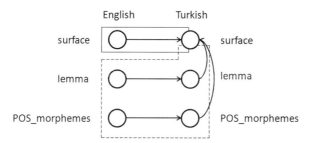

Fig. 10.6 Alternative path model for decoding

parallel words in English and Turkish. The first baseline uses the representation in the first row, while the second baseline uses the second factored representation.[23]

Moses lets word alignment to align over any of the factors. We aligned our training sets using only the root factor to conflate statistics from different forms of the same root. The rest of the factors are then automatically assumed to be aligned, based on the root alignment. Furthermore, in factored models, we can employ different language models for different factors. For the initial set of experiments we used 3-gram language models for all the factors.

For factored decoding, we employed a model whereby we let the decoder translate a surface form directly, but if/when that fails, the decoder can back-off with a generation model that builds a target word from independent translations of the root and tags. This alternative path model is illustrated in Fig. 10.6.

The results of our baseline models are given in top two rows of Table 10.5. As expected, the word-based baseline performs worse than the factored baseline. We believe that the use of multiple language models (some much less sparse than the surface language model) in the factored baseline is the main reason for the improvement.

---

[23]Note that for Turkish, this representation is equivalent to surface words in that the surface is unique given this representation.

Table 10.5 BLEU scores for a variety of transformation combinations

Experiment	Ave	STD	Max	Min
Baseline	17.08	0.60	17.99	15.97
Factored baseline	18.61	0.76	19.41	16.80
Noun+Adj	21.33	0.62	22.27	20.05
Verb	19.41	0.62	20.19	17.99
Adv	18.62	0.58	19.24	17.30
Verb+Adv	19.42	0.59	20.17	18.13
Noun+Adj+Verb+Adv	21.67	0.72	22.66	20.38
Noun+Adj+Verb+Adv+PostP	21.96	0.72	22.91	20.67

10.4.2.2 Applying Syntax-to-Morphology Mapping Transformations

To gauge the effects of transformations separately, we first performed them in batches on the English side. These batches were:

1. transformations involving nouns and adjectives, labeled as *Noun+Adj*
2. transformations involving verbs, labeled as *Verb*,
3. transformations involving adverbs, labeled *Adv*,
4. transformations involving both verbs and adverbs, labeled *Verb+Adv*.

We also performed one set of transformations on the Turkish side. In general, English prepositions translate as case markers on Turkish nouns. However, there are quite a number of lexical *postpositions* in Turkish which also correspond to English prepositions. To normalize these with the handling of case-markers, we treated these postpositions as if they were case-markers and attached them to the immediately preceding noun, and then aligned the resulting training data (labeled *PostP*).[24]

The results of these experiments are presented in Table 10.5. We can observe that the combined syntax-to-morphology transformations on the source side provide a substantial improvement by themselves and a simple target side transformation on top of those provides a further boost to 21.96 BLEU which represents a 28.57% relative improvement over the word-based baseline and a 18.00% relative improvement over the factored baseline.

We can see that every transformation improves the baseline system and the highest performance is attained when all transformations are performed. However when we take a closer look at the individual transformations performed on English side, we observe that not all of them have the same effect. While *Noun+Adj* transformations give us an increase of 2.73 BLEU points, *Verbs* improve the result by only 0.8 points and improvement with *Adverbs* is even lower. To understand why we get such a difference, we investigated the correlation of the decrease in

---

[24]Note that in this case, the translations would be generated in the same format, but we then split such postpositions from the words they are attached to, during decoding, and *then* evaluate the BLEU score.

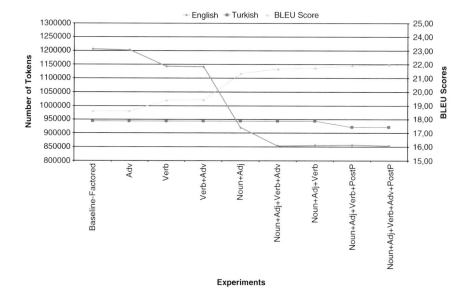

Fig. 10.7 BLEU scores vs number of tokens in the training sets

Table 10.6 Details of word, root, and morphology BLEU scores

		1-gram	2-gram	3-gram	4-gram
BLEU	21.96	55.73	27.86	16.61	10.68
BLEU-R	27.63	68.60	35.49	21.08	13.47
BLEU-M	27.93	67.41	37.27	21.40	13.41

the number of tokens on both sides of the parallel data, with the change in BLEU scores. The graph in Fig. 10.7 plots the BLEU scores and the number of tokens in the two sides of the training data as the data is modified with transformations. We can see that as the number of tokens in English decreases, the BLEU score increases. In order to measure the relationship between these two variables statistically, we performed a correlation analysis and found that there is a strong negative correlation of −0.99 between the BLEU score and the number of English tokens. We can also note that the largest reduction in the number of tokens comes with the application of the *Noun+Adj* group of transformations, which correlates with the largest increase in BLEU score.

It is also interesting to look at the *n*-gram precision components of the BLEU scores (again averaged). In Table 10.6, we list these for words (actual BLEU), roots (BLEU-R) to see how effective we are in getting the root words right, and morphological tags (BLEU-M) to see how effective we are in getting just the morphosyntax right. It seems we are getting almost 69% of the root words and 68% of the morphological tags correct, but not necessarily getting the combination equally as good, since only about 56% of the full word forms are correct.

Table 10.7 Details of word, root, and morphology BLEU scores, with 8-gram tag LM and 3/4-gram root LMs

	1-gram	2-gram	3-gram	4-gram
3-gram root LM				
BLEU	22.61	55.85	28.21	17.16	11.36
BLEU-R	28.21	68.67	35.80	21.55	14.07
BLEU-M	28.68	67.50	37.59	22.02	14.22
4-gram root LM				
BLEU	22.80	55.85	28.39	17.34	11.54
BLEU-R	28.48	68.68	35.97	21.79	14.35
BLEU-M	28.82	67.49	37.63	22.17	14.40

10.4.2.3 Experiments with Higher-Order Language Models

Factor phrased-based functionality in Moses allows the use of multiple language models for the target side, for different factors during decoding. Since the number of possible distinct morphological tags (the full morphological tag vocabulary size) in our training data is small (about 3700) compared to distinct number of surface forms (about 52K) and distinct roots (about 15K including numbers), it makes sense to investigate the contribution of higher order n-gram language models for the morphological tag factor on the target side, to see if we can address the observation in the previous section.

Using the data transformed with *Noun+Adj+Verb+Adv+PostP* transformations which previously gave us the best results overall, we experimented with using higher order models (4- to 9-gram) during decoding, for the morphological tag factor models, keeping the surface and root models at 3-gram. We observed that for all the ten data sets, the improvements were consistent for up to 8-gram. The BLEU with the 8-gram *for only the morphological tag factor* averaged over the ten data sets was **22.61** (max: 23.66, min: 21.37, std: 0.72) compared to the 21.96 in Table 10.6 with a 3-gram language model for the morphological tag factor.

Using a 4-gram *root* language model, considerably less sparse than word forms but more sparse than tags, we get a BLEU score of **22.80** (max: 24.07, min: 21.57, std: 0.85). The details of the various BLEU scores are shown in the two halves of Table 10.7. It seems that larger n-gram LMs contribute to the larger n-gram precisions contributing to the BLEU but not to the unigram precision.

In order to alleviate the lack of large-scale parallel corpora for the English–Turkish language pair, we also experimented with augmenting the training data with reliable phrase pairs obtained from a previous alignment, as we did for the first approach earlier. This augmentation was applied to all ten data sets and the new models were trained. The resulting BLEU score was **23.78** averaged over ten data sets (max: 24.52, min: 22.25, std: 0.71).

10.4.3   Experiments with Constituent Reordering

The transformations in the previous section *do not perform any constituent level reordering*, but rather *eliminate* certain English function words as tokens in the text and fold them into complex syntactic tags. That is, no transformation reorders the English SVO order to Turkish SOV, for instance, or move post-nominal prepositional phrase modifiers in English, to prenominal phrasal modifiers in Turkish. Now that we have the parses of the English side, we also investigated a more comprehensive set of reordering transformations which perform the following constituent reordering to bring English constituent order more in line with the Turkish constituent order at the top and embedded phrase levels:

- Object reordering (*ObjR*), in which the objects and their dependents are moved in front of the verb but after the subject and its dependents.
- Adverbial phrase reordering (*AdvR*), which involves moving post-verbal adverbial phrases in front of the verb (and object if there was one).
- Passive sentence agent reordering (*PassAgR*), in which any post-verbal agents marked by *by* are moved in front of the verb (which would already have the complex tag marking the passivization).
- Subordinate clause reordering (*SubCR*) which involves moving post-nominal relative clauses or prepositional phrase modifiers in front of any modifiers of the head noun. Similarly any prepositional phrases attached to verbs are moved to in front of the verb.

An example English sentence with multiple reordering applied is presented in Fig. 10.8.[25] The first part of the figure contains an English sentence with its Turkish translation. The second part of Fig. 10.8 presents the constituent partitions of the English and Turkish sentences and the alignment between these constituent parts. The link between the aligned subordinate clauses causes a crossing with the other alignment links. Performing a subordinate clause reordering removes this crossing and returns aligned clauses. In addition to this top level reordering, there is also another misalignment within the subordinate clause as seen in the third part of Fig. 10.8. Performing an object reordering within the subordinate clause returns a *monotonic* alignment between English and Turkish constituents.

These reorderings were performed on top of the data obtained with the *Noun+Adj +Verb+Adv+PostP* transformations and used the same decoder parameters. Table 10.8 shows the performance obtained after various combination of reordering operations over the ten data sets. Although there were some improvements for certain cases, none of reordering gave consistent improvements for all the data sets. A cursory examination of the alignments produced after these reordering transformations indicated that the resulting root alignments were not necessarily that close to being monotonic as we would have expected.

---

[25]In order to provide a simple and clear representation, the example sentences contain the surface form of the words as opposed to the morphemic representation used earlier.

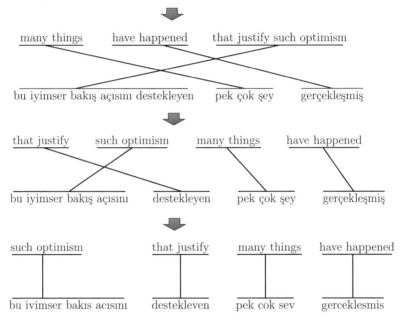

Fig. 10.8 An English sentence with multiple constituent reorderings applied

Table 10.8 BLEU scores after reordering transformations

Experiment	Ave.	STD	Max.	Min.
Baseline	21.96	0.72	22.91	20.67
ObjR	21.94	0.71	23.12	20.56
ObjR+AdvR	21.73	0.50	22.44	20.69
ObjR+PassAgR	21.88	0.73	23.03	20.51
ObjR+SubCR	21.88	0.61	22.77	20.92

Constituent reordering does not provide a significant improvement in the test scores but one wonders if it improves the alignment quality by producing more monotonic alignments. Therefore, we looked at the alignment files that were created by GIZA during the training. Two metrics were used in order to analyze the monotonicity of these alignments; (1) *absolute distance* metric which finds the absolute distance between the positions of two tokens of an alignment and (2) *crossing alignments* metric which is the number of times the links of alignments cross each other. In case of a monotonic alignment both of these metrics will return a value close to zero since the positions of aligned words in sentences will be close to each other with no other alignment link crossing the other alignment's link.

The frequency of the applied constituent reordering and the absolute distance and crossing alignments values are presented in Table 10.9. According to the table

Table 10.9 Average number of crossings and absolute distance

Experiment	Frequency	Crossings	Distance
No reordering	–	3.45	5.56
Obj	15,804	3.40	5.54
Obj+Adv	20,882	3.40	5.54
Obj+Passive	18,582	3.39	5.54
Obj+Subord	19,285	3.37	5.50

on average an alignment link can be crossed with 3 or 4 other alignment links. Furthermore, distance information indicates that on the average the alignment of the ith word is most probably somewhere close to $(i + 6)$th or $(i − 6)$th position of the translated sentence.

As seen in Table 10.9, performing different types of constituent reordering is slightly reducing both metrics. Due to the high frequency of object reordering, the reduction with object reordering is more than the others. Similarly the change with subordinate clause reordering is more than the change with adverb or passive reordering. The reason for this may be the length differences of these phrases. In subordinate clause, we move a whole clause while in adverb or passive, reordering is just limited with a couple of words.

10.5 Conclusions

This chapter presented an overview of statistical machine translation research on translating from English to Turkish. The two approaches described above both attempt to handle issues related to Turkish morphology but in two very different ways. The first one splits Turkish words into morphemes and uses standard techniques to transfer English into a sequence of Turkish morphemes which are then composed into Turkish words. While this looks intuitive, the decoder usually gets some of the morphology wrong as it does not distinguish between generating constituents and generating morphemes in a word.

The second approach essentially would like to generate whole Turkish words to avoid generating malformed sequence of morphemes, from relatively larger chunks of English words related syntactically in limited ways: it tries to identify English syntactic constructions that can map wholesale into Turkish words with complex morphology. In retrospect our impression is that while we identified quite many low-hanging alignable such patterns, it is not clear that our recall was high: the transformation rule base could have been extended by including many more rules but this would have been a very large manual text mining effort and any automatic ways of inducing rules through machine learning was not very obvious. It is possible this may be less effective if the available data is much larger, but we have reasons to believe that they may still be effective then also. The reduction in size of the source language side of the training corpus seems to be definitely effective and there is no reason why such a reduction (if not more) will not be observed in larger data. Also, the preprocessing of English prepositional phrases and many adverbial phrases usually involve rather long distance relations in the source side syntactic structure[26] and when such structures are coded as complex tags on the nominal or verbal heads, such long distance syntax is effectively "localized" and thus can be better captured with the limited window size used for phrase extraction.

Since these works, there have been some improvements in a number of directions. Nowadays there is significantly more parallel data involving Turkish, though not uniformly of high-quality. Among these we can point to the following:

- EUbookshop: Corpus of documents from the EU bookshop collected within the project LetsMT! available at opus.lingfil.uu.se/EUbookshop.php (Accessed Sept. 14, 2017). project.letsmt.eu (Accessed Sept. 14, 2017) includes small parallel corpora (varying in size from 1.1K to 23.7K) between Turkish and 31 languages.
- KDE: A parallel corpus of KDE4 localization files with more than 90 languages available at opus.lingfil.uu.se/KDE4.php (Accessed Sept. 14, 2017). For Turkish, there is parallel corpora available for several languages but the largest one is a Turkish–English parallel corpus with 153K sentence pairs and 1.38M words.
- SETIMES: A parallel corpus of news articles in the Balkan languages (including Turkish–English with 200K sentence pairs and 9.57M words) available at opus.lingfil.uu.se/SETIMES2.php (Accessed Sept. 14, 2017), originally extracted from www.eurasiareview.com/author/setimes/ (Accessed Sept. 14, 2017).
- OpenSubtitles: Corpus of subtitles for several movies and series available at opus.lingfil.uu.se/OpenSubtitles2013.php (Accessed Sept. 14, 2017), originally from www.opensubtitles.org (Accessed Sept. 14, 2017).
- PHP: A small corpora translated versions of the PHP manual available at opus.lingfil.uu.se/PHP.php (Accessed Sept. 14, 2017).
- Tanzil: Quran translations compiled by Tanzil Project, available at opus.lingfil.uu.se/Tanzil.php (Accessed Sept. 14, 2017).
- Tatoeba: Approximately 200K translated sentences by Tatoeba, available at opus.lingfil.uu.se/Tatoeba.php (Accessed Sept. 14, 2017).
- Ubuntu: A parallel corpus of Ubuntu localization documents, available at opus.lingfil.uu.se/Ubuntu.php (Accessed Sept. 14, 2017).
- GNOME: A small corpus of GNOME localization documents, available at opus.lingfil.uu.se/GNOME.php (Accessed Sept. 14, 2017).
- WIT: Web Inventory of Transcribed and Translated Talks (WIT) corpus is multilingual transcriptions of TED talks, available at wit3.fbk.eu (Accessed Sept. 14, 2017).

More recent works on Turkish have explored hierarchical or syntax-based methods for English–Turkish translation along with unsupervised segmentation approaches (Durgar-El Kahlout et al. 2012; Yılmaz and Durgar-El Kahlout 2014;

---

[26]For instance, consider the example in Fig. 10.4 involving *if* with some additional modifiers added to the intervening noun phrase.

Yılmaz et al. 2013). These and other related techniques will become effective, as more robust and wider-coverage language processing tools for Turkish become available.

References

Bisazza A, Federico M (2009) Morphological pre-processing for Turkish to English statistical machine translation. In: Proceedings of IWSLT, Tokyo, pp 129–135

Carpuat M (2009) Toward using morphology in French-English phrase-based SMT. In: Proceedings of WMT, Athens, pp 150–154

Chung T, Gildea D (2009) Unsupervised tokenization for machine translation. In: Proceedings of EMNLP, Singapore, pp 718–726

Corston-Oliver S, Gamon M (2004) Normalizing German and English inflectional morphology to improve statistical word alignment. In: Proceedings of AMTA, Washington, DC

Durgar-El Kahlout İ (2009) A prototype English-Turkish statistical machine translation system. PhD thesis, Sabancı University, Istanbul

Durgar-El Kahlout İ, Oflazer K (2006) Initial explorations in English to Turkish statistical machine translation. In: Proceedings of WMT, New York, NY, pp 7–14

Durgar-El Kahlout İ, Oflazer K (2010) Exploiting morphology and local word reordering in English-to-Turkish phrase-based statistical machine translation. IEEE Trans Audio Speech Lang Process 18(6):1313–1322

Durgar-El Kahlout İ, Mermer C, Doğan MU (2012) Recent improvements in statistical machine translation between Turkish and English. In: Vertan C, von Hahn W (eds) Multilingual processing in Eastern and Southern EU languages: low-resourced technologies and translation. Cambridge Scholars Publishing, Cambridge

Eyigöz E, Gildea D, Oflazer K (2013a) Multi-rate HMMs for word alignment. In: Proceedings of WMT, Sofia, pp 494–502

Eyigöz E, Gildea D, Oflazer K (2013b) Simultaneous word-morpheme alignment for statistical machine translation. In: Proceedings of NAACL-HLT, Atlanta, GA, pp 32–40

Goldwater S, McClosky D (2005) Improving statistical MT through morphological analysis. In: Proceedings of EMNLP, Vancouver, BC, pp 676–683

Koehn P, Och FJ, Marcu D (2003) Statistical phrase-based translation. In: Proceedings of NAACL-HLT, Edmonton, AB, pp 127–133

Koehn P, Hoang H, Birch A, Callison-Burch C, Federico M, Bertoldi N, Cowan B, Shen W, Moran C, Zens R, Dyer C, Bojar O, Constantin A, Herbst E (2007) Moses: open source toolkit for statistical machine translation. In: Proceedings of ACL, Prague, pp 177–180

Lee YS (2004) Morphological analysis for statistical machine translation. In: Proceedings of NAACL-HLT, Boston, MA, pp 57–60

Luong MT, Nakov P, Kan MY (2010) A hybrid morpheme-word representation for machine translation of morphologically rich languages. In: Proceedings of EMNLP, Cambridge, MA, pp 148–157

Mermer C, Akın AA (2010) Unsupervised search for the optimal segmentation for statistical machine translation. In: Proceedings of the ACL student research workshop, Uppsala, pp 31–36

Minkov E, Toutanova K, Suzuki H (2007) Generating complex morphology for machine translation. In: Proceedings of ACL, Prague, pp 128–135

Naradowsky J, Toutanova K (2011) Unsupervised bilingual morpheme segmentation and alignment with context-rich hidden semi-markov models. In: Proceedings of ACL-HLT, Portland, OR, pp 895–904

Nguyen T, Vogel S, Smith NA (2010) Nonparametric word segmentation for machine translation. In: Proceedings of COLING, Beijing, pp 815–823

Niessen S, Ney H (2004) Statistical machine translation with scarce resources using morpho-syntatic information. Comput Linguist 30(2):181–204

Nivre J, Hall J, Nilsson J, Chanev A, Eryiğit G, Kübler S, Marinov S, Marsi E (2007) MaltParser: a language-independent system for data-driven dependency parsing. Nat Lang Eng 13(2):95–135

Oflazer K (1994) Two-level description of Turkish morphology. Lit Linguist Comput 9(2):137–148

Oflazer K (1996) Error-tolerant finite-state recognition with applications to morphological analysis and spelling correction. Comput Linguist 22(1):73–99

Oflazer K, Durgar-El Kahlout İ (2007) Exploring different representational units in English-to-Turkish statistical machine translation. In: Proceedings of WMT, Prague, pp 25–32

Papineni K, Roukos S, Ward T, Zhu WJ (2002) BLEU: a method for automatic evaluation of machine translation. In: Proceedings of ACL, Philadelphia, PA, pp 311–318

Popovic M, Ney H (2004) Towards the use of word stems and suffixes for statistical machine translation. In: Proceedings of LREC, Lisbon, pp 1585–1588

Sadat F, Habash N (2006) Combination of Arabic preprocessing schemes for statistical machine translation. In: Proceedings of COLING-ACL, Sydney, pp 1–8

Schmid H (1994) Probabilistic part-of-speech tagging using decision trees. In: Proceedings of the international conference on new methods in language processing, Manchester

Stolcke A (2002) SRILM – an extensible language modeling toolkit. In: Proceedings of ICSLP, Denver, CO, vol 2, pp 901–904

Talbot D, Osborne M (2006) Modelling lexical redundancy for machine translation. In: Proceedings of COLING-ACL, Sydney, pp 969–976

Tantuğ AC, Oflazer K, Durgar-El Kahlout İ (2008) BLEU+: a tool for fine-grained BLEU computation. In: Proceedings of LREC, Marrakesh, pp 1493–1499

Toutanova K, Klein D, Manning CD, Singer Y (2003) Feature-rich part-of-speech tagging with a cyclic dependency network. In: Proceedings of NAACL-HLT, Edmonton, AB, pp 252–259

Yang M, Kirchhoff K (2006) Phrase-based backoff models for machine translation of highly inflected languages. In: Proceedings of EACL, Trento, pp 41–48

Yeniterzi R (2009) Syntax-to-morphology alignment and constituent reordering in factored phrase-based statistical machine translation from English to Turkish. Master's thesis, Sabancı University, Istanbul

Yeniterzi R, Oflazer K (2010) Syntax-to-morphology mapping in factored phrase-based statistical machine translation from English to Turkish. In: Proceedings of ACL, Uppsala, pp 454–464

Yılmaz E, Durgar-El Kahlout İ (2014) The use of recurrent neural networks language model in Turkish-English machine translation. In: Proceedings of IEEE signal processing and communications applications conference, Trabzon, pp 1247–1250

Yılmaz E, Durgar-El Kahlout İ, Aydın B, Özil ZS (2013) TÜBİTAK Turkish-English submissions for IWSLT 2013. In: Proceedings of IWSLT, Heidelberg, pp 152–159

Yuret D, Türe F (2006) Learning morphological disambiguation rules for Turkish. In: Proceedings of NAACL-HLT, New York, NY, pp 328–334

Zollmann A, Venugopal A, Vogel S (2006) Bridging the inflection morphology gap for Arabic statistical machine translation. In: Proceedings of NAACL-HLT, New York, NY, pp 201–204

Chapter 11
Machine Translation Between Turkic Languages

A. Cüneyd Tantuğ and Eşref Adalı

Abstract Turkish belongs to the Turkic family of languages and these languages exhibit tremendous similarity when it comes to morphological and grammatical structure but have somewhat different lexicons owing to various historical, geographical, and cultural interactions with neighboring languages. In this chapter we briefly cover the similarities and differences of these languages and introduce a machine translation methodology that exploits the similarities among these languages. This methodology relies on rule-based and statistical components and can be applicable for not only Turkic languages but also any other cognate language pairs.

11.1 Introduction

Contrary to machine translation between unrelated languages (such as say Turkish and English), machine translation between closely-related languages would conceivably be relatively easier given substantial similarities between their lexical stock and their morphological and syntactic structures. Usually, even a rule-based word level machine translation system between cognate languages can produce high-quality outputs by taking advantage of the linguistic similarities in morphology and syntactic structures.

There have been a number of studies on machine translation between related languages like Czech-Russian (Hajič 1987), Czech-Slovak (Hajič et al. 2000, 2003), Spanish-Portuguese (Garrido-Alenda et al. 2003; Corbi-Bellot et al. 2005; Homola and Kuboň 2008), Catalan-Aranese Occitan (Oller and Forcada 2006), Irish-Scottish Gaelic (Scannell 2006), Czech-Lower Serbian and Macedonian (Dvořák et al. 2006). Additionally, recent studies have tried to incorporate the advantages of language similarities in SMT (Tiedemann 2009; Nakov and Tiedemann 2012).

A. C. Tantuğ (✉) · E. Adalı
Istanbul Technical University, Istanbul, Turkey
e-mail: tantug@itu.edu.tr; adali@itu.edu.tr

© Springer International Publishing AG, part of Springer Nature 2018   237
K. Oflazer, M. Saraçlar (eds.), *Turkish Natural Language Processing*,
Theory and Applications of Natural Language Processing,
https://doi.org/10.1007/978-3-319-90165-7_11

There are also a number of studies on machine translation between Turkic languages. Hamzaoğlu (1993) and Fatullayev and Shagavatov (2008) have developed Azerbaijani-Turkish systems. Altıntaş (2000) and Altıntaş and Güvenir (2003) describe Crimean Tatar-Turkish machine translation systems. Tantuğ et al. (2007); Durgar-El Kahlout (2008) have developed Turkmen-Turkish machine translation system which we describe later in the chapter. Finally we note the recent work by Tyers et al. (2012) on Tatar-Bashkir machine translation and Salimzyanov et al. (2013) on Kazakh-Tatar machine translation.

Most of these systems are direct translation systems as when parallel sentences in these languages exhibit predominantly monotonic alignments. Thus a rule-based lexical transfer approach with additional rules for some exception handling and complementary statistical language modeling often works.

In this chapter, after a short overview of Turkish languages, we describe a machine translation system from a resource poor Turkic language, Turkmen, to Turkish. The ideas and resources used in this system can certainly be adapted to other Turkic languages. For more detailed information about Turkic languages, including aspects not necessarily related to machine translation, we refer the reader to one of the widely-available resources.[1]

11.2   Turkic Languages

As a sub-family of Altaic language family, Turkic Language Family comprises 34 languages in total. Figure 11.1 shows the family tree of Turkic languages, where the major Turkic languages with more than 5 million speakers are shown in boldface, whereas the extinct languages (with less than 5000 speakers) are shown in italic face.[2] Table 11.1 lists the major Turkic languages along with the number of native speakers and regions they are spoken.

11.2.1   Similarities and Differences of Turkic Languages

Although the Turkic languages are to a large extent similar at many linguistic levels, the differences between these languages are not negligible. Throughout the history, these languages have been influenced by other non-Turkic languages due to religious, political, and economical influences. For instance, an intense Russian influence can easily be observed for most of the Turkic languages as the regions they were spoken were part of the USSR (and are still parts of Russia), while Arabic, Greek, French, and English have had more influence on Turkish in Turkey, Middle East, and Eastern Europe.

---

[1]For example, en.wikipedia.org/wiki/Turkiclanguages (Accessed Sept. 14, 2017).

[2]Note that the minor variants of these languages are not shown for the sake of clarity.

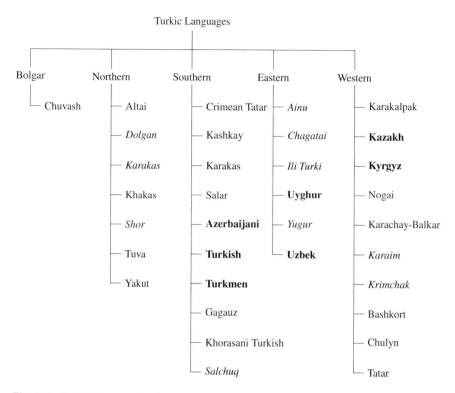

Fig. 11.1   Turkish language family

Table 11.1   Major Turkic languages

Language	Region	Scripts	# of speakers
Turkish	Turkey, Cyprus, Greece, Macedonia	Latin	70.8M
Azerbaijani	Azerbaijan, Iran	Latin, Arabic	24.2M
Uzbek	Uzbekistan, Afghanistan, Tajikistan	Latin, Cyrilic, Arabic	21.9M
Uyghur	China, Kazakhstan	Arabic	8.7M
Kazakh	Kazakhstan	Cyrilic, Latin, Arabic	8.0M
Turkmen	Turkmenistan, Azerbaijan, Iran	Latin	7.5M
Kyrgyz	Kyrgyzstan, Tajikistan, Afghanistan, China	Cyrilic, Arabic	2.9M

The oldest alphabet of Turkic peoples is known as the Göktürk Alphabet. This alphabet can be seen on the eighth century monuments of Orhon, Yenisey, and Talas which are presently in Mongolia. After the waning of the Göktürk state, Uyghurs produced a new alphabet named Uyghur. Over time, Turkic peoples adopted Arabic, Cyrillic, and Latin alphabets depending on religious or political reasons.

As Table 11.1 shows, languages in the Turkic family are written with various scripts like Latin, Arabic, and Cyrilic mostly due to geographical and political reasons. Although different scripts are used for Turkic languages, the morphological structures are very close. All Turkic languages have a very productive inflectional

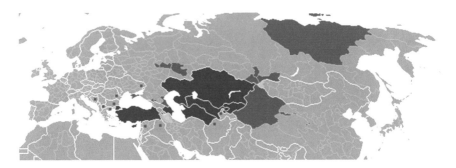

Fig. 11.2 Geographical map of Turkic language countries (Source: Wikipedia)

Table 11.2 Past definitive tense

	1Sg	2Sg	3Sg	1Pl	2Pl	3Pl	Negative
TUR	gel-di-m	gel-di-n	gel-di	gel-di-k	gel-di-niz	gel-di-ler	gel-me-di-m
	oku-du-m	oku-du-n	oku-du	oku-du-k	oku-du-nuz	oku-du-lar	oku-ma-du-m
AZE	gəl-di-m	gəl-di-n	gəl-di	gəl-di-k	gəl-di-niz	gəl-di-lər	gəl-mə-di-m
	oxu-du-m	oxu-du-n	oxu-du	oxu-du-q	oxu-du-nuz	oxu-du-lar	oxu-ma-dı-m
TKM	gel-di-m	gel-di-ň	gel-di	gel-di-k	gel-di-ňiz	gel-di-ler	gel-me-di-m
	oka-dy-m	oka-dy-ň	oka-dy	oka-dy-k	oka-dy-ňyz	oka-dy-lar	oka-ma-dy-m
UZB	kel-di-m	kel-di-ng	kel-di	kel-di-k	kel-di-ngiz	kel-ishdi-lar	kel-ma-di-m
	oqi-di-m	oqi-di-ng	oqi-di	oqi-di-k	oqi-di-ngiz	oqi-shdi-lar	oqi-ma-di-m
KAZ	gel-di-m	kel-di-ň	kel-di	kel-di-k	kel-di-ňder	kel-di	kel-me-di-m
		kel-di-ňiz			kel-di-ňizder		
	oqı-dı-m	oqı- dı-ň	oqı-dı	oqı-dı-k	oqı-dı-ňdar	oqı-dı	oqı-ma-dı-m
		oqı-dı-ňız			oqı-dı-ňızdar		
KYR	kel-di-m	kel-di-ň	kel-di	kel-di-k	kel-di-ňiz(der)	kel-iş-ti	kel-be-di-m
	oqu-du-m	oqu-du-ň	oqu-du	oqu-du-k	oqu-du-ňuz(dar)	oqu-ş-tu	oqu-ba-dı-m
UYG	kel-di-m	kel-di-ň	kel-di	kel-di-k	kel-di-ňlar	kel-di	kel-mi-di-m
	oqu-dı-m	oqu-dı-ň	oqu-dı	oqu-dı-q	oqu-dı-ňlar	oqu-dı	oqı-mi-dy-m
TAT	kil-de-m	kil-de-ň	kil-de	kil-de-k	kil-de-gez	kil-de-lär	kil-mä-de-m
	ukı-dı-m	ukı-dı-ň	ukı-dı	ukı-dı-k	ukı-dı-gız	ukı-dı-lar	ukı-ma-dı-m

and derivational morphology where suffixes are affixed to a root word or to another suffix. While suffixes can be different among Turkic languages, the morphophonology and morphotactics rules are nearly same for all Turkic languages. For example, all Turkic languages have some kind of vowel harmony and consonant mutation rules and have almost the same order of morphemes. There is no lexical gender. In syntax, the unmarked constituent order is Subject Object Verb (SOV), those all permutations are essentially possible (Fig. 11.2).

In order to show the morphological similarities between these languages, we present in Table 11.2, a few forms of the basic past tense forms of two verbs gel- (come) and oku- (read) in eight Turkic languages.

Table 11.3 Example sentences in Turkic languages

TUR	Dağ	dağa	kavuşmaz,	insan	insana	kavuşur.
AZE	Dağ	dağa	qovuşmaz,	insan	insana	qovuşar.
TKM	Dag	daga	duşmaz,	adama	adama	duşar.
KAZ	Taw	tawğa	qosılmas,	adam	adamğa	qosıladı.
KIR	Too	too	menen körüşpöyt,	adam	adam	körüşöt.
TAT	Taw	tawga	kilmäs,	adäm	adämgä	oçrar.
UYG	Tagh	tagh bilen	tipishalmas,	insan	insan bilen	tipishar.
UZB	Tog'ning	ko'rki	tosh bilan,	odamning	ko'rki	bosh bilan.
ENG	mountain	mountain	meet	human	human	meet
GLOSS	+Nom	+Dat	+Neg+Aor,	+Nom	+Dat	+Pos+Aor.

TUR	Ağır	kazan	geç	kaynar.
AZE	Ağır	qazan	gec	qaynayar.
TKM	Agyr	gazan	giç	qaýnar.
KAZ	Awur	qazan	keş	qaynaydı
KIR	Oor	kazan	keç	kaynayt.
TAT	Avır	kazan	ozak	kaynıy.
UYG	Eghir	qazan	waqche	qaynaydu.
UZB	Teran	daryo	tinch	oqar.
ENG	heavy/big	cauldron	late	boils.
GLOSS				

Fig. 11.3 An alignment example between Turkish and Turkmen parallel sentences

Table 11.3 shows two example sentences in six different Turkic languages. While there exist some lexical and morphological differences, this example properly demonstrates that the sentences display significant similarities.

From the point of view of syntactical structure, an almost one-to-one mapping can be observed between Turkic languages. However, word-by-word correspondence fails in many situations because of one-to-many or many-to-one mappings stem from multi-word expressions (MWEs). For example, in many cases with adjective participles, it is inevitable to change the position of some morphemes among other words in the adjectival phrase between Turkmen and Turkish. A sample alignment is given in Fig. 11.3.

In this alignment, one can readily see the replacement of Turkmen +*iñiz*[3] morpheme with its Turkish equivalent +*iniz*, and also the positional change of the

---

[3] Second person plural possessive agreement suffix.

morpheme from the noun to the participle adjectival form. Additionally, a typical instance of a case where SL MWE is aligned with a single TL word occurs in the end of the example sentence. These and many other examples show that in spite of the syntactical similarities, word-by-word translation is not sufficient solely, and additional sentence level processing must be employed.

Since the origins of the Turkic languages are same, their lexicons also share considerable amount of root words, sometimes with only minor variations. Most of the variations are observed in orthography whereas spoken languages have more common patterns. Personal pronouns, date/time expressions, organ names, main color names, numbers are nearly same for all Turkic languages. However, due to the strong Russian and Arabic influence on some of the Turkic languages, root words borrowed from these are widely in use, specifically for the technical and political terms.

11.3 Machine Translation Between Turkic Languages

In this section we describe a machine translation system architecture between Turkic languages. It is also possible to use this methodology to build a machine translation system between any closely-related language pair. The proposed system relies on morphological and lexical transfer and have both rule-based and statistical components.

Basically, the proposed system can be considered as a direct translation system that is capable of translating the morphological structures as well as the root words. For Turkic languages, the system relies on morphological pre-processing and post-processing steps.

A direct translation is straightforward except for the ambiguities generated. Our general approach is to perform (ambiguous) morphological analysis on the source side, where minimally we expect that a morphological analyzer exists and then ambiguously encode into a lattice without necessarily resolving any ambiguities (if we do not have a morphological disambiguator for the source language) and then transfer them to the target language (Turkish in this case) by translating the root words again ambiguously. We then use target side statistical models to help disambiguate. Similarities in word orders mean that almost no word order rearrangements need to made, but, of course there are quite many exceptions that we also handle with rules during the transfer.

The proposed system architecture is presented in Fig. 11.4.

11.3.1 Preprocessing

The first two components of the system execute standard procedures to split the input text into sentences and tokenize the elements of the input sentences. Although the sentence splitting and tokenization tools developed for the source language are

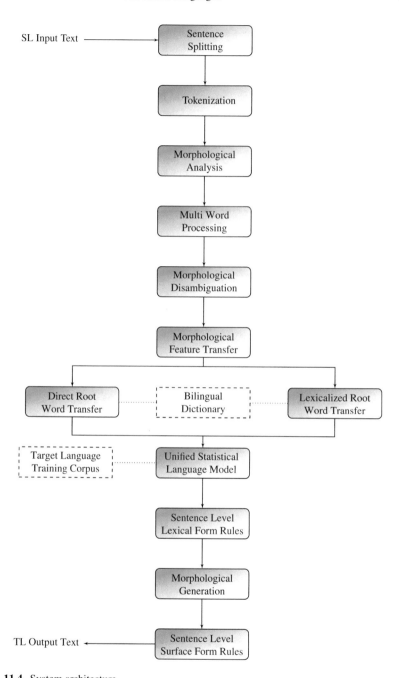

Fig. 11.4 System architecture

preferable, language independent sentence segmentation and tokenization modules can also be used if no language specific tools are available.

The next module in the pipeline is the source language morphological analyzer. This is a compulsory component in this architecture that transduces input tokens to possible morphological parses. So, in order to build a machine translation system between any Turkic language pair, a source language morphological analyzer must be acquired in the minimum configuration.

Another important module in this architecture is the multi-word expression identification module. A multi-word expression is a sequence of words forming a phrase that is not entirely predictable on the basis of standard grammar rules and lexical components. MWE processing in agglutinative languages is a quite difficult task owing to unlimited number of surface forms. Quick look-up lists for MWEs are of limited use for Turkic languages since the components of MWEs can suffer a derivational and/or inflectional process. The types of MWEs and identification strategies are well-studied for Turkish (Oflazer et al. 2004) (see also Chap. 2). The same approach has also been used for Turkmen (Tantuğ et al. 2007). These studies show that the MWEs in other Turkic languages can be processed in the similar fashion. Oflazer classified Turkish MWEs into three different categories:

1. Lexicalized MWEs
2. Semi-Lexicalized MWEs
3. Non-Lexicalized MWEs

Identification of a lexicalized MWE is a trivial task since the lexical items constituting the MWE are fixed. A simple lookup list will be sufficient to determine this type of MWEs. However, semi-lexicalized collocations can undergo any kind of morphological process. For example, Turkish equivalents of Turkmen noun *gürüm-jürüm* and verb *bol-* are *gizli (hidden)* and *ol- (to be)*, respectively. On the other hand, their combination *"gürüm-jürüm bol-"* should be translated as *kaybol- (to lose)*. But, the phrase *"gürüm-jürüm bol-"* occurs usually in inflected and/or derived forms along the running text:

gürüm-jürüm bolypdyrsyñ	kaybolmuşsundur
	(you must have been lost)
gürüm-jürüm boldy	kayboldu
	((he) is lost)

In non-lexicalized type of MWEs, morphosyntactic patterns are used to construct the collocation. Some modal formations are expressed by the combination of two or more words, one or both of which is a verb. For instance, when placed after another verb ending +*yp*/+*ip*, the verb *ber- (to give (to))* indicates that an action is performed for the benefit of someone else, while the same formation with *al- (to take (from))* indicates an action performed for oneself as seen.

Source language root word and morphological features which are not enclosed by the upper dotted box are variable parts of this type of collocation. While combining

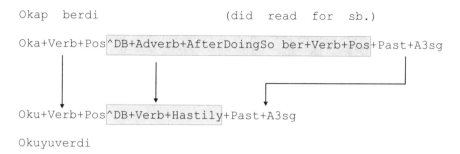

Fig. 11.5 MWE translation example

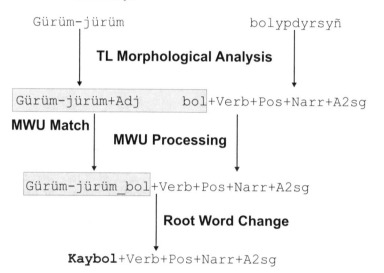

Fig. 11.6 MWE translation example

these two words, MWE recognizer also transforms the inside of the box into an appropriate TL structure.

Determining MWEs in the source side improves the translation performance since the alignment in Turkic languages are sometimes one-to-many. Figures 11.5 and 11.6 show two examples of the translation process of multi-word expressions from Turkmen as the source language to Turkish.

11.3.2 Morphological Disambiguation

The outputs of any Turkic language morphological analyzer are usually ambiguous; thus, a morphological disambiguation module should be integrated after the analyzer. In practice, such a disambiguation tool for any Turkic language other than Turkish is not available. For these cases, an appropriate strategy might be preserving the input side morphological ambiguities through the translation process and applying statistical techniques on the target side for resolving these ambiguities. So, while having a morphological disambiguation module is beneficial, the proposed architecture is still valid even in the absence of the disambiguation component.

11.3.3 Morphological Feature Transfer

The first step of the actual translation starts with the transferring of morphological features. Although the morphological structures are similar for Turkic languages, a module is required for morpheme translations and occasional reorderings. This module may be based on hand-crafted rules prepared using the contrastive knowledge between languages. This translation step is generally one-to-one, so the output of the morphological transferring module is usually not ambiguous.

11.3.4 Lexical Transfer

The second step of the translation involves the translation of root words. The root word transfer module takes all possible morphological analyses of input source language words and replaces the root of the parse with one or multiple target word(s) selected from the bilingual root word transfer dictionary. Therefore, one-to-many mappings taking place in this process cause another type of ambiguity, namely *lexical ambiguity*. The part-of-speech of the root word may be taken into account when performing this mapping, so that spurious mappings based on just the written form can be eliminated.

The following example root transfer rules contain mapping of the Turkmen root *boz-* to different Turkish root words based on its part-of-speech: it maps to the Turkish adjective *gri (gray)* as an adjective and to the Turkish verb *sil (erase)* as a verb. Some entries from such a bilingual dictionary implemented with Xerox Regular Expression Language (Karttunen et al. 1997) are given below:

```
define AdjDict      gri <- boz \/ _ +Adj
                    ...
define VerbDict     sil <- boz \/ _ +Verb
                    söyle <- geple \/ _ +Verb
                    konuş <- geple \/ _ +Verb
                    ...
```

The second set of rules also shows that the verb *geple* has two corresponding Turkish entries, *söyle (say)* and *konuş (talk)*, producing ambiguous outputs, whereas in the former case, ambiguity is resolved by the help of POS information.

For some words, a simple replacement of the root word cannot produce a legal word form in the generation stage. This stems usually from the productive structure of the Turkic languages. For the sake of clarity, consider the following Turkmen word *ulumsylyk (vanity)* and its morphological analysis.

```
ulumsylyk ulumsy+Adj^DB+Noun+Ness+A3sg+Pnon+Nom
```

In the bilingual dictionary, *kibirli (arrogant)* is the corresponding Turkish root word for Turkmen root word *ulumsy*. However, a direct replacement of Turkmen root with its Turkish counterpart causes a failure in generation stage, because *kibirli* is not a legal root word in Turkish. In fact, *kibirli* is actually derived from the original root *kibir (arrogance)* with the suffix *+li (with)*. So, to produce the right word form, a special lexicalized rule is required which replaces the *ulumsy+Adj* structure with the proper morphological representation of *kibirli*:

```
kibir+Noun+A3sg+Pnon+Nom^DB+Adj+With <- ulumsy+Adj
```

This type of lexicalized rules are not necessary for the source words which are derived by the suffixes that have direct equivalents in TL with same semantics, such as *+lyk* (Turkmen) and *+lık* (Turkish) suffix pair.

A sample of root word and morphological transfer process is given in Fig. 11.7.

In some rare cases, mapping the available morphological features and translating the root word are not sufficient to generate a legal Turkish lexical structure as

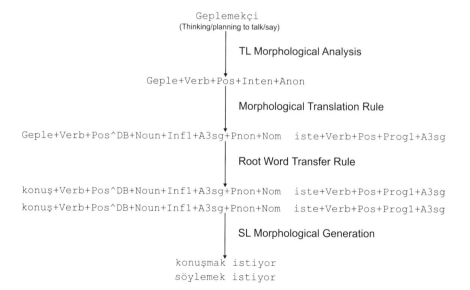

Fig. 11.7 Morphological transfer example

sometimes some required feature on the target side may not be explicitly available on the source word. For such a case, we use rules that look at much wider context, mostly using additional heuristics to infer such features.

11.3.5   Statistical Disambiguation Module

To resolve both source side morphological ambiguities and lexical transfer ambiguities, we employ statistical language models (LM) on the target language side. A LM is normally generated by using surface forms; but this causes serious data sparseness problems for Turkic languages due to the agglutinative structure as the vocabulary size is quite large. Instead of building a single LM to model the full word forms, we have employed a unified language modeling concept where different granularity LMs are utilized to model different parts of the language. As a first step, target training corpora are morphologically analyzed and disambiguated to build various types of LMs. For example, one type of LM which is trained on only disambiguated root words can play an effective role in solving lexical ambiguity problems. In Fig. 11.8, we show roots in Turkmen sentence with their Turkish root translations in a lattice. The transition probabilities come from the LM probabilities. The most probable path (corresponding to the most probable root word translation) can be found by standard algorithms. Table   11.4 shows the decoding outputs of the example sentence in Fig. 11.8, where the bold sentence indicates the right translation. The bigram LM achieves to resolve the ambiguities so that the right translation gained the first rank and is selected as the output.

Similarly, source language morphological ambiguities can be resolved by LMs trained on other morphological features (for instance, the last set of inflectional groups in the analyses or full morphological features except the root word). Input of this statistical processing module is an ordered bag of all possible translations of the input sentence, including all kinds of ambiguities. LM based disambiguation

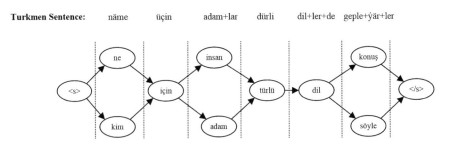

English Equivalent : *Why do people talk in various languages ?*

Fig. 11.8 Language model for decoding of translation

Table 11.4 Decoding results for the sample sentence

LM order	Rank	Most probable sentence
Unigram	1	ne için insanlar türlü dillerde söylüyorlar
	2	**ne için insanlar türlü dillerde konuşuyorlar**
	3	ne için adamlar türlü dillerde söylüyorlar
Bigram	1	**ne için insanlar türlü dillerde konuşuyorlar**
	2	ne için insanlar türlü dillerde söylüyorlar
	3	ne için adamlar türlü dillerde söylüyorlar
Trigram	1	**ne için insanlar türlü dillerde konuşuyorlar**
	2	ne için insanlar türlü dillerde söylüyorlar
	3	ne için adamlar türlü dillerde söylüyorlar

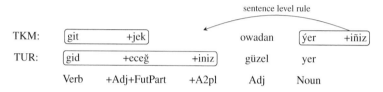

Fig. 11.9 Sentence level processing of a participle phrase

module would then try to find the contextually most appropriate morphological interpretations.

11.3.6 Sentence Level Rules

Sentence level rules are the rules that implement some modifications to overcome the problems of word-by-word transfer paradigm. These rules do some sentence level work such as identifying participle phrases and making morpheme arrangements within the phrase. For this, instead of a parser, we employ some chunking rules just for finding certain phrase patterns. In the following example, a morpheme of an adjective phrase discovered in the original sentence is replaced and relocated to its expected grammatical place in Turkish (Fig. 11.9).

Certainly, these kinds of rules require the morphological representation of the sentence. However, a small set of sentence level rules require the surface realizations of the words. Mostly, this set of rules deal with orthographic changes caused by the phonetic interactions between words. Note that the set of rules operating on surface forms must be performed after the target language morphological generation stage.

11.3.7   Morphological Generation

Morphological representation of the translated target language sentence is transformed into the final representation by generating surface forms of the words using a target language morphological generator. We have used a well-known wide-coverage Turkish morphological analyzer (Oflazer 1994) in reverse (generation) direction to synthesize resulting word forms. Except for very few cases, each Turkish morphological parse maps to only one surface form, hence no ambiguity arises during this generation phase.

11.4   Machine Translation Evaluation on Turkic Languages

The usability of a machine translation system is usually dependent on the quality of the output of the system. Machine translation evaluation is still an active research area with efforts to discover better metrics. Generally, automatic machine translation performance evaluation metrics aim to measure both the level of adequacy and the fluency of the translations. However, the majority of the current automatic evaluation metrics such as BLEU, NIST, METEOR, WER, TER are based on matching of surface forms or root words (ignoring morphological structure) which causes serious drawbacks for the morphologically rich languages. A considerable amount of semantic information conveyed in the morphology is not evaluated in these metrics. So, many alternative evaluation methods are suggested to alleviate these shortcomings for languages with complex morphology (El Kholy and Habash 2011; Bouamor et al. 2014; Kos and Bojar 2009).

BLEU+ is a machine translation evaluation metric designed for agglutinative languages and also suitable for Turkic languages (Durgar-El Kahlout 2008). Although BLEU+ is based on the well-known BLEU metric, BLEU+ suggests using a similarity score between 0 and 1 for word matching instead of using straight match and no match for words (and also bigrams, trigrams, etc.). A 0 similarity score means completely different words, and 1 means completely same wordforms. Design of such a re-modified precision function requires the morphological representation of the word forms. This representation must contain a root word and, if any, other suffix(es). Table 11.5 shows an example machine translation output and reference sentence both in surface form representation and morphemic representation. Note that the suffixes are normalized to simplify the morphophonological variations.

Given a candidate translation and a set of reference sentences in morphemic representation, a re-modified precision score is computed by a novel matching metric which compares the morphological representation of two words, and produces a similarity score.

$$similarity(w_i, w_j) = similarity_{root}(w_i, w_j) \cdot similarity_{suffix}(w_i, w_j) \qquad (11.1)$$

Table 11.5 An example machine translation candidate output and reference translation

Output:	iki aile arasındaki husumet ve kavga uzun yıllardır sürüyordu.
	iki aile ara+sh+nda+ki husumet ve kavga uzun yıl+lar+dhr sür+hyor+dh.
Ref.:	iki aile arasında düşmanlık ve çatışma uzun senelerdir sürmekteydi.
	iki aile ara+sh+nda düşmanlık ve çatışma uzun sene+lar+dhr sür+makta+ydh.
	(The hostility and fight between two families was lasting for many years.)

This similarity metric makes use of the morphological units of the word forms: both the root word and suffix(es). The two word forms in comparison must have *compatible* root words to have a non-zero similarity score. Another interpretation of this is that if the root words are not similar, other morphological similarities will be ignored and the result will be no match.

11.4.1 Root Matching

Word choice freedom in translation process usually triggers a problematic issue in automatic machine translation evaluation. If machine translation output contains a synonym of the reference word, this cannot be recognized as a match and causes a reduction in the overall evaluation score. BLEU counts on multiple references to attack this problem by the assumption that reference translations may contain the possible synonymous words. Unfortunately, current BLEU based machine translation evaluations are generally carried out with only one reference. So, we opted for using WordNET as a solution for finding synonyms, similar to the METEOR metric with a minor change. We defined a penalty value $penalty_s$ for this kind of root match cases. In the example above, *yıl* and *sene* are synonyms of each other and also same suffix combination is affixed after each root word. The resulting word forms *yıllardır* and *senelerdir* have exactly the same meaning. Although they cannot contribute to BLEU score, very close hypernyms of the reference root words sometimes occur in the candidate, and still maintain the reference meaning. BLEU+ classifies this sort of hypernyms as compatible root words whereas the semantic difference is penalized with a penalty weight. The longer the hypernym relation distance is between the root words, the harsher they are penalized.

The general formulation of root matching strategy is given below:

$$similarity_{root}(w_i, w_j) = \begin{cases} 1 & \text{if } r_i = r_j \\ penalty_s & \text{if } r_i \text{ is synonym of } r_j \\ penalty_{hk} & \text{if } r_i \text{ is a hypernym of } r_j \text{ at with a distance of k} \\ & \text{or vice versa} \\ 0 & \text{otherwise} \end{cases}$$

$$(11.2)$$

11.4.2 Feasible Suffix Pairs

When the morphology of a language is rich, some of the different morphemes can convey the same or very close morphosemantics, even some subsets of them can be used interchangeably. We name this kind of morphemes as *feasible morpheme pairs*. A list of feasible morpheme pairs is generated with the linguistic knowledge. This list is a collection of ternary tuples, (m_{i1}, m_{i2}, p_i) where m_{i1} and m_{i2} are feasible morphemes of the ith item in the list and p_i is the semantic similarity ratio on a scale of 0–1. Table 11.6 shows a portion of the Turkish feasible morpheme pairs.

To compare two word forms whose root words are compatible, the following formula is used (Table 11.7):

$$similarity_{suffix}(w_i, w_j) = \prod_i^{n_1 \text{ or } n_2} c(s_{1i}, s_{2i}) \tag{11.3}$$

$$c(m_1, m_2) = \begin{cases} 1 & \text{if } m_1 = m_2 \\ p_i & \text{if } (m_1, m_2) \text{ is the } i\text{th element in the list} \\ 0 & \text{otherwise} \end{cases} \tag{11.4}$$

$$similarity(\text{gelebilmekteydi, gelebiliyormuş}) = 1.00 \times 1.00 \times 0.95 \times 0.70 = 0.665 \tag{11.5}$$

Table 11.6 Feasible morpheme pairs in Turkish

Morpheme 1	Example 1	Morpheme 2	Example 2	Similarity score
+hyor	gel+hyor	+makta	gel+makta	0.95
	geliyor		gelmekte	
	he is coming		he is (in a state of) coming	
+yacak	gel+yacak	+yacak+dhr	gel+yacak+dhr	0.90
	gelecek		gelecektir	
	he will come		he *will* come	
+dh	gel+dh	+mhs	gel+mis	0.70
	geldi		gelmiş	
	he came (past tense)		he came (hearsay past tense)	

Table 11.7 Feasible morpheme pairs in Turkish

gelebilmekteydi	gel	+abil	+makta	+ydh
	↓	↓	↓	↓
	1.00	1.00	0.95	0.70
	↓	↓	↓	↓
gelebiliyormuş	gel	+abil	+hyor	+muş

Since some morphemes may have same spelling with different morphosemantics, we resort to limiting contexts of the feasible pairs by specifying the applicable part-of-speech tags explicitly. However, owing to the lack of POS information in morphemic representations, an external source like WordNET is required.

11.5 Conclusions

The morphological and syntactical similarities shared by Turkic languages make machine translation between these languages relatively easier. Our solution to translation between these language relies on a methodology that takes into account the lack of NLP tools and corpora for many of these languages. This hybrid approach employs finite-state fashioned rule-based components and statistical language model components with most of the disambiguation being done on the Turkish side where more tools and resources are available. Additionally, we introduced BLEU+, a machine translation evaluation metric for Turkic languages. This metric is similar to BLEU, but it incorporates both root similarities and morphological similarities, which are crucial for evaluating translations also at the morphological level.

References

Altıntaş K (2000) Turkish to Crimean-Tatar machine translation system. Master's thesis, Bilkent University, Ankara

Altıntaş K, Güvenir HA (2003) Learning translation templates for closely related languages. In: Proceedings of knowledge-based intelligent information and engineering systems, Oxford, pp 756–762

Bouamor H, Alshikhabobakr H, Mohit B, Oflazer K (2014) A human judgement corpus and a metric for Arabic MT evaluation. In: Proceedings of EMNLP, Doha, pp 207–213

Corbi-Bellot AM, Forcada ML, Ortíz-Rojas S, Pérez-Ortiz JA, Ramírez-Sánchez G, Sánchez-Martínez F, Alegria I, Mayor A, Sarasola K (2005) An open-source shallow-transfer machine translation engine for the romance languages of Spain. In: Proceedings of EAMT, Budapest

Dvořák B, Homola P, Kuboň V (2006) Exploiting similarity in the MT into a minority language. In: Proceedings of the workshop on minority languages: "strategies for developing machine translation for minority languages", Genoa, pp 59–64

El Kholy A, Habash N (2011) Automatic error analysis for morphologically rich languages. In: Proceedings of the MT summit, Xiamen, pp 225–232

Fatullayev A, Shagavatov S (2008) Turkish-Azerbaijani translation module of Dilmanc MT system. In: Proceedings of the second international conference on problems of cybernetics and informatics, Baku, Azerbaijan, pp 150–153

Garrido-Alenda A, Gilabert-Zarco P, Pérez-Ortíz JA, Pertusa-Ibáñez A, Ramírez-Sánchez G, Sánchez-Martínez F, Scalco MA, Forcada ML (2003) Shallow parsing for Portuguese-Spanish machine translation. In: Proceedings of the workshop on tagging and shallow processing of Portuguese, Lisbon, pp 135–144

Hajič J (1987) RUSLAN - an MT system between closely related languages. In: Proceedings of EACL, Copenhagen, pp 113–117

Hajič J, Hric J, Kuboň V (2000) Machine translation of very close languages. In: Proceedings of NAACL-ANLP, Seattle, WA, pp 7–12

Hajič J, Homola P, Kuboň V (2003) A simple multilingual machine translation system. In: Proceedings of the MT summit, New Orleans, LA

Hamzaoğlu İ (1993) Machine translation from Turkish to other Turkic languages and an implementation for the Azeri language. Master's thesis, Boğaziçi University, Istanbul

Homola P, Kuboň V (2008) Improving machine translation between closely related romance languages. In: Proceedings of EAMT, Hamburg, pp 72–77

Karttunen L, Gaal T, Kempe A (1997) Xerox finite-state tool. Technical report, Xerox Research Centre Europe, Meylan

Kos K, Bojar O (2009) Evaluation of machine translation metrics for Czech as the target language. Prague Bull Math Ling 92:135–148

Nakov P, Tiedemann J (2012) Combining word-level and character-level models for machine translation between closely-related languages. In: Proceedings of ACL, Jeju, pp 301–305

Oflazer K (1994) Two-level description of Turkish morphology. Lit Ling Comput 9(2):137–148

Oflazer K, Çetinoğlu Ö, Say B (2004) Integrating morphology with multi-word expression processing in Turkish. In: Proceedings of the ACL workshop on multiword expressions: integrating processing, Barcelona, pp 64–71

Oller CA, Forcada ML (2006) Open-source machine translation between small languages: Catalan and Aranese Occitan. In: Proceedings of the workshop on minority languages: "strategies for developing machine translation for minority languages", Genoa, pp 51–54

Salimzyanov I, Washington JN, Tyers FM (2013) A free/open-source Kazakh-Tatar machine translation system. In: Proceedings of the MT summit, Nice, pp 174–182

Scannell KP (2006) Machine translation for closely related language pairs. In: Proceedings of the workshop on minority languages: "strategies for developing machine translation for minority languages", Genoa, pp 103–109

Tantuğ AC, Adalı E, Oflazer K (2007) A MT system from Turkmen to Turkish employing finite state and statistical methods. In: Proceedings of the MT summit, Copenhagen, pp 459–465

Tantuğ AC, Oflazer K, Durgar-El Kahlout İ (2008) BLEU+: a tool for fine-grained BLEU computation. In: Proceedings of LREC, Marrakesh, pp 1493–1499

Tiedemann J (2009) Character-based PSMT for closely related languages. In: Proceedings of EAMT, Barcelona, pp 12–19

Tyers FM, Washington JN, Salimzyanov I, Batalov R (2012) A prototype machine translation system for Tatar and Bashkir based on free/open-source components. In: Proceedings of the workshop on language resources and technologies for Turkic languages, Istanbul, pp 11–14

Chapter 12
Sentiment Analysis in Turkish

Gizem Gezici and Berrin Yanıkoğlu

Abstract In this chapter, we give an overview of sentiment analysis problem and present a system to estimate the sentiment of movie reviews in Turkish. Our approach combines supervised learning and lexicon-based approaches, making use of a recently constructed Turkish polarity lexicon called SentiTurkNet. For performance evaluation, we investigate the contribution of different feature sets, as well as the effect of lexicon size on the overall classification performance.

12.1 Introduction

Sentiment analysis aims to identify the *polarity* and strength of the opinions indicated in a given text, that together define its *semantic orientation*. The polarity can be indicated categorically as *positive*, *objective*, or *negative*, or numerically, indicating the strength of the opinion on a canonical scale.

Automatic extraction of the sentiment can be very useful in analyzing what people think about specific issues or items, by analyzing large collections of textual data sources such as personal blogs, product review sites, and social media. Commercial interest to this problem has been strong, with companies showing interest to public opinion about their products and financial companies offering advice on the general economic trend by following the sentiment in social media (Pang and Lee 2008). In the remainder of this chapter, we use the terms "document," "review," and "text" interchangeably, to refer to a text whose sentiment polarity or opinion strength is to be estimated.

G. Gezici · B. Yanıkoğlu (✉)
Sabancı University, Istanbul, Turkey
e-mail: gizemgezici@sabanciuniv.edu; berrin@sabanciuniv.edu

© Springer International Publishing AG, part of Springer Nature 2018   255
K. Oflazer, M. Saraçlar (eds.), *Turkish Natural Language Processing*,
Theory and Applications of Natural Language Processing,
https://doi.org/10.1007/978-3-319-90165-7_12

12.1.1 Approaches

There exist two fundamental approaches for sentiment analysis in the state-of-the-art: (1) linguistic or lexicon-based (Turney 2002) and (2) statistical or based on *supervised learning* (Pang et al. 2002). The first approach has the advantage of being simple, while the second approach is typically more successful since it learns from samples of documents with known sentiment in the given domain, without necessarily relying on specially compiled lexicons.

A *polarity lexicon* contains the sentiment polarity of words or phrases. Senti-Wordnet (Esuli and Sebastiani 2006) and SenticNet (Poria et al. 2012) are two of the most commonly used domain-independent polarity lexicons, for sentiment analysis. The lexicon-based approach obtains the polarities of the words or phrases in a document from a polarity lexicon, for the goal of determining the semantic orientation of the document (Turney 2002). The approach may be as simple as estimating the document polarity from that of the average polarity of the constituent words, or can be more complex where different properties of the text can be exploited with the hope of obtaining a more accurate semantic orientation. For instance, the number of subjective words in the document, or the *purity* of the constituent words may be considered (Taboada et al. 2011). The distinctive aspect of lexicon-based approaches is that they do not involve any domain-specific learning.

Supervised learning approaches *learn from data*. While different learning techniques vary on how they use the available labelled data, called the *training data*, the common approach represents each review in the training data using a vector of features (e.g., length, average word polarity, etc.), and a model is learned to associate a feature vector representation with the desired output. The problem can be approached as a *classification* problem (e.g., a review is classified as positive or negative) or *regression* problem (e.g., number of stars given by a review is predicted.) Furthermore, classification can be *binary* (positive/negative) or *ternary* (positive/negative/neutral). The problem gets more difficult as the number of classes increases.

The model that is learned using the training data is tested on a separate *test* data, in order to gauge the generalization performance of the system. However, if there is no designated test set, the available data is split into two datasets as *train* and *validation*, such that the success of the model trained using the training portion is evaluated using the validation portion. For evaluation, the estimated labels (class labels or regression values) are compared with the true labels assigned in the validation/test data.

In case the available data is not very large, instead of splitting the data as training and test, one can make use of a technique called *cross-validation*. Cross-validation is a model validation technique for assessing how well the results of a predictive model would be generalized to an independent dataset. In k-fold cross-validation, the whole data is divided into k equal-sized subsets, where $k - 1$ subsets are used for training and one is used for testing. To reduce variability, this train-test cycle is repeated for multiple rounds and at each round different partitions are used for training and test, and the results obtained with the each test data are averaged at the end.

In the basic approach to sentiment analysis, a given text is viewed as a bag-of-words (BoW); that is, it is represented as a set of words ignoring any word order information (Pang et al. 2002). With this representation, the sentence "A is better than B" is the same as "B is better than A." However, in many cases, the loss of word order information may not necessarily have such a drastic impact (e.g., in "excellent movie was"). Alternatives to the bag-of-words approach are also possible, where word polarities of sentences at significant locations (e.g., first and last sentences) are taken into consideration (Zhao et al. 2008; Gezici et al. 2012).

Some of the supervised learning methods require a polarity lexicon in addition to the training corpus, in order to extract features of the given text (e.g., average word polarity, length, and the number of negative words, etc.) that are later exploited in the learning algorithm. In the Latent Dirichlet Allocation (LDA) approach, one of the most successful supervised learning approaches, the probability distributions of topic and word occurrences in different categories (e.g., positive and negative reviews) are learned by using a training corpus and the classification of a new text is done based on the likelihood estimated from these different distributions (Bespalov et al. 2011, 2012). Deep learning approaches found to be successful in many different pattern recognition problems in recent years are also applied to sentiment analysis with good results (Severyn and Moschitti 2015; Tang et al. 2014; Socher et al. 2013).

12.1.2 Data Type

There are two main types of data for which automatic sentiment analysis is of interest. In *reviews*, the text is generally longer and writers express their opinions on different aspects of the product (e.g., a movie, a hotel, a cell phone). In contrast, with *tweets*, writers express opinions using a very short text which brings in additional complications during automatic processing.

In the hotel domain, the TripAdvisor dataset is well-known, consisting of reviews that are crawled from the TripAdvisor website.[1] In the movie domain, there is a dataset of reviews from IMDB website.[2] For product reviews, researchers often rely on online product reviews from various websites such as Amazon.[3] Tweets in various topics can be collected using keyword searches but are generally more difficult to analyze for sentiment, owing to their short length, prevalence of spelling error and/or abbreviations, though, there have been significant developments on this

---

[1] www.tripadvisor.com (Accessed Sept. 14, 2017).

[2] www.imdb.com (Accessed Sept. 14, 2017).

[3] www.amazon.com (Accessed Sept. 14, 2017).

through a yearly evaluation campaigns organized at relevant conferences (Rosenthal et al. 2014).

12.1.3 Domain Dependence

Words may have different meanings in different domains. For instance, the word "small" has a negative connotation in the hotel domain whereas it is in general positive in the cellphone domain. Since domain-independent lexicons such as SentiWordNet (Esuli and Sebastiani 2006) and SenticNet (Poria et al. 2012) do not contain homonyms (a word that has diverse meanings in different contexts), they may mislead the sentiment analysis system. Hence, one may need a domain-specific lexicon which can be constructed by using a corpus of labeled reviews in a specific domain.

In the rest of the chapter, after a brief discussion of the related work, we discuss difficulties encountered in Turkish sentiment analysis and describe a Turkish sentiment analysis framework and give experimental results on a movie dataset.

12.2 Related Work

Research in sentiment analysis has been active for the last fifteen plus years in line with increasing academic and commercial interest in the topic. Pang and Lee (2008) present a detailed survey of the previous work. Here we only summarize research about the fundamental issues in sentiment analysis and discuss issues in Turkish sentiment analysis and describe and evaluate a system for Turkish.

In their seminal work, Pang et al. (2002) evaluate several features with three different machine learning methods, on a dataset collected from movie reviews crawled from the IMDB internet movie database. A Support Vector Machine (SVM) classifier taking as input the occurrence counts of unigrams and bigrams gives the best performance (82.9% accuracy on 1400 movie reviews).

As mentioned earlier in earlier work a document was typically viewed as a bag-of-words and its sentiment orientation was estimated from features extracted from this "bag" (e.g., words and their frequencies, etc.) (Hatzivassiloglou and McKeown 1997; Pang et al. 2002; Pang and Lee 2004; Mao and Lebanon 2006). As word-order information was no longer available, researchers explored different methods so that they can analyze phrases and sentences. Wilson et al. (2004) represented the document data in a tree structure and then generated features for displaying the relations in the tree with the help of boosting and rule-based methods. Gezici et al. (2012) analyzed sentence-level features in order to bridge the large gap between word and document-level sentiment analysis.

While exploring features at different levels, researchers observed that one of the most important review properties that is highly relevant to sentiment analysis is *subjectivity*. They found that identifying subjective parts within the text first may help to estimate the overall sentiment more accurately. Wiebe (2000) investigated the impact of adjective orientation and gradability on sentence subjectivity. The aim of the approach is to understand whether a given sentence is subjective or not, by looking at the adjectives in that sentence. Wiebe et al. (2004) presented a broad subjectivity analysis along with a comprehensive survey of subjectivity recognition using various features and clues. One of the first datasets generated for the classification of subjectivity consists of 5000 subjective movie review snippets and 5000 movie plot sentences which are assumed to be objective (Pang and Lee 2004). Using this dataset, Pang and Lee built a two-layer algorithm for classification, where the first layer differentiated subjective sentences from objective ones and classified subjective sentences as positive or negative. The two-layer classification process increased the overall result by 4% from 82.9% accuracy (Pang et al. 2002) to 86.9% accuracy (Pang and Lee 2004).

Since 2007, SemEval (Semantic Evaluation) evaluation campaigns have promoted and benchmarked progress in sentiment analysis, receiving a large participation from around the world. Best systems in SemEval 2015 consist of deep learning techniques and ensemble methods (Severyn and Moschitti 2015; Hagen et al. 2015) and achieve around 65% accuracy in ternary classification of tweets.

Most sentiment analysis research in literature is for English and most resources for sentiment analysis (e.g., polarity lexicons, parsers) are for English as well. However research on sentiment analysis of non-English texts has picked interest in recent years. For instance, Ghorbel and Jacot (2011) formulated a method to classify French movie reviews by using supervised learning and linguistic features that are extracted with part-of-speech tagging and chunking, using the semantic orientation information of words from SentiWordNet (Esuli and Sebastiani 2006). French words in the reviews are translated to English so as to obtain their semantic orientation from SentiWordNet.

Sentiment analysis of texts in Turkish has also attracted research interest in recent years and there is still a lot to do in the field. Eroğul (2009) is one of the first studies on Turkish sentiment analysis and develops an SVM classifier for Turkish movie reviews, crawled from the website BeyazPerde.[4] It uses n-grams as features and studies the effect of part-of-speech tagging, spelling-checking and stemming on the overall result. The classifier achieves an 85% of accuracy on the binary sentiment classification of Turkish movie reviews.

More recently, Vural et al. (2013) have proposed a lexicon-based sentiment analysis system using SentiStrength, a lexicon-based sentiment analysis library developed by Thelwall et al. (2010). The library generates a positive or a negative sentiment score for each word in a given text. The authors evaluated their unsupervised Turkish sentiment analysis framework on the same dataset that was already used in Eroğul (2009) and reported an accuracy of 76% for positive/negative classification.

---

[4] www.beyazperde.com (Accessed Sept. 14, 2017).

Türkmenoğlu and Tantuğ (2014) present a lexicon-based framework similar to the systems described in Thelwall et al. (2010) and Vural et al. (2013), with additional handling of simple negation and multi-word expressions. They report achieving accuracy of 79.0% and 75.2% on movie review and tweets datasets, respectively.

As a related problem, *emotion analysis* has also attracted attention from researchers. Boynukalın (2012) presents an emotion classification framework for Turkish and comparative experimental results of several classifiers, for a new dataset for Turkish emotion analysis, with an investigation of the effects of newly added features which are compatible with the morphological characteristics of Turkish language. In another study on emotion analysis of Turkish texts, Çakmak et al. (2012) investigate the feasibility of using a fuzzy-logic representation of Turkish emotion-related words and indicate that there is a strong connection between emotions expressed by word roots and sentences in Turkish.

Analysis of Turkish political news and tweets has also attracted researcher interest. Kaya et al. (2012) investigate the performance of supervised machine learning algorithms such as Naive Bayes, maximum entropy, SVM, and the character based n-gram language models. They observe that maximum entropy and the n-gram models outperform the SVM and Naive Bayes approaches, and report 76–77% accuracy using different features. Kaya (2013) describes an improved version of this system by implementing transfer learning into the existing framework. This system accomplishes a significant improvement over the previous systems, and with all of the three machine learning approaches above, reports accuracy values over 90% for the sentiment classification of Turkish political opinion columns.

12.3 Main Difficulties for Turkish Sentiment Analysis

The motivation behind building a sentiment analysis framework specific to Turkish, rather than utilizing an already established system for English and translating it into Turkish, is due to certain important differences between these two languages. These main differences can be summarized in three categories as follows:

1. *Morphology of Turkish*: Turkish is an agglutinative language (see Chap. 2) which allows the generation of many inflected and derived variant forms of a word adding suffixes to a root word. These derivational and inflectional suffixes may change the part-of-speech and semantic orientation of the word (e.g., "beğendim" (I liked it), "beğenmedim" (I didn't like it)).

   The practical effect of the agglutinative morphology in sentiment analysis is that it makes it infeasible to build a (polarity) lexicon that would contain all variants of Turkish words. Hence, sentiment analysis systems for agglutinative languages like Turkish face some extra challenges compared to those for which a reasonable size lexicon (e.g., 30,000 as for English) is sufficient for many applications.

2. *Turkish character set*: Turkish has several characters that do not exist in the English alphabet: "ç", "ğ", "ı", "ö", "ş", "ü". In informal writing, people tend to substitute the closest ASCII characters for some of these letters (e.g., "ç" is written as "c"). While as such these words are readable in context by humans, they cause complications for robust determination of the words in a sentence. A preprocessing step known as "deasciification" (i.e., converting the ASCII English characters to their Turkish equivalents) to find the words and obtain their polarities from the lexicon) is thus needed before sentiment analysis so that the input is in proper Turkish.
3. *Complexity of negation*: In Turkish, there are several ways a word may be negated, which in turn changes the expressed sentiment:

   • The suffixes *me/ma* negate a verb (e.g., "beğen<u>me</u>dim" (I did not like it)).
   • The productive derivational suffixes *siz/sız/suz/süz* derive an adjective whose meaning adds "without" to the noun they are attached to (e.g., "başarı<u>sız</u>" (without success—unsuccessful).
   • The word "değil" (is/are not) negates nominal or adjectival predications (e.g., "güzel değil" (is not beautiful))
   • The word"yok" (does not exist) indicates nonexistence or unavailability (e.g., "konusu yok" (does not have a topic))

12.4 Practical Sentiment Analysis for Turkish

In this section, we present a basic sentiment analysis system for Turkish, starting with a basic baseline approach and showing how subsequent steps utilizing simple natural language processing (NLP) steps and increasing the lexicon size improve the basic results.

Our evaluation procedure is composed of two main parts: first, we report the effectiveness of different sets of features in classifying movie reviews in Turkish. We then investigate the influence of lexicon size on detecting the overall sentiment of the reviews in the same dataset.

12.4.1 Resources

12.4.1.1 Polarity Lexicon

Our polarity lexicon is the first comprehensive Turkish polarity lexicon, *Senti-TurkNet*, described in Dehkharghani et al. (2016) and developed using several resources both in English and in Turkish. In building this lexicon, authors did not translate SentiWordNet to Turkish as was done in Türkmenoğlu and Tantuğ (2014), but rather they compiled the lexicon using various NLP techniques and

Table 12.1 Sample entries from SentiTurkNet

Synset	POS tag	Negative	Objective	Positive
mükemmel, kusursuz (excellent)	a	0.000	0.000	1.100
kötü (bad)	a	0.946	0.018	0.036
çekici, güzel (beautiful, attractive)	a	0.000	0.000	1.000
şaka, latife (joke)	n	0.060	0.397	0.543
gülmek (to laugh)	v	0.095	0.095	0.810
fiilen, gerçekten (really)	b	0.060	0.872	0.068

additional available resources such as Turkish WordNet (Bilgin et al. 2004), English WordNet (Fellbaum 1998), SentiWordNet (Esuli and Sebastiani 2006), and SenticNet (Cambria et al. 2010).

SentiTurkNet consists of 15,000 synsets with their part-of-speech tags[5] and three associated polarity values—positive, negative, and neutral/objective. The polarity scores stand for the measurement of negativity, objectivity, and positivity, and sum up to 1. Some sample entries from SentiTurkNet are provided in Table 12.1.

Note that a given word may belong to different synsets with different sentiment polarity and hence, one needs to find the correct synset to obtain the correct sentiment polarity. If this is not feasible, then sentiment polarity values across different synsets corresponding to the given word may be averaged. We took the latter approach in this work.

12.4.1.2 Seed Words

Seed words are highly sentiment-bearing words (e.g., "excellent," "horrible") that are expected to be strong indicators of a review's sentiment. Seed word sets have been commonly used for sentiment analysis (Hu and Liu 2004; Qiu et al. 2011).

An advantage of such sentiment-bearing words is that their sentiment polarity does not change much across different domains. For instance, *muhteşem* (excellent) is a positive word in Turkish and *berbat* (awful) is a negative word, independent of the context. Hence they may be helpful in domain-independent tasks.

Another important quality of seed words is that they are often not used in negated form, simplifying the analysis of the sentiment they carry. For instance, while one may use "not very good," where the sentiment polarity of the word "good" is reversed, it is not common to use highly sentiment-bearing words in negated form (as in "it was not excellent").

We use a positive seed word list of 34 words and a negative seed word list of 93 words in this work. A sample of ten positive and ten negative seed words from this list is shown in Table 12.2.

---

[5]We label these as follows in this chapter: a—adjective, n—noun, v—verb, and b—adverb.

Table 12.2 Sample seed words

Positive words	Type	Negative words	Type
muhteşem (magnificent)	a	fiyasko (failure)	n
güzel (beautiful)	a	berbat (awful)	a
eğlenceli (enjoyable)	a	hayalkırıklığı (disappointment)	n
harika (awesome)	a	sıradan (average)	a
şahane (fantastic)	a	sıkıcı (boring)	a
etkileyici (fascinating)	a	olumsuz (negative)	a
başyapıt (masterpiece)	n	vasat (mediocre)	a
kaliteli (good quality)	a	felaket (terrible)	a
kusursuz (perfect)	a	beğenmedim (I did not like)	v
inanılmaz (incredible)	a	değmez (not worth it)	v

Table 12.3 Booster words

Word	POS tag
en (most)	b
gerçekten (really)	b
çok (very)	b
bayağı (too many/much)	b

12.4.1.3 Booster Word List

Booster words are adverbs that accentuate the sentiment polarity of the words that follow: For instance, the words "very" or "really," as in "it was a really good movie," are examples of such words. The boosting effect has already been investigated for Turkish (Türkmenoğlu and Tantuğ 2014).

We have a very small list of four commonly used booster words shown in Table 12.3. Strengthening is done by shifting the polarity value of the corresponding adjective towards its sentiment pole, i.e., positive or negative. We chose a value of 0.4 for shifting.

12.4.2 Methodology

Our approach combines supervised learning and lexicon-based approaches. In the baseline approach, we simply compute the average polarity of the words (adjectives, verbs, and nouns) in the review and train a classifier (Naive Bayes or SVM) to classify the reviews as positive or negative just based on this average polarity feature. Then we measure the effectiveness of more complex processing techniques or additional features such as handling negation, considering the effects of booster words and using additional features derived from seed words. In all of these approaches, the document is viewed as a bag-of-words.

Table 12.4 Sample preprocessing

Input	Preprocessing step	Output	Lexicon search
hoslanmadim	Deasciification	hoşlanmadım	Not successful
hoşlanmadım	Root extraction	hoşlan	Not successful
hoşlan	Adding infinitive suffixes	hoşlanmak	Successful

12.4.2.1 Preprocessing

Before feature extraction, several steps are necessary as preprocessing steps, in order to obtain the corresponding polarity values of the word in a review. These polarity values form the basis of the features used in this work. As an initial step, we tokenize the given text into words and then we use Zemberek (Akın and Akın 2007) for deasciification.

While obtaining the polarity value for each word in the document, the following procedure is used: We first search the word itself in the lexicon (SentiTurkNet) with the part-of-speech tag information. If the word is not found, then we identify the root with Zemberek and search for the root in the lexicon. If we still cannot find it and the part-of-speech tag of the word is verb, then we search the root of the word by adding the infinitive suffixes (mek/mak in Turkish) to the end of it. If none of these help to find the polarity values for the word, this means that the word does not exist in the lexicon, therefore its polarity values are set to 0. The process is illustrated in Table 12.4 for the sample word "hoslanmadim (I did not like it)," written using only ASCII versions of some of the characters.

A word in the lexicon may have multiple synset entries. In order to get the correct polarity values, it is important to find the correct synset, or as a lesser alternative, to compute an average of the polarity values of all corresponding synsets. We take the latter approach in this work for simplicity.

12.4.2.2 Basic Approach

In the basic approach, we only use the average polarity of the constituent words to estimate the document polarity. The overall average sentiment polarity is computed by averaging the polarity of all potentially sentiment-bearing words in the document (adjectives, verbs, and nouns), while adverbs affect the overall polarity indirectly if they are in the booster list shown in Table 12.3. The average polarity of a given text is computed as follows:

$$F_1 = \frac{1}{N} \sum_{w_i} pol(w_i) \qquad (12.1)$$

where w_i are the corresponding words in the document, N is the total number of sentiment-bearing words, and $pol(w_i)$ is computed from the polarity values obtained from SentiTurkNet. The *average polarity* of a word w, denoted by $pol(w)$, is calculated as:

$$pol(w) = (pol^+ - pol^-)/2 \tag{12.2}$$

where pol^+ and pol^- represent the positive and negative polarity values assigned to the word w in the polarity lexicon. For simplicity, we do not take into account of the neutral/objective polarity value of the word in this work. An alternative to using the average polarity is to use the *dominant polarity* of a word (Demiröz et al. 2012).

12.4.2.3 Handling Negation

In comparison to English, negation handling is quite complicated for Turkish. For instance, in English the word *not* is used for negation purposes, while in Turkish negation can happen in several different forms as described earlier. We take into consideration all words or suffixes that signal negation except *yok* because it requires the negation analysis at the sentence-level, instead of the word-level.

For each negated word, we negate its polarity $pol(w_i)$ as defined in Eq. (12.2) and recompute the average polarity to give feature F_2.

12.4.2.4 Booster Effect

As mentioned in earlier booster words strengthen the meaning of adjectives that they modify. In order to take them into account, we compute the average review polarity by considering the effects of booster words shown in Table 12.3, by shifting the polarity of affected words. Booster word handling is performed after negation handling, to obtain feature F_3.

12.4.2.5 Seed Words

We have chosen a positive seed word list of 34 words and a negative seed word list of 93 words, as discussed earlier. The seed word enables us to estimate sentiment that is less error-prone in comparison to using a large polarity lexicon that may contain errors. The corresponding features F_4 and F_5 are the positive and negative seed word counts in a review. Feature F_4 is computed as:

$$F_4 = \sum_{w_i} PositiveSeed(w_i)$$

Table 12.5 Sample sentences

Sample input	Relevant words	F_1	F_2	F_3
Hata-larla dolu (full of errors)	hata (n; pol = −0.47)	−0.47	−0.47	−0.47
Hiç sev-me-dim (I did not like at all)	sev (v; pol = 0.37), -me (negation)	0.37	−0.37	−0.37
Cok guzel-di (was very beautiful)	çok (booster), güzel (a; pol = 0.5)	0.50	0.50	0.90

where w_i are the sentiment-bearing words that are adjectives, adverbs, verb and nouns in the document and $PositiveSeed(w_i)$ returns 1 if the word w_i is a positive seed word and zero otherwise. Similarly, feature F_5 captures the number of negative seed words in the review.

12.4.2.6 Sample Analysis

Table 12.5 shows the feature values for three separate sentences. The first example has a single sentiment bearing word, no booster words nor any negation suffix. Hence all three features have the same value. In the second example there is a negation suffix (-me), which is considered in determining the average polarity in F_2. As there are no booster words, F_3 is the same as F_2. The third example contains a booster word, therefore the average polarity of the following adjective is shifted by 0.4 towards the positive end in determining F_3.

12.4.2.7 Classifier Training

We randomly split the available data into train and test sets containing a balanced number of positive and negative reviews in each. Then, the system is trained using a Naive Bayes or SVM classifier using only the training set and tested on the test set. Our system is implemented in Java and uses WEKA (Hall et al. 2009) for classifier training and testing. WEKA is a commonly used machine learning toolbox that provides many supervised as well as unsupervised algorithms (Hall et al. 2009).

We used the LibSVM package which is implemented in WEKA for parameter optimization, training and testing stages. Before the actual training with the SVM classifier, we performed parameter optimization. For optimization, we performed a 5-fold cross-validation on the training data and found the best parameter values as 10.0 and 10.0, for the cost and gamma parameters. We then re-trained the system with all of the training data.

12.5 Experimental Evaluation

12.5.1 Data

We evaluated the proposed approach and features using the Turkish movie reviews dataset that was compiled by Demirtaş and Pechenizkiy (2013) from a well-known movie site called Beyazperde.[6] The star ratings of reviews (1–5 stars) are used as ground-truth labels for evaluation. Since we only address the binary classification problem, 4 or 5-star reviews are considered as positive reviews while 1 or 2-star reviews are considered negative. We excluded reviews with 3-stars from the study, as often done in binary classification evaluations. As a result, we obtained a total set of 5331 positive and 5330 negative movie reviews. Some sample positive and negative movie reviews from the database are shown in Tables 12.6 and 12.7, respectively. The data is split into train and test sets with equal proportion of positive and negative reviews in each.

12.5.2 Results

We report results obtained with both Naive Bayes or SVM classifiers, using the features in increasing complexity. Table 12.8 presents correct classification accuracies with basic and more complex features, while Table 12.9 presents the effect of the lexicon size in overall accuracy.

Table 12.6 Positive movie reviews

Review	Gloss
"gerçek bir başyapıt"	" a true masterpiece"
"gelmiş geçmiş en iyi 10 filmden biri"	"it's one of the top 10 movies ever"
"tek kelimeyle kusursuz"	"in one word: perfect"

Table 12.7 Negative movie reviews

Review	Gloss
"benim için sadece büyük bir hayalkırıklığı"	"for me it's just a big disappointment"
"hiç beğenmedim bu filmi"	"I didn't like this movie at all"
"berbat bir film"	"it's a terrible movie"

---

[6]Reviewers on Beyazperde rate movies star ratings of 1–5 scale, in addition to the review they enter.

Table 12.8 Classification accuracy with different features

Features	Accuracy (NB)	Accuracy (SVM)
F_1 (Basic)	67.49%	67.61%
F_2 (w/ neg. handling)	69.29%	69.42%
F_3 (w/ neg. + booster handling)	68.22%	68.19%
F_2, F_4, F_5 (neg. handling + seed words)	75.16%	73.70%

Table 12.9 The effect of lexicon size on the classification performance

Lexicon size (number of words)	Accuracy (NB)	Accuracy (SVM)
100	51.27%	51.29%
1000	51.85%	51.88%
5000	52.07%	53.28%
(All) 15,000	67.49%	67.61%

12.5.2.1 Feature Efficacy

While the basic baseline approach only uses the raw sentiment polarities of words in estimating the average polarity of a given document, the second approach extends it with negation handling to compute the average document polarity more accurately, and the third approach includes both negation and booster word handling. As we see in Table 12.8, the basic approach obtains 67.49% accuracy with the Naive Bayes classifier and 67.61% with the SVM classifier, while the best results are obtained with negation handling and seed words, achieving 75.16% with the Naive Bayes and 73.70% with the SVM classifiers, respectively. Somewhat surprisingly, booster effect handling does not improve accuracy, while considering seed words does.

12.5.2.2 Lexicon Effect

The second part of our evaluation investigates the effect of the lexicon size on obtaining the overall sentiment of a given review. Increasing the lexicon size generally improves the classification performance, since with a larger lexicon, the system knows about the semantic orientations of more words.

To generate lexicons of various sizes, we started with the polarity values of the seed words, obtained from the SentiTurkNet (Dehkharghani et al. 2016), and the rest of the new lexicon was filled by randomly choosing the necessary number of synsets from the lexicon. To obtain more robust results, we randomly chose the rest of the words in the new lexicon five times and obtained results and computed an average over these.

This process was repeated until the lexicon size reached that of SentiTurkNet which contains 15,000 synsets. In investigating the effect of lexicon size, we only used our basic feature, F_1. Results are displayed in Table 12.9 where the last row corresponds to the basic approach given in Table 12.8. As can be seen, a larger lexicon always brings better classification performance as the added words help estimate the review polarity more accurately.

12.6   Conclusions

Interest in sentiment analysis is growing rapidly thanks to its use in collecting public's opinion in several different application areas. Various approaches described in literature range from simple approaches based on the use of domain-independent polarity lexicons to deep learning techniques that can capture long-term interactions between words in a review. Suitability and success of different approaches depend upon many factors, including the availability polarity resources in the given language and domain, availability of labeled data, the length of the review to be analyzed. Through a simple system, we have demonstrated the effects of some of the necessary natural language processing steps (i.e., negation handling and booster word handling), along with the effect of seed words and lexicon size. We achieve 75% accuracy on binary classification of movie reviews in Turkish. Our results show that having even a small set of domain-dependent seed words and a large domain-independent polarity lexicon affects recognition accuracy the most. Future work in this area seems to be geared towards building resources in new languages, as well as machine learning techniques such as deep learning that leverage large amounts of unlabeled data, in addition to labeled data and sentiment resources.

References

Akın AA, Akın MD (2007) Zemberek, an open source NLP framework for Turkic languages. Structure 10:1–5

Bespalov D, Bai B, Qi Y, Shokoufandeh A (2011) Sentiment classification based on supervised latent n-gram analysis. In: Proceedings of the ACM international conference on information and knowledge management, Glasgow, pp 375–382

Bespalov D, Qi Y, Bai B, Shokoufandeh A (2012) Sentiment classification with supervised sequence embedding. In: Proceedings of conference on machine learning and knowledge discovery in databases, Bristol, pp 159–174

Bilgin O, Çetinoğlu Ö, Oflazer K (2004) Building a Wordnet for Turkish. Rom J Inf Sci Technol 7(1–2):163–172

Boynukalın Z (2012) Emotion analysis of Turkish texts by using machine learning methods. Master's thesis, Middle East Technical University, Ankara

Çakmak O, Kazemzadeh A, Yıldırım S, Narayanan S (2012) Using interval type-2 fuzzy logic to analyze Turkish emotion words. In: Proceedings of the annual summit and conference of signal information processing association, Los Angeles, CA, pp 1–4

Cambria E, Speer R, Havasi C, Hussain A (2010) Senticnet: a publicly available semantic resource for opinion mining. In: Proceedings of AAAI fall symposium: commonsense knowledge, Arlington, VA, vol 10, p 02

Dehkharghani R, Saygın Y, Yanıkoğlu B, Oflazer K (2016) SentiTurkNet: a Turkish polarity lexicon for sentiment analysis. Lang Resour Eval 50(3):667–685

Demiröz G, Yanıkoğlu B, Tapucu D, Saygın Y (2012) Learning domain-specific polarity lexicons. In: Proceedings of the workshop on sentiment elicitation from natural text for information retrieval and extraction, Brussels, pp 674–679

Demirtaş E, Pechenizkiy M (2013) Cross-lingual polarity detection with machine translation. In: Proceedings of the international workshop on issues of sentiment discovery and opinion mining, Chicago, IL, pp 9:1–9:8

Eroğul U (2009) Sentiment analysis in Turkish. Master's thesis, Middle East Technical University, Ankara

Esuli A, Sebastiani F (2006) Sentiwordnet: a publicly available lexical resource for opinion mining. In: Proceedings of LREC, Genoa, vol 6, pp 417–422

Fellbaum C (1998) WordNet: an electronic lexical database. MIT Press, Cambridge, MA

Gezici G, Yanıkoğlu B, Tapucu D, Saygın Y (2012) New features for sentiment analysis: do sentences matter? In: Proceedings of the International Workshop on Sentiment Discovery from Affective Data, Bristol, pp 5–15

Ghorbel H, Jacot D (2011) Sentiment analysis of French movie reviews. In: Advances in distributed agent-based retrieval tools. Springer, Berlin

Hagen M, Potthast M, Büchner M, Stein B (2015) Webis: an ensemble for Twitter sentiment detection. In: Proceedings of SEMEVAL, Denver, CO, pp 582–589

Hall M, Frank E, Holmes G, Pfahringer B, Reutemann P, Witten IH (2009) The WEKA data mining software: an update. ACM SIGKDD Explor Newsl 11(1):10–18

Hatzivassiloglou V, McKeown KR (1997) Predicting the semantic orientation of adjectives. In: Proceedings of ACL-EACL, Madrid, pp 174–181

Hu M, Liu B (2004) Mining and summarizing customer reviews. In: Proceedings of the 10th ACM SIGKDD international conference on knowledge discovery and data mining, Seattle, WA, pp 168–177

Kaya M (2013) Sentiment analysis of Turkish political columns with transfer learning. PhD thesis, Middle East Technical University, Ankara

Kaya M, Fidan G, Toroslu İH (2012) Sentiment analysis of Turkish political news. In: Proceedings of the 2012 IEEE/WIC/ACM international joint conferences on web intelligence and intelligent agent technology, Macau, pp 174–180

Mao Y, Lebanon G (2006) Isotonic conditional random fields and local sentiment flow. In: Proceedings of NIPS, Vancouver, pp 961–968

Pang B, Lee L (2004) A sentimental education: sentiment analysis using subjectivity summarization based on minimum cuts. In: Proceedings of ACL, Barcelona, pp 271–278

Pang B, Lee L (2008) Opinion mining and sentiment analysis. Found Trends Inf Retr 2(1–2):1–135

Pang B, Lee L, Vaithyanathan S (2002) Thumbs up?: sentiment classification using machine learning techniques. In: Proceedings of EMNLP, Philadelphia, PA, pp 79–86

Poria S, Gelbukh A, Cambria E, Das D, Bandyopadhyay S (2012) Enriching SenticNet polarity scores through semi-supervised fuzzy clustering. In: Proceedings of the workshop on sentiment elicitation from natural text for information retrieval and extraction, Brussels, pp 709–716

Qiu G, Liu B, Bu J, Chen C (2011) Opinion word expansion and target extraction through double propagation. Comput Linguist 37(1):9–27

Rosenthal S, Ritter A, Nakov P, Stoyanov V (2014) Semeval-2014 task 9: sentiment analysis in twitter. In: Proceedings of SEMEVAL, Dublin, pp 73–80

Severyn A, Moschitti A (2015) UNITN: training deep convolutional neural network for Twitter sentiment classification. In: Proceedings of SEMEVAL, Denver, CO, pp 464–469

Socher R, Perelygin A, Wu J, Chuang J, Manning CD, Ng AY, Potts C (2013) Recursive deep models for semantic compositionality over a sentiment treebank. In: Proceedings of EMNLP, Seattle, WA, pp 1631–1642

Taboada M, Brooke J, Tofiloski M, Voll K, Stede M (2011) Lexicon-based methods for sentiment analysis. Comput Linguist 37(2):267–307

Tang D, Wei F, Qin B, Liu T, Zhou M (2014) Coooolll: a deep learning system for twitter sentiment classification. In: Proceedings of SEMEVAL, Dublin, pp 208–212

Thelwall M, Buckley K, Paltoglou G, Cai D, Kappas A (2010) Sentiment strength detection in short informal text. J Am Soc Inf Sci Technol 61(12):2544–2558

Türkmenoğlu C, Tantuğ AC (2014) Sentiment analysis in Turkish media. Technical report, Istanbul Technical University, Istanbul

Turney PD (2002) Thumbs up or thumbs down?: semantic orientation applied to unsupervised classification of reviews. In: Proceedings of ACL, Philadelphia, PA, pp 417–424

Vural AG, Cambazoğlu BB, Şenkul P, Tokgöz ZÖ (2013) A framework for sentiment analysis in Turkish: application to polarity detection of movie reviews in Turkish. In: Proceedings of ISCIS, Paris, pp 437–445

Wiebe J (2000) Learning subjective adjectives from corpora. In: Proceedings of AAAI, Austin, TX, pp 735–740

Wiebe J, Wilson T, Bruce R, Bell M, Martin M (2004) Learning subjective language. Comput Linguist 30(3):277–308

Wilson T, Wiebe J, Hwa R (2004) Just how mad are you? Finding strong and weak opinion clauses. In: Proceedings of AAAI, San Jose, CA, pp 761–769

Zhao J, Liu K, Wang G (2008) Adding redundant features for CRF-based sentence sentiment classification. In: Proceedings of EMNLP, Honolulu, HI, pp 117–126

Chapter 13
The Turkish Treebank

Gülşen Eryiğit, Kemal Oflazer, and Umut Sulubacak

Abstract In the last three decades, treebanks have become a crucial resource for building and evaluating natural language processing tools and applications. In this chapter, we review the essential aspects of the first treebank for Turkish that was built in early 2000s and its evolution and extensions since then.

13.1 Introduction

In the last three decades, treebanks such as the Penn Treebank (Marcus et al. 1993) have become a crucial resource for building and evaluating natural language processing tools and applications. Although the compilation of such structurally annotated corpora is time-consuming and expensive, the eventual benefits outweigh the initial cost. Around 2000, with a set of future applications in mind, we undertook the design of a treebank corpus architecture for Turkish, which we believe encodes the lexical and structural information relevant to Turkish, and developed a modest sized treebank. In this chapter, we present the issues that we have encountered in designing a treebank for Turkish along with the rationale for the representation choices we have made, and the evolution of the treebank over the years in response to various developments.

In the resulting representation, the information encoded in the complex agglutinative word structures is represented as a sequence of inflectional groups separated by derivational boundaries. A tagset reduction is not attempted as any such reduction leads to the removal of potentially useful syntactic markers, especially in the encoding of derived forms. At the syntactic level, we opted to just represent

G. Eryiğit · U. Sulubacak
Istanbul Technical University, Istanbul, Turkey
e-mail: gulsen.cebiroglu@itu.edu.tr; sulubacak@itu.edu.tr

K. Oflazer (✉)
Carnegie Mellon University Qatar, Doha-Education City, Qatar
e-mail: ko@cs.cmu.edu

© Springer International Publishing AG, part of Springer Nature 2018
K. Oflazer, M. Saraçlar (eds.), *Turkish Natural Language Processing*,
Theory and Applications of Natural Language Processing,
https://doi.org/10.1007/978-3-319-90165-7_13

relationships between lexical items (or rather, inflectional groups) as dependency relations. The representation is extensible so that relations between lexical items can be further refined by augmenting syntactic relations by finer distinctions which are more semantic in nature.

On the syntax side, although Turkish has an unmarked SOV constituent order, it is considered as a free constituent order language, as all constituents, including the verb, can move freely as demanded by the discourse context with very few syntactic constraints (Erguvanlı 1979). Case marking on nominal constituents usually indicates their syntactic role. Constituent order in embedded clauses is substantially more constrained but deviations from the default order, however infrequent, can still be found. Turkish is also a pro-drop language, as the subject, if necessary, can be elided and recovered from the agreement markers on the verb. Within noun phrases, there is a loose order with specifiers preceding modifiers, but within each group, order (e.g., between cardinal and attributive modifiers) is mainly determined by the aspect that is to be emphasized. For instance, the Turkish equivalents of *two young men* and *young two men* are both possible: the former being the neutral case or the case where youth is emphasized, while the latter is the case where the cardinality is emphasized. A further but relatively minor complication is that Turkish will allow for discontinuous constituents. For example, various verbal adjuncts may intervene in well-defined positions within NPs causing discontinuous constituents.

13.2 What Information Needs To Be Represented?

We expected this treebank to be used by a wide variety "consumers," ranging from linguists investigating morphological structure and distributions, syntactic structure, constituent order variations, to computational linguists extracting language models or training and evaluating parsers, etc. We therefore employed an extendable multi-tier representation, so that any future extensions can be easily incorporated if/when necessary.

13.2.1 Representing Morphological Information

At the lowest level, we would like to represent three main aspects of a lexical item:

- The word itself, e.g., `evimdekiler` (those in my house).
- The lexical structure, as a sequence of free and bound morphemes.[1]

    `ev+Hm+DA+ki+lAr`

---

[1] See Chap. 2 for conventions for morphophonological symbols.

- The morphological features encoded by the word as a sequence of morphological and POS feature values all of which except the root are symbolic, e.g.,

```
ev+Noun+A3sg+P1sg+Loc^DB+Adj+Rel^DB+Noun+Zero+A3pl+Pnon+Nom
```

A point to note about this representation is that information that is conveyed covertly by zero-morphemes that is not explicit in the lexical representation is represented here (e.g., if a plural marker is not present, then the noun is singular hence +A3sg is the feature supplied even though there is no overt morpheme).[2]

The first two components of the morphological information do not deserve any more details for the purposes of this presentation. The third component with its relation to lexical tag information will be detailed later.

The prevalence of productive derivational word forms brings a challenge to representing such information using a finite (and possibly reduced) tagset. The usual approaches to tagset design typically assume that the morphological information associated with a word form can be encoded using a finite number of cryptically coded symbols from some set whose size ranges from a few tens [e.g., Penn Treebank tag set (Marcus et al. 1993)] to hundreds or even thousands [e.g., Prague Treebank tag set (Hajič 1998)]. But, such a finite tagset approach for languages like Turkish inevitably leads to loss of information. The reason for this is that the morphological features of intermediate derivations can contain markers for syntactic relationships. Leaving out this information within a fixed-tagset scheme may prevent important syntactic information from being represented.

For these reasons we have decided not to compress in any way the morphological information associated with a Turkish word and represent such words as a sequence of *inflectional groups* (IGs), separated by ^DBs denoting derivation boundaries. Thus, a word is represented in the following general form:

$$root+Infl_1\text{^}DB+Infl_2\text{^}DB+\cdots\text{^}DB+Infl_n$$

where Infl$_i$ denote relevant inflectional features including the part-of-speech for the root or any of the forms. For instance, the derived modifier sağlamlaş-tırdığımızdaki (at the time we caused …to become strong) would be represented by the five IGs:

```
1)   sağlam+Adj^DB
2)   +Verb+Become^DB
3)   +Verb+Caus+Pos^DB
4)   +Noun+PastPart+A3sg+Pnon+Loc^DB
5)   +Adj+Rel
```

Note that it is possible to come up with a finite (but large) inventory of IGs which can be compactly coded, but we feel that apart from saving storage such an encoding serves no real purpose while the resulting opaqueness prevents facilitated

---

[2]Refer to Chap. 2 for a list of morphological features.

Table 13.1 Parse and IG statistics from a Turkish corpus

	All tokens	All but high freq function words and and punctuation
Morph. parses per token	1.76	1.93
IGs per parse	1.38	1.48
% Tokens with a single parse	55	45
% Parses with 1 IGs	72	65
% Parses with 2 IGs	18	23
% Parses with 3 IGs	7	9
% Parses with > 3 IGs	3	3
Max number of IGs in a parse	7	7
Unique IGs ignoring roots	2448	

access to component features. The discussions about what a syntactic unit should be in morphologically rich languages such as Turkish, Japanese, and others and at what granularity level are still active in the Turkish computational linguistics community (Çöltekin 2016) and in the Universal Dependencies Project.[3]

Although we have presented a novel way of looking at the lexical structure, the reader may have received the impression that words in Turkish have overly complicated structures with many IGs per word. Various statistics actually indicate that this is really not the case. For instance, the statistics presented in Table 13.1, compiled from about 850,000 word corpus of Turkish news text indicate that on average the number of IG's per word is less than 2. Thus, for instance, modeling each word uniformly with 2 IGs each may be a very good approximation for statistical modeling (Hakkani-Tür et al. 2002).

Turkish is also very rich in lexicalized and non-lexicalized collocations (see Chap. 2). These also need to be represented for proper capture of syntactic relations.

13.2.2 Representing Syntactic Relations

We would like to represent the syntactic relations between lexical items (actually between inflectional groups as we will see in a moment) using a simple dependency framework. Our arguments for this choice essentially parallel those of recent works on this topic (Hajič 1998; Skut et al. 1997; Lepage et al. 1998). Free constituent ordering and discontinuous phrases make constituent-based representations rather difficult and unnatural to employ. It is however possible to use constituency where it makes sense and bracket sequences of tokens to mark segments in the texts whose

---

[3]The Universal Dependencies project [universaldependencies.org (Accessed Sept. 14, 2017)] is an international collaborative project to make cross-linguistically consistent treebanks available for a wide variety of languages.

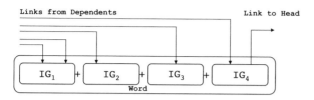

Fig. 13.1 Inflectional groups and dependency links

internal dependency structure would be of little interest. For instance, collocations, date–time expressions, or multiword proper names[4] are examples whose internal structure is of little syntactic concern, and can be bracketed a priori as chunks and then related to other words. If necessary, a constituent-based representation can be extracted from the dependency representation (Lin 1998).

An interesting observation that we can make about Turkish is that when a word is considered as a sequence of IGs, syntactic relation links are between the last IG of a (dependent) word, and one of the IGs of the (head) word almost always somewhere to the right, as exemplified in Fig. 13.1. A second observation is that (again with minor exceptions) the dependency links between the IGs, when drawn above the IG sequence, do not cross (although this is not a concern here). Figure 13.2 shows an example Turkish sentence, where the +'s indicate the derivation and hence IG boundaries. Note for instance that, for the word *büyümesi* the previous two words link to its first (verbal) IG, while its second IG (infinitive nominal) links to the final verb as subject.

The syntactic relations that we have currently opted to encode in our syntactic representation are the following[5]:

1.	SENTENCE	2.	DETERMINER	3.	QUESTION.PARTICLE
4.	INTENSIFIER	5.	RELATIVIZER	6.	CLASSIFIER
7.	POSSESSOR	8.	NEGATIVE.PARTICLE	9.	OBJECT
10.	MODIFIER	11.	DATIVE.ADJUNCT	12.	FOCUS.PARTICLE
13.	SUBJECT	14.	ABLATIVE.ADJUNCT	15.	INSTRUMENTAL.ADJUNCT
16.	ETOL	17.	LOCATIVE.ADJUNCT	18.	COORDINATION
19.	S.MODIFIER	20.	EQU.ADJUNCT	21.	APPOSITION
22.	VOCATIVE	23.	COLLOCATION	24.	ROOT

Some of the relations above perhaps require some more clarification. *Object* is used to mark objects of verbs and the nominal complements of postpositions. A *classifier* is a nominal modifier in nominative case (as in *book cover*) while a *possessor* is a genitive case-marked nominal modifier. For verbal adjuncts, we

---

[4]Which incidentally do not follow Turkish noun phrase rules so have to be treated specially anyway.

[5]ETOL encodes light verb constructions involving the Turkish verbs *et-* (do) and *ol-* (be).

Bu   eski   bahçe-de+ki
bu+Det   eski+Adj   bahçe+A3sg+Pnon+Loc^DB+Adj+Rel

gül-ün   böyle
gül+Noun+A3sg+Pnon+Gen   böyle+Adv

büyü+me-si
büyü+Verb+Pos^DB+Noun+Inf+A3sg+P3sg+Nom

herkes-i   çok
herkes+Pron+A3sg+Pnon+Acc çok+Adv

etkile-di.
etkile+Verb+Pos+Past+A3sg

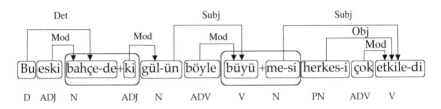

Fig. 13.2 An example Turkish sentence and its dependency representation

indicate the syntactic relation with a marker paralleling the case marking, though the semantic relation they encode is not only determined by the case marking but also the lexical semantics of the head noun and the verb it is attached to. For instance, a dative adjunct can be a *goal*, a *destination*, a *beneficiary* or a *value carrier* in a transaction, or a *theme*, while an ablative adjunct may be *reason*, a *source*, or a *theme*.

13.2.3 Example of a Treebank Sentence

In this section we present the detailed representation of a Turkish sentence in the treebank. Each sentence is represented by a sequence of attribute lists of the words involved, bracketed with tags <S> and </S>.[6] Figure 13.3 shows the treebank encoding for the sentence given earlier. Each word is bracketed by <W> and </W> tags. The rest of the symbols denote the following:

1. IX denotes the position index of the word in the sentence.
2. LEM denotes the morphological root of the word as delivered by the morphological analyzer.

---

[6]Words in this context may also be a lexicalized or non-lexicalized collocations.

```
<S>
<W IX=1 LEM="bu" MORPH="bu" IG=[(1, "bu+Det")]
REL=[(3,1,(DETERMINER)]> Bu </W>

<W IX=2 LEM="eski"' MORPH="eski" IG=[(1, "eski+Adj")]
REL=[3,1,(MODIFIER)]> eski> </W>

<W IX=3 LEM="bahçe" MORPH="bahçe+DA+ki" IG=[(1,
"bahçe+A3sg+Pnon+Loc") (2, "+Adj+Rel")] REL=[4,1,(MODIFIER)]>
bahçedeki </W>

<W IX=4 LEM="gül" MORPH="gül+nHn" IG=[(1,"gül+Noun+A3sg+Pnon+Gen")]
REL=[6,1,(SUBJECT)]> gülün </W>

<W IX=5 LEM="böyle" MORPH="böyle" IG=[(1,"böyle+Adv")]
REL=[6,1,(MODIFIER)]> böyle </W>

<W IX=6 LEM="büyü" MORPH="büyü+mA+sH" IG=[(1,"büyü+Verb+Pos") (2,
"+Noun+Inf+A3sg+P3sg+Nom")] REL=[9,1,(SUBJECT)]> büyümesi </W>

<W IX=7 LEM="herkes" MORPH="herkes+yH" IG=[(1,"herkes+Pron+A3sg+Pnon+Acc")]
REL=[9,1,(OBJECT)]> herkesi </W>

<W IX=8 LEM="çok" MORPH="çok" IG=[(1,"çok+Adv")]
REL=[9,1,(MODIFIER)]> çok </W>

<W IX=9 LEM="etkile" MORPH="etkile+DH" IG=[(1,
"etkile+Verb+Pos+Past+A3sg")] REL=[]> etkiledi </W>

</S>
```

Fig. 13.3 Treebank encoding of a Turkish sentence in the treebank

3. MORPH indicates the morphological structure of the word as a sequence of morphemes, essentially corresponding to the lexical form. The morphemes may involve meta-symbols (mentioned earlier) for indicating any phonological classes of symbols.[7]

4. IG is a list of pairs of an inflectional group number and the symbolic representation of that inflectional group.

5. REL encodes the relationship of this word, as indicated by its last inflection group, to an inflectional group of some other word. The first component of REL is the index of a word, the second component is the number of the inflection group in that word that this word's last IG is linked to, and the third component is a list of relation labels for any possible syntactic (e.g., dative adjunct) or semantic relationships (e.g., destination) between the IGs involved.

   For example, the fourth and fifth words in the sentence are the subject and the adverbial modifier, respectively, of the verb in the first IG of the sixth word, while the second IG of the same word (6) is the subject of the main verb in word 9.

---

[7]In the initial version of the treebank, this field was left empty, with the expectation that it would be provided in future versions.

A collocation would be represented by coalescing the information of individual components. For instance, the non-lexicalized collocation *gelir gelmez* (as soon as ...comes)[8] and its adjunct

ev-e
```
ev+Noun+A3sg+Pnon+Dat
```

gel-ir   gel-me-z
```
gel+Verb+Pos+Aor+A3sg        gel+Verb+Neg+Aor+A3sg
```
would be represented as

```
      . . .
   <W IX=5 LEM="ev" MORPH="ev+yA" IG=[(1,"ev+A3sg+Pnon+Dat")]],
      REL=[6,1,(DATIVE-ADJ,DEST)]> eve </W>

   <W IX=6 LEM="gelmek" MORPH="gel+Hr gel+mA+z"
      IG=[(1, "gel+Verb+Pos")(2, "+Adv+AsSoonAs")],
      REL=[...]> gelir gelmez   </W>
   . . .
```

where the non-lexicalized collocation has been treated as a derivational process and an adverbial IG +Adv+AsSoonAs has been created.

13.3   Evolution of the Turkish Treebank

The original METU-Sabancı Turkish Dependency Treebank gave birth to several resources in recent years. Many studies, especially on parsing (Eryiğit et al. 2011; Sulubacak and Eryiğit 2013) but also others (Eryiğit et al. 2015; Adalı et al. 2016; Pamay et al. 2015; Eryiğit et al. 2015; Sulubacak and Eryiğit 2013; Sulubacak et al. 2016b) introduced their own versions of the treebank, each with a slightly altered morphosyntactic representation or an augmented semantic representation. These studies collectively paved the way for a rich, multi-faceted treebank, featuring a wide variety of annotations on the same basic sentences for some contested linguistic phenomena such as coordination structures, inflectional groups, and multiword expressions.

13.3.1   The CoNLL Format

The Turkish Treebank was used in dependency parsing competitions on multilingual dependency parsing in 2006 CoNLL (Buchholz and Marsi 2006) and 2007 CoNLL (Nilsson et al. 2007). However, the representation of the treebank needed to

---

[8]See Chap. 2.

Table 13.2 CoNLL-X representation of a Turkish treebank sentence

ID	FORM	LEMMA	CPOSTAG	POSTAG	FEATS	HEAD	DEPREL
1	Çocuk	çocuk	*Noun*	*Noun*	*A3sg | Pnon | Nom*	*6*	SUBJECT
2	_	gül	*Verb*	*Verb*	*Pos | Aor | A3sg*	*3*	DERIV
3	gülerken	_	*Adverb*	*While*	_	*6*	MODIFIER
4	çok	çok	*Adverb*	*Adverb*	_	*5*	MODIFIER
5	_	güzel	*Noun*	*NAdj*	*A3sg | Pnon | Nom*	*6*	DERIV
6	güzeldi	_	*Verb*	*Zero*	*Past | A3sg*	*0*	PREDICATE
7	.	.	*Punc*	*Punc*	_	*6*	PUNCTUATION

be adapted to the CoNLL sentence format, which was not necessarily ready to accommodate the IG-based syntactic representation we had employed.

The widely recognized CoNLL-X format organizes sentences on a grid, where each row corresponds to an elementary parsing unit, and the parametric data associated with each unit are arranged under the columns as shown in Table 13.2. Although not as expressive, the CoNLL format is more human-readable and has gained popularity. Though the XML format has remained in use, it was eventually largely replaced by the CoNLL format.

In line with the original XML representation, the CoNLL format considers parsing units to be IGs rather than orthographic words, and thus has a separate line and ID for every IG in the sentence. The **FORM** and **LEMMA** fields contain, respectively, the surface form and the root for each unit. For words with multiple IGs, the stem of the word is given under the **LEMMA** field of the first IG and the surface form is given under the **FORM** field of the last IG, while the rest of the **FORM** and **LEMMA** fields for the other IGs are omitted by use of an underscore.

Morphological information is distributed over the **CPOSTAG**, **POSTAG**, and **FEATS** fields. The fields **CPOSTAG** and **POSTAG** provide fields, respectively, for a coarse-grained and fine-grained part-of-speech tag. It is just as viable to use **POSTAG** for the regular POS tags and **CPOSTAG** for a more general category, as using **CPOSTAG** for the regular POS tags and reserving **POSTAG** for when sub-parts-of-speech are present, though the latter is exercised more frequently. Usually in the absence of different coarse- and fine-grained POS tags, **CPOSTAG** and **POSTAG** are assigned the same value. The **FEATS** field contains the rest of the inflectional features after the POS tag, separated by vertical bars.

Dependency information is likewise collected under the **HEAD** and **DEPREL** fields, where the former contains the ID of the head unit and the latter lists the dependency relation. Since IGs are represented as separate units, an exclusive field to indicate the IG inside the head word is not required. Note that the CoNLL format is sometimes extended with two more fields, **PHEAD** and **PDEPREL**, to denote a projective head and dependency relation for the unit, for when a projective version of the sentence has to be additionally provided.

The CoNLL format has a drawback in its representation of IGs in that it is unable to properly represent the form and root for individual derivation steps. Sulubacak

Table 13.3 Augmented CoNLL representation supporting lexical information on IGs—FORM and LEMMA

ID	FORM	LEMMA	CPOSTAG	POSTAG	FEATS	HEAD	DEPREL	IG
1	Çocuk	çocuk	*Noun*	*Noun*	*A3sg | Pnon | Nom*	*6*	SUBJECT	0
2	güler	gül	*Verb*	*Verb*	*Pos | Aor | A3sg*	*3*	DERIV	1
3	gülerken	güler	*Adverb*	*While*	_	*6*	MODIFIER	0
4	çok	çok	*Adverb*	*Adverb*	_	*5*	MODIFIER	0
5	güzel	güzel	*Adj*	*Adj*	*A3sg | Pnon | Nom*	*6*	DERIV	1
6	güzeldi	güzel	*Verb*	*Zero*	*Past | A3sg*	*0*	PREDICATE	0
7	.	.	*Punc*	*Punc*	_	*6*	PUNCTUATION	0

and Eryiğit (2013) tried to address this problem and proposed an alternative means for distinguishing IGs that did not require the intermediate forms and lemmata to be left out, using the additional `IG` column as shown in Table 13.3. Although this variant is not in common use at the time of writing, it has been shown to be quite feasible and effective.

13.3.2 Branches of the Turkish Treebank

Sulubacak et al. (2016b) critically analyzed and discussed the annotation framework of the METU-Sabancı Treebank, and then proposed a revised framework based on the original annotation framework of the treebank. Following this revised framework, the ITU-METU-Sabancı Treebank, also referred to as the IMST,[9] was created as a reannotation of the original treebank using the same raw tokenized sentences and building upon the original morphological analyses. Despite still being a fledgling treebank, the IMST has also had some measure of success and is already well recognized.

Another design decision for the IMST was to add *deep* dependencies. Deep dependencies are secondary dependencies of tokens to other logical heads, often with different dependency relations, in addition to their regular *surface* dependencies. Although the annotation of these dependencies is favored often because they function as cues for semantic parsers, it also violates the restriction of each constituent having a single head, which is a common requirement of most parsers and sentence formats.

It is also not possible to annotate multiple heads for a single constituent under the standard CoNLL format due to its grid system. A representation to allow multiple dependency arcs from the same dependent was considered first in preparation for the IMST, and the augmented CoNLL format as shown in Table 13.4 was

---

[9]All treebanks described in this section are available at ITU Turkish Natural Language Processing Pipeline: tools.nlp.itu.edu.tr (Accessed Sept. 14, 2017).

Table 13.4 Augmented CoNLL representation supporting deep dependencies

ID	FORM	LEMMA	CPOSTAG	POSTAG	FEATS	HEAD	DEPREL
1	Çocuk	çocuk	*Noun*	*Noun*	*A3sg | Pnon | Nom*	2	SUBJECT
1	Çocuk	çocuk	*Noun*	*Noun*	*A3sg | Pnon | Nom*	6	SUBJECT
2	_	gül	*Verb*	*Verb*	*Pos | Aor | A3sg*	3	DERIV
3	gülerken	_	*Adverb*	*While*	_	6	MODIFIER
4	çok	çok	*Adverb*	*Adverb*	_	5	MODIFIER
5	_	güzel	*Noun*	*NAdj*	*A3sg | Pnon | Nom*	6	DERIV
6	güzeldi	_	*Verb*	*Zero*	*Past | A3sg*	0	PREDICATE
7	.	.	*Punc*	*Punc*	_	6	PUNCTUATION

proposed. In the augmented format, there are as many rows for a dependent as there are dependencies from it, each differing only in **HEAD** and **DEPREL** to represent the secondary dependencies and having the same values for all other fields. Secondary dependencies were only annotated to represent coreference links from zero pronouns and to mark shared modifiers for tokens in coordination in the IMST, and no real hierarchy was enacted between *deep* and *surface* dependencies in these cases.

The ITU Validation Set (IVS) (Eryiğit 2007; Eryiğit and Pamay 2014) is another small treebank containing 300 sentences. This treebank follows the same annotation conventions as the original treebank, so it could also be considered a progeny of the original METU-Sabancı Treebank. Though the ITU Validation Set was first introduced in 2007 to be used as a test set in the CoNLL XI Shared Task (Eryiğit 2007), it has been reannotated by Eryiğit and Pamay (2014) with the revised annotation framework following the IMST.

13.4 The ITU Web Treebank

The revised annotation framework originally designed for the IMST was also used to annotate the ITU Web Treebank (IWT) (Pamay et al. 2015), a fairly comprehensive, brand new treebank composed of user-generated content appearing with Web 2.0. The IWT is the first of its kind in being a web treebank of Turkish sentences, following similar efforts such as the French Social Media Bank (Seddah et al. 2012) and the English Web Treebank (Petrov and McDonald 2012).

With its 5009 sentences and 47,245 tokens, the IWT is almost as large as the original Turkish Treebank. The sentences in the IWT were compiled from a wide variety of websites from five main domains (news story comments, personal blog comments, customer product reviews, social network posts, and discussion forum posts), where common users participate in creating the content. These sentences were first manually tokenized and normalized before morphosyntactic annotation. On top of this layer, the annotation of the IWT is very similar to the IMST (and therefore the original METU-Sabancı Treebank), including the annotation of deep dependencies, except for certain extensions to the morphological tag set that were needed to properly represent web-specific entities. Nonetheless, the non-canonical content of the IWT is largely complementary to the well-edited sentences of its predecessors.

13.5   The Annotation Tool

Annotation platforms are very useful for the human annotation process of the treebanks and easy-to-use interfaces were shown to increase the speed and the quality of the annotations. The need for an annotation platform tailored for Turkish's specific representation schemes (introduced above), the technological development allowing web-based applications and the need to annotate the non-canonical language coming with Web 2.0 yielded the development of different annotation platforms for Turkish Treebank development over the recent years (Oflazer et al. 2003; Eryiğit 2007; Pamay et al. 2015). The common properties of these platforms are that they allow the selection of the correct morphological analysis from the possible morphological analyses produced by a morphological analyzer and the annotation of syntactic relations between inflectional groups. These platforms mainly consist of three levels of annotation and may be used to produce results for each of these, namely, morphological analysis, morphological disambiguation, and syntactic analysis stages. Eryiğit (2007) also allows semi-automatic annotation for different layers by the use of NLP plugins, transforming the annotation process from a manual procedure (starting from scratch) into a check-and-correct procedure. Pamay et al. (2015) introduced a web-based application supporting annotation for the normalization layer in addition to the morphology and syntax layers, and allowing concurrent operation by multiple annotators on the same data as well as arbitration and vetting by an annotation supervisor. The platform comes with a set of changes in the annotation interfaces in compliance with the changes in the annotation methodologies for web data compatibility.

Figures 13.4 and 13.5 show the two annotation screens from Pamay et al. (2015) sampling the annotations for the input sentence "Rahat et Müşfik Kenter" (*Rest in peace Müşfik Kenter*). The platform generates morphological analyses for certain orthographically tagged tokens such as web entities in addition to the output fetched from a morphological analyzer, to be later disambiguated manually. In the first screen, the language expert is asked to first normalize the instances and then select the relevant morphological analyses from the automatically produced results. In the second screen, the annotator continues with the dependency analysis by first selecting the source and target IGs and then the relation type from a dynamic combo box appearing on top of the dependency arc. The screen shows the dependency tree along with the dependency relation table for the sentence being processed. The dependency annotation interface supports the specification of multiple head tokens for a given constituent, allowing the annotation of deep dependencies on the tool

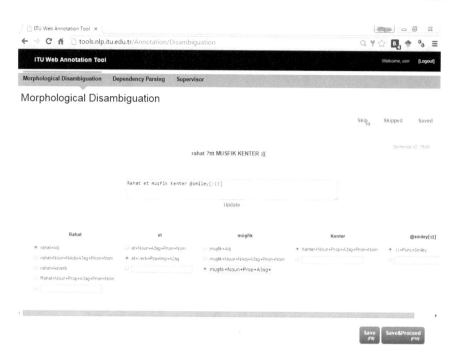

Fig. 13.4 Treebank annotation tool—normalization and morphology layer

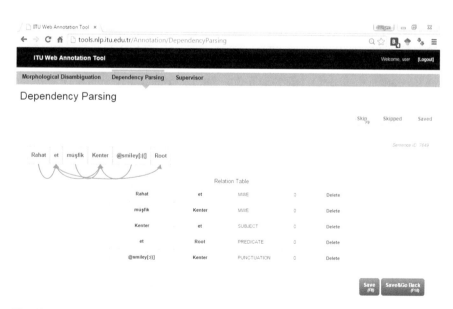

Fig. 13.5 Treebank annotation tool—syntactic layer

while still enforcing at least one head for each dependent. The interface also displays the root node as a separate token and allows regular dependencies to the root node. The platform has been used during the development of IMST, IVS, and IWT.

13.6   The Turkish Universal Dependencies Treebank

The Turkish Universal Dependencies Treebank[10] (Sulubacak et al. 2016a), also called the IMST-UD, is a Universal Dependencies (UD) treebank published in UD version 1.3. UD is an international project to unify treebanks for a wide variety of languages under a common annotation framework, with the aim of investigating the cross-linguistic efficiencies of various language processing tasks such as parsing. This is achieved by a number of annotation principles for the unified representation, in order to make it as consistent as possible with all the involved languages. In accordance with these principles, UD treebanks are required to use a set of universal annotation guidelines based on the Google Universal Part-of-Speech Tagset (Petrov et al. 2012) for parts-of-speech, the Interset framework (Zeman 2008) for morphological features, and Stanford Dependencies (de Marneffe et al. 2006, 2014; Tsarfaty 2013) for dependency relations. Moreover, word segmentation (as in IGs) is disallowed in most cases, but conversely required in others, such as for bound enclitics like copulas and a restricted set of derivations such as relativizers.

Although the Turkish UD Treebank follows an entirely different annotation framework from the previously discussed treebanks, it is effectively the most recent offspring of the METU-Sabancı Treebank, since it was automatically adapted from the IMST. IMST-UD was created through an automatic conversion procedure from the IMST, successively adjusting the typology, the morphological representation, the dependency relations, and finally the sentence format, in compliance with the UD guidelines.

The CoNLL-U format was specifically designed for the UD project, and is the most recent variant of the CoNLL format. It is largely identical to the original CoNLL sentence format as shown in Table 13.5, with a few notable exceptions. The distinction between **CPOSTAG** and **POSTAG** was eliminated in favor of a single **UPOSTAG** column for the universal POS tag, along with an **XPOSTAG** column for optionally including language-specific parts-of-speech. Since the IG representation is no longer compatible with the CoNLL-U format, segments are instead represented by a separate header column, which is not considered a token of the sentence. The format also introduces a standard way for specifying deep dependencies, where all secondary dependencies are arranged in a **HEAD:DEPREL** format and specified in the exclusive **DEPS** column, delimited by vertical bars. The final precautionary column is reserved for any other annotation.

---

[10]Available at universaldependencies.org (Accessed Sept. 14. 2017).

Table 13.5 The CoNLL-U representation

ID	FORM	LEMMA	UPOSTAG	XPOSTAG	FEATS	HEAD	DEPREL	DEPS	MISC
1	Çocuk	çocuk	NOUN	Noun	Case=Nom | Number=Sing | Person=3	4	nsubj	2:nsubj	–
2	gülerken	gül	VERB	Verb	Aspect=Perf | Mood=Ind | Negative=Pos | Number=Sing | Person=3 | Tense=Aor | VerbForm=Trans	4	advcl	–	–
3	çok	çok	ADV	Adverb	–	4	det	–	–
4-5	güzeldi								
4	güzel	güzel	ADJ	NAdj	–	0	root	–	–
5	di	i	AUX	Zero	Aspect=Perf | Mood=Ind | Negative=Pos | Number=Sing | Person=3 | Tense=Past	4	cop	–	–
6	.	.	PUNCT	Punc	–	4	punct	–	–

13.7   Conclusions

This chapter has described the rationale for the representational choices for encoding the syntactic structure of Turkish sentences in a dependency treebank resources, along with the description of newer versions of the treebank employing recent standardized representational frameworks and extensions of the treebank covering more casual texts found on the web. Over the years, the Turkish treebank has been instrumental in developing other tools and applications for Turkish. We strongly encourage interested researchers to explore the various treebanks offered at the URLs mentioned in this chapter and experiment with them.

Acknowledgements   The development of the Turkish Treebank and its extensions were supported by grants 199E026 and 112E276 from TÜBİTAK (Turkish Scientific and Technological Research Council) and by the ICT COST Action IC1207 PARSEME (PARSing and Multi-word Expressions).

References

Adalı K, Dinç T, Gökırmak M, Eryiğit G (2016) Comprehensive annotation of multiword expressions for Turkish. In: Proceedings of TurCLing 2016, the first international conference on Turkic computational linguistics, Konya, pp 60–66

Buchholz S, Marsi E (2006) CoNLL-X shared task on multilingual dependency parsing. In: Proceedings of CONLL, New York, NY, pp 149–164

Çöltekin Ç (2016) (When) do we need inflectional groups? In: Proceedings of TurCLing 2016, the first international conference on Turkic computational linguistics, pp 38–43

de Marneffe MC, MacCartney B, Manning C (2006) Generating typed dependency parses from phrase structure parses. In: Proceedings of LREC, Genoa, pp 449–454

de Marneffe MC, Dozat T, Silveira N, Haverinen K, Ginter F, Nivre J, Manning CD (2014) Universal Stanford Dependencies: a cross-linguistic typology. In: Proceedings of LREC, Reykjavík, pp 4585–4592

Erguvanlı EE (1979) The function of word order in Turkish grammar. PhD thesis, UCLA, Los Angeles, CA

Eryiğit G (2007) ITU treebank annotation tool. In: Proceedings of the linguistic annotation workshop, Prague, pp 117–120

Eryiğit G (2007) ITU validation set for METU-Sabancı Turkish Treebank. www.web.itu.edu.tr/ gulsenc/papers/validationset.pdf. Accessed 14 Sept 2017

Eryiğit G, Pamay T (2014) ITU validation set. Türkiye Bilişim Vakfı Bilgisayar Bilimleri ve Mühendisliği Dergisi 7(1):103–106

Eryiğit G, İlbay T, Can OA (2011) Multiword expressions in statistical dependency parsing. In: Proceedings of the workshop on statistical parsing of morphologically rich languages, Dublin, pp 45–55

Eryiğit G, Adalı K, Torunoğlu-Selamet D, Sulubacak U, Pamay T (2015) Annotation and extraction of multiword expressions in Turkish treebanks. In: Proceedings of the workshop on multiword expressions, Denver, CO, pp 70–76

Hajič J (1998) Building a syntactically annotated corpus: the Prague Dependency Treebank. In: Hajicova E (ed) Issues in valency and meaning: studies in honour of Jarmila Panenova. Charles University Press, Prague

Hakkani-Tür DZ, Oflazer K, Tür G (2002) Statistical morphological disambiguation for agglutinative languages. Comput Hum 36(4):381–410

Lepage Y, Shin-Ichi A, Susumu A, Hitoshi I (1998) An annotated corpus in Japanese using Tesniere's structural syntax. In: Proceedings of the workshop on the processing of dependency-based grammars, Montreal, pp 109–115

Lin D (1998) A dependency-based method for evaluating broad-coverage parsers. Nat Lang Eng 4(02):97–114

Marcus M, Marcinkiewicz M, Santorini B (1993) Building a large annotated corpus of English: the Penn Treebank. Comput Linguist 19(2):313–330

Nilsson J, Riedel S, Yuret D (2007) The CoNLL 2007 shared task on dependency parsing. In: Proceedings of CoNLL, Prague, pp 915–932

Oflazer K, Say B, Hakkani-Tür DZ, Tür G (2003) Building a Turkish Treebank. In: Treebanks: building and using parsed corpora. Kluwer Academic, Berlin

Pamay T, Sulubacak U, Torunoğlu-Selamet D, Eryiğit G (2015) The annotation process of the ITU Web Treebank. In: Proceedings of the linguistic annotation workshop, Denver, CO, pp 95–101

Petrov S, McDonald R (2012) Overview of the 2012 shared task on parsing the web. In: Notes of the first workshop on syntactic analysis of non-canonical language

Petrov S, Das D, McDonald R (2012) A universal part-of-speech tagset. In: Proceedings of LREC, Istanbul, pp 2089–2096

Seddah D, Sagot B, Candito M, Mouilleron V, Combet V (2012) The French Social Media Bank: a treebank of noisy user generated content. In: Proceedings of COLING, Mumbai, pp 2441–2457

Skut W, Krenn B, Brants T, Uszkoreit H (1997) An annotation scheme for free word order languages. In: Proceedings of the conference on applied natural language processing, Washington, DC, pp 88–95

Sulubacak U, Eryiğit G (2013) Representation of morphosyntactic units and coordination structures in the Turkish dependency treebank. In: Proceedings of the workshop on statistical parsing of morphologically rich languages, Seattle, WA, pp 129–134

Sulubacak U, Gökırmak M, Tyers F, Çöltekin Ç, Nivre J, Eryiğit G (2016a) Universal dependencies for Turkish. In: Proceedings of COLING, Osaka, pp 3444–3454

Sulubacak U, Pamay T, Eryiğit G (2016b) IMST: a revisited Turkish dependency treebank. In: Proceedings of TurCLing 2016, the first international conference on Turkic computational linguistics, Konya, pp 1–6

Tsarfaty R (2013) A unified morpho-syntactic scheme of Stanford Dependencies. In: Proceedings of ACL, Sofia, pp 578–584

Zeman D (2008) Reusable tagset conversion using tagset drivers. In: Proceedings of LREC, Marrakesh, pp 213–218

Chapter 14
Linguistic Corpora: A View from Turkish

Mustafa Aksan and Yeşim Aksan

Abstract Usage-based linguistic studies have gained new insights as corpus-based and corpus-driven analyses have advanced in recent years. Linguists working in different domains have turned to corpora as a major source in their study of language at all levels of representation. Currently, corpus linguistics is evolving into a sophisticated methodology in extracting and analyzing data. Building and using corpora in Turkish linguistics is a recent undertaking, initially motivated by work on natural language processing (NLP) research. The number of available corpora is increasing and linguistic research has come to test hypotheses on attested data, or uncover more lexical and grammatical patterns of use that have gone unnoticed in the absence of corpus data. Advances in NLP research and tools provided for corpus building and annotation further contribute to corpus studies in Turkish linguistics.

14.1 Introduction

In his comment on the state of American linguistics in the mid-1950s, Newmeyer (1986, p. 2) defines the period as "a period of optimism." The common understanding among the linguists of the time was that the field had achieved a level of sophistication in which major problems were solved and all that was left to do was to provide details: "…punch the data into the computer and out would come the grammar!" The success of linguistics was beyond doubt as other social sciences were imitating linguistics in adopting its methods in their research. However, the introduction of generative grammar in the late 1950s marked the end of classical structuralism and changed the course of the field. This new revolution in linguistics also "curtailed" early corpus-based theoretical frameworks and introduced "idealizations and abstractions" which had little to do with the empirical methodologies of corpus studies (Barlow 2011).

M. Aksan (✉) · Y. Aksan
Mersin University, Mersin, Turkey

© Springer International Publishing AG, part of Springer Nature 2018
K. Oflazer, M. Saraçlar (eds.), *Turkish Natural Language Processing*,
Theory and Applications of Natural Language Processing,
https://doi.org/10.1007/978-3-319-90165-7_14

Slowly but steadily, empirical linguistics made an impressive comeback, especially after the early 1990s, "... when computational linguistics embraced corpora as the automated analysis of large quantities of text data started to make serious impact in the development of speech recognition, machine-aided translation, and other natural language processing tasks ..." (Leech 2011, p. 157). The approach of linguists toward usage-based studies and the recognition of the role of frequencies and patterns determined via corpus-analytic tools resulted in a significant increase in the number of linguistic corpora. The intricate relationship between data, theory, and methodology is now being discussed in a new perspective motivated by the extensive use of corpus data in all fields of linguistics.

The use of corpora in Turkish linguistics studies is a comparatively recent enterprise. One major reason for this late involvement is the fact that no linguistic corpora was available until the early 2000s. As is well known to many involved in the process, corpus building is a labor-intensive and time-consuming activity that requires committed institutional backing. The small number of linguistics departments in Turkey and the lack of appreciation and funding of such work were the main reasons for lack of such corpora.

In this chapter, we present a brief review of the available Turkish linguistic corpora and corpus-based and corpus-driven linguistics: We will review the evolution of linguistic corpora in general, and corpora as a source in linguistic analysis and corpus linguistics as a method in linguistics along with arguments concerning the nature of the object, addressing the relationship between research questions and the typology of corpora. Then we will discuss the use of corpora in different fields of linguistics and the defining standards of representative and balanced corpora. The final section will give a brief review of the available linguistic corpora in Turkish, some linguistic work using corpus data, and their evaluation.

14.2   Brief History of Corpus Linguistics

It is by now customary to distinguish between the preelectronic and post-electronic eras in the development of corpus linguistics. Svartvik (2007), for example, notes that the initials BC, for a corpus linguist, stands for Before Computers. The preelectronic period refers to corpus studies that were predecessors of contemporary work and which were mostly done before the 1960s. For some, the early studies go back to the thirteenth century indexing work on the Bible and for others, to recent times as recently as the beginnings of the twentieth century work of American Structuralism in collecting textual samples of language use (Leech 1992).

Advances in computer technology, such as increase in storage capacities and the sophistication of available software, had a major impact on the progress of corpus linguistics. In fact, it is such advances that have empowered corpus linguistics to achieve its status today. Equally, we may say that linguistics also provided a strong impetus in developing many practical applications in computing in general because it demanded new types of software in processing natural language for its complex manifestations at different levels.

Apart from the concordances derived from data stored on punch cards that appeared in the late 1950s, Francis and Kučera (1964) constructed the first ever electronic corpus of written English at Brown University in 1961. The Brown Corpus set the standards for corpus design with a size of one million words. The developments following the Brown Corpus are described as five phases or stages in Renouf (2007, p. 28). The stages are determined on the basis of the periods in which a specific corpus was constructed as well as the "types, styles and design" of the corpora of the time.

1. 1960s onwards: the one-million-word (or less) Small Corpus (standard, general and specialized, sampled, multimodal, multidimensional)
2. 1980s onwards: the multimillion-word Large Corpus (standard, general and specialized, sampled, multimodal, multidimensional)
3. 1990s onwards: the 'Modern Diachronic' Corpus (dynamic, open-ended, chronological data flow)
4. 1998 onwards: the Web as corpus (Web texts as sources of linguistic information)
5. 2005 onwards: the Grid (pathway to distributed corpora, consolidation of existing corpus types)

The impact of computer science and computer technology became more significant in the second and third stages. The development of desktop computing freed many corpus developers and corpus users from mainframes and the rapid growth and expansion of the Internet and data storage capacities helped to store and share data efficiently, and a new generation of scanners increased the capabilities of data entry processes. In linguistics, it became clear that some questions of lexis and grammar cannot be pursued properly in small-size corpora. Thus, the demands of corpus-based analyses and the appealing developments in computer technologies gave way to corpora of a different generation. Some of the resulting multimillion-word super-corpora of these two stages include the *Birmingham Corpus* (1980–1986, 20 million words), the *Bank of English* (1980 onwards, 650 million words as of 2012), and the *British National Corpus* (1991–1995, 100 million words).

In the next stage of development, after the advance of super corpora, almost the same motivations and emerging technological potentials contributed not only to size, but also to types of corpus construction. The development of the *monitor corpus*, or *modern diachronic corpus*, as recalled by Renouf (2007), goes back to 1982. It was observed that language is changing and language change can be captured and observed in corpora. The idea that innovations, variations, and changes in lexis and grammar can be followed in "dynamic corpus unbroken chronological text" resulted in a distinct type of corpus, continuously adding texts from the *Times* (starting in 1988), followed by the monitor corpora of the *Independent* and the *Guardian* journalistic texts.

The expansion of the Web in the 1990s led to a new era in corpus linguistics research and introduced new and improved corpus tools. The World Wide Web itself has become an online corpus in that some of the texts stored on the Web appear only in electronic form and never in any other format, more varied language is manifested on the Web, and there are citations of new and rare lexical items and patterns that are not found in ordinary available corpora. Furthermore, the Web provided a cheap and easy means of building corpora with huge amounts of accessible texts, representing present-day language, updated continuously. The advantages that the Web as corpus presented were soon overshadowed by the problems observed by researchers. It was argued that the data on the Web is too heterogeneous and unstructured ("cheap and dirty") to derive any reliable conclusions when corpus linguistics methods are put to use.

14.3   Linguistic Corpora and Corpus Linguistics

A *Glossary of Corpus Linguistics* (Baker et al. 2006, p. 48) defines corpus as

> **corpus** The word *corpus* is Latin for body (plural *corpora*). In linguistics, a corpus is a collection of texts (a 'body' of language) stored in an electronic database. Corpora are usually large bodies of machine-readable text containing thousands or millions of words. A corpus is different from an archive in that often (but not always) the texts have been selected so that they can be said to be representative of a particular language variety or genre, therefore acting as a standard reference. Corpora are often annotated with additional information such as part-of-speech tags or to denote prosodic features associated with speech. Individual texts within a corpus usually receive some form of meta-encoding in a header, giving information about their genre, the author, date and place of publication, etc.

A review of other existing definitions suggests that there are rarely disagreements among researchers in the field. This consensus is captured by McEnery et al. (2006, p. 5):

> …a corpus is a collection of (1) machine readable (2) authentic texts (including transcripts of spoken data) which is (3) sampled to be (4) representative of a particular language or language variety.

In the same *Glossary*, (Baker et al. 2006, p. 50) defines corpus linguistics as

> **corpus linguistics** A scholarly enterprise concerned with the compilation and analysis of corpora (Kennedy 1998, p. 1). According to McEnery and Wilson (1996, p. 1) it is the 'study of language based on examples of "real life" language use' and 'a methodology rather than an aspect of language requiring explanation or description.'

While linguists do not diverge in defining corpus, they disagree in defining the field itself. With respect to the corpora themselves, the arguments commonly concern the typology of corpora, the methods by which they are designed and constructed, and the extent to which they should meet the now-standard criteria to count as linguistic corpora. The major disagreement concerns the very nature of the field. Put simply, a group of corpus linguists conceptualize their enterprise as a "methodology" in doing any type of linguistic analysis in which corpus tools provide special qualitative and quantitative methods for the questions at hand.

For another group of linguists, the so-called "neo-Firthians," corpus linguistics is a "theory". A neo-Firthian corpus linguist asserts that corpus linguistics is "a theoretical approach to the study of language" (Teubert 2005).

The ever-increasing use of corpora in linguistic research introduced new concepts and methods that helped uncover many aspects of language structure and language use, which ultimately lead to new theories of language. In a recent introduction to the field, (McEnery and Hardie 2012) argue that corpus linguistics "... is not directly about the study of any particular aspect of language. Rather, it is an area which focuses upon a set of procedures, or methods, for studying language." Accordingly, they also argue that

> The procedures themselves are still developing, and remain an unclearly delineated set— though some of them, such as concordancing, are well established and are viewed as central to the approach. Given these procedures, we can take a corpus-based approach to many areas of linguistics. ... it may refine and redefine a range of theories of language. It may also enable us to use theories of language which were at best difficult to explore prior to the development of corpora of suitable size and machines of sufficient power to exploit them. Importantly, the development of corpus linguistics has also spawned, or at least facilitated the exploration of, new theories of language—theories which draw their inspiration from attested language use and the findings drawn from it (McEnery and Hardie 2012, p. 1).

A linguistic corpus is designed by a set of *external* and *internal* criteria. External criteria (*situational*) relate to the selection of texts on the bases of registers, genres, and time span, among others. Internal criteria (*linguistic*) are concerned with the distribution of linguistic features across texts that make up the corpus. It is evident that external criteria do not take into account the linguistic characteristics. Internal criteria, on the other hand, present a problematic situation in which a corpus builder decides in advance which linguistic features are to be represented in the corpus. However, it is helpful in selecting text types with different linguistic features to be added next to the corpus.

The defining features that stand out as the most significant in measuring a corpus as a reliable source for linguistic analysis are *representativeness* and *balance*. In an earlier study on corpus representativeness, Biber (1993, p. 243) explains the standards:

> Some of the first considerations in constructing a corpus concern the overall design: for example, the *kinds* of texts included, the *number* of texts, the *selection* of particular texts, the selection of text samples from within texts, and the *length* of text samples. Each of these involves a sampling decision, either conscious or not. [emphasis added]

Representativeness is a much-debated feature that sets a linguistic corpus apart from an archive or collection of texts. In other words, representativeness makes a corpus a reliable source for any linguistic analysis to derive valid conclusions on language structure or use. Despite its importance in corpus design, there exists little agreement about representativeness. Leech (2007) indicates that for some researchers, if a corpus lacks representativeness, any conclusion derived from such a corpus will be confined to that particular corpus only, and cannot be extended or generalized to language.

Balance is another hard-to-define requirement for linguistic corpora. Leech (2007, pp. 136–138) indicates

> An obvious way forward is to say that a corpus is 'balanced' when the size of its subcorpora (representing particular genres or registers) is proportional to the relative frequency of occurrence of those genres in the language's textual universe as a whole. In other words, balancedness equates with proportionality. ... There is one rule of thumb that few are likely to dissent from. It is that in general, the larger a corpus is, and the more diverse it is in terms of genres and other language varieties, the more balanced and representative it will be.

It is expected that a balanced corpus covers as much variety of text categories as possible to represent the language. At present, there is no concrete measure to judge the balance of a corpus other than informed and intuitive judgments. The research interests and their extent determine the type of the corpus to be built. A common typology of corpora include the following:

- *General Corpora*: The driving force in the construction of a general corpus is to produce a *reference corpus* of language use that would be balanced and representative. A general corpus may contain written or spoken texts or may contain texts from both media. The major aim is to represent texts from different genres, domains, and types in a balanced manner so that the conclusions drawn from quantitative and qualitative analyses of corpus data will hold true for language use in general. The British National Corpus (BNC) is one such general reference corpus of modern English having 100 million words and comprising 4048 written texts and ten million words of transcribed spoken data. It is a balanced and representative corpus of modern English as it includes texts sampled from national and regional newspapers and journals, popular and academic books, university essays, e-mail samples, unpublished letters, and reports from different ages, institutions, and readerships. The success of the BNC as a representative and balanced general corpus led others to adopt its basic design principles, including the American National Corpus, the Korean National Corpus, the Polish National Corpus, and recently, the Turkish National Corpus.
- *Specialized Corpora*: Relatively small-sized and specialized in terms of genre or domain, these types of corpora are more varied and available in greater numbers. The current tendency in specialized corpus creation is mostly observed in professional and academic domains. Some representatives of such specialized corpora include the Corpus of Professional Spoken American English (CPSA)[1] and the Michigan Corpus of Academic Spoken English (MICASE).[2] A specialized corpus can also be created by extracting relevant text data from a larger general corpus.
- *Written Corpora*: The Brown Corpus is not only the first corpus, but it is at the same time the first written corpus of English in modern times. The texts that make up the corpus data are collected from written media, sampled from

---

[1] Athelstan Corpus of Spoken, Professional American-English: www.athel.com/cpsa.html (Accessed Sept. 14, 2017).

[2] quod.lib.umich.edu/m/micase (Accessed Sept. 14, 2017).

15 categories. A counterpart of the Brown Corpus, the Lancaster-Oslo-Bergen Corpus of British English (LOB), is constructed following the same principles of the Brown Corpus, and thus they have collectively come to represent varieties of the same language, providing a reliable means of comparison between two varieties of English. Later, in the early 1990s, the Freiburg-LOB Corpus of British English (FLOB) and the Freiburg-Brown Corpus of American English (Frown) were developed to represent written American and British varieties of English. Furthermore, comparisons of these two Freiburg corpora with previous Brown/LOB corpora revealed data on language change in the time span between the 60s and the 90s.

- *Spoken Corpora*: Compared to general or written corpora, it is harder to construct and annotate the spoken corpora of a language. Only recently, we witnessed an increase in the number of spoken corpora due to improvements in recording technologies and automated transcription software. Pioneering corpora for spoken English were built in the late 1960s, such as the London-Lund Corpus (LLC) (Greenbaum and Svartvik 1990), followed by others, including the Lancaster/IBM Spoken English Corpus (SEC),[3] the Cambridge and Nottingham Corpus of Discourse in English (CANCODE) (Carter and McCarthy 2004), the Santa Barbara Corpus of Spoken American English (SBCSAE) (Du Bois et al. 2005), and the Wellington Corpus of Spoken New Zealand English (WSC) (Holmes et al. 1998). The only existing and linguistically reliable new-generation spoken corpus of Turkish is the Spoken Turkish Corpus (STC) (Ruhi et al. 2010a). The Turkish National Corpus (TNC) (Aksan et al. 2012) also has a spoken component of one million words as a reflection of its adherence to the design principles of the BNC.

- *Synchronic Corpora*: Linguists build synchronic corpora in order to observe language change and language variation in corpus data, primarily for the purpose of providing a "snapshot" of language use at a certain point or period of time. In such corpora, all the texts should be selected from the same time period to account for varieties of the language synchronically present. The International Corpus of English (ICE) is built for the synchronic analysis of the English spoken in Britain, the USA, Australia, Canada, and other countries where English is the first language (Greenbaum 1991). It consists of twenty corpora of one million words each, with samples of both written and spoken English.

- *Diachronic Corpora*: Corpora that are constructed for a linguistic account of language in time commonly contain texts representing language use during different periods of the language under investigation. Given the recent history of sound recording technologies, diachronic corpora represent written language over time, for example, the Helsinki Diachronic Corpus of English Texts (Rissanen et al. 1991)

---

[3]ICAME Corpus Collection: Information: clu.uni.no/icame/lanspeks.html (Accessed Sept. 14, 2017).

- *Learner Corpora*: Corpus use in the language classroom has found its place in teaching and learning contexts. For example, the International Corpus of Learner English (ICLE) (Granger 2003) and as its sub-corpus Turkish International Corpus of Learner English (TICLE) (Kilimci and Can 2009) have been a source of research in teaching contexts in recent years.
- *Monitor Corpora*: A monitor corpus is different from the (static) others presented here in the sense that it is constantly growing (dynamic) with the addition of new material. The Bank of English (BoE)[4] and Corpus of Contemporary American English (COCA) (Davis 2008) are well-known corpora of this type for English.

14.4   Use of Corpora in Linguistics

A corpus is constructed primarily to represent language use in a balanced manner in order to study language empirically on the basis of real data. The role and function of corpora in linguistic analyses can be viewed from different perspectives, depending on the research questions at hand. Lüdeling and Kytö (2008, p. ix) summarize the use of corpora in linguistic analyses for three major purposes: (1) empirical support, (2) frequency information, and (3) meta-information.

The corpus query tools help researchers in finding examples of real language use that are relevant to their questions, that is what they now have as an example is a citation of actual language use rather than the alternative—a made-up example or a sample derived by chance and most often de-contextualized. Providing evidence for language structure and use from corpora is not limited to a specific level of linguistic analysis but works at all levels, from sound to form and to function. The data in corpora are tagged and annotated and thus provide the exact type of sampling that empirically supports the hypotheses. As a repository of real language samples, a corpus query returns citations of language use that had not been envisaged before. Additionally, the empirical nature of corpora makes it possible to replicate the analysis conducted, which is not possible with data based on introspection.

Citations retrieved from a corpus do not simply represent a particular linguistic manifestation, but also provide quantificational information. The occurrences in the data and the patterns in which they occur also provide evidence for their distribution. Depending on the level of analysis and particular research questions at hand, linguists may derive various conclusions regarding different aspects of natural language use. The frequency information concerning distribution of units and patterns may have practical as well as theoretical implications in linguistics.

The language use captured in linguistic corpora further incorporates "meta" information for its users in terms of major participants or components of a communication event. These include the gender of the participants, their age as well as their dialectical background, the medium of the text and its specific genre, among

---

[4]Titania, The Bank of English: www.titania.bham.ac.uk (Accessed Sept. 14, 2017).

others, all of which provide significant information to a linguist in an analysis of natural language use in context.

When we narrow down the actual corpus linguistic work conducted over the years, we observe that they cover major areas. Meyer (2004) lists these general areas which further include many other subfields of linguistics: Grammatical studies of specific linguistic constructions, lexicography, language variation, historical linguistics, contrastive analysis and translation theory, natural language processing, language acquisition, and language pedagogy.

The ever-growing number of publications and the appearance of special journals in the field clearly underline the increasing importance of corpora in linguistics. It is evident that linguists with different interests will continue to build and use corpora in the future. As before, contributions from neighboring disciplines like computational linguistics and natural language processing research will continue to play a significant role in the future of corpus linguistics. As observed by Sampson (2013), there is currently a rising trend in linguistic analyses to adopt empirical approaches.

14.5 Turkish Linguistic Corpora

We may argue that there are at least three different kinds of corpora in Turkish today: (1) large-sized general linguistic corpora that are constructed and made available for users with proper corpus tools, (2) small-sized specialized corpora that are constructed for the study of specific research questions and are confined to the builders only, and (3) NLP corpora built with no linguistic criteria in mind, but rather as tools for testing algorithms devised for different applications.

We cannot say Turkish is a well-studied language when compared to other languages, for which there are well-documented histories and grammars. In other words, there exist catalogs of constructions or structures that have been collected and documented; however, the number of grammars or general descriptions of Turkish at different levels of representations are quite limited in number. Most linguistic works in current Turkish studies concentrate on a small number of fields like discourse analysis, pragmatics, or syntax. Rarely do we find works on semantics or lexicology or in any other domains, probably because they require enriched datasets. A well-balanced and representative corpus of Turkish is thus a necessity in studying the language where the accumulated and documented potentials of the language and its representative datasets are relatively small in number.

What may be called preelectronic corpora of Turkish are, in fact, not collections of texts, but rather compilations of lexemes. As early as the tenth century, we find the very first dictionary of Turkic languages, namely, the *Compendium of the Languages of the Turks*,[5] compiled by Mahmud al-Kashgari in 1072. Two major undertakings

---

[5] Divânu Lügati't-Türk.

of the Turkish Language Institute (TDK) in the early 1930s may also be considered early examples of data compilations. The monumental *Derleme Sözlüğü* (Dictionary of Compilations), motivated by the Turkification of lexis during the early years of the newly founded Republic, aimed at compiling vocabulary from the existing dialects of the time. From printed material, a number of themes were listed and then collectors were recruited from village intelligentsia to record samples of lexis. Initially printed in four volumes, this huge dictionary reached its current twelve volumes over the years 1963–1982. The second dictionary, *Tarama Sözlüğü*, also aimed at finding and revitalizing native lexical stock, was published in eight volumes in 1977. The dictionary compiled lexical items of Turkish origin from about 160 different historical texts starting as early as the thirteenth century. In both cases, however, the linguistic material is not extracted from a specially constructed corpus.

The pioneering work and current studies suggest that the role of data seems to be well appreciated in Turkish linguistic work. Apart from very few theoretical studies, almost all linguistic analyses are empirical and data-based. A typical research in Turkish linguistics gathers a "data base" or a "data set" in the analysis of the question at hand. We may say that there are very small-sized special corpora employed in almost all usage-based empirical research. However, these are severely confined in their form and size, they are not available for other researchers, and the data was collected with a specific problem at hand. Such work does not preprocess the data or use corpus-analytic tools.

Work in computational linguistics in Turkish has a longer history than Turkish corpus linguistic studies. The early beginning of corpus research in Turkish was in fact prompted by NLP research and computational linguistics analyses. In computational linguistics and in NLP, large-scale corpora are constructed for "practical" purposes. In a very reductionist manner, it is possible to say that researchers in these domains built corpora first and foremost to evaluate the algorithms that they had developed and to use corpora as a testing ground.

A comprehensive history of computational linguistics in Turkey has yet to be written; however, there are occasional references to earlier work in the field. The first known electronic corpus for linguistic analysis was constructed by Köksal (1976) for "automatic morphological analysis." Köksal tested and evaluated his algorithm over a corpus of 1534-word text sample randomly selected from daily newspapers. Even at this very early stage, some degree of representativeness and balance was sought: "…materials have been selected from the most important six daily newspapers representing different political views and linguistic trends." (Köksal 1976). Köksal's work recognizes the rich morphology of Turkish and possible morpheme combinations, and also points to major challenges further ahead, noting potential fields of application, urging building larger corpora for automated language analyses.

14.5.1   METU-Turkish Corpus

The first electronic linguistic corpus designed and compiled to represent modern Turkish is the Middle East Technical University (METU) Turkish Corpus. The developers of the METU Turkish Corpus (hereafter MTC) note this fact and state that the basic aim was to design a balanced written corpus on Turkish with the hope that it will prove useful to descriptive and theoretical studies alike (Say et al. 2004).

The MTC is also a mother corpus from which two subcorpora are derived. The first one is a morphologically and syntactically annotated treebank of Turkish, namely, the METU-Sabancı Turkish Treebank (Oflazer et al. 2003) (see also Chap. 13), which contains almost 7260 sentences and 65,000 words, and syntactic annotation is realized in a dependency-based XML-compliant format. The genre distribution in the treebank follows the MTC. The METU-Sabancı Turkish Treebank has served as a significant electronic resource for many studies for a long time (see, e.g., Kırkıcı (2009) for realizations of nominal compounds; Çetinoğlu (2014) for developing morphological disambiguators on the basis of the Turkish Treebank). The METU-Turkish Discourse Bank (METU-TDB) (Zeyrek et al. 2013) (see also Chap. 16), which is the first attempt to develop discourse annotation procedures in Turkish, is the second sub-corpus. In order to build an annotated discourse resource for Turkish, an approximately 400,000-word sub-corpus was extracted from MTC datasets, and discourse connectors (i.e., coordinating conjunctions, subordinating conjunctions, discourse adverbials, and phrasal expressions) were annotated manually, sharing the same principles as the Penn Discourse Treebank (Zeyrek et al. 2009). The METU-TDB project has so far developed the sub-corpus, the annotation tool, and the TDB query browser as its products that are freely distributed to academic users.[6]

In introducing the design decisions and principles of the MTC and the processes that led to its construction, the builders are not only confronted with issues facing "trailblazers" in general, but also are faced with many standard problems that corpus builders had to tolerate during construction. The constant reference to "limited resources" by the builders in presenting their construction process and its effects on the final product can be observed in a number of places as we will note below.

The MTC is a two million-word general corpus, composed of post-1990 written texts representing different genres. It includes texts from ten different genres and consists of 520 sample texts from 291 different sources (Table 14.1). The corpus does not have a spoken component, the lack of which is explained by the limitation of resources and experience required to process spoken language data at the time of the design process (Say et al. 2004).

As for representativeness of the corpus, the developers suggest that they preferred an "opportunistic" approach. It appears that within the severely limited prospects of accessing and digitizing the data sources (restricted permissions granted by the

---

[6]www.medid.ii.metu.edu.tr/ (Accessed Sept. 14, 2017).

Table 14.1 Genre distribution of the MTC (Say 2006) (reprinted with permission)

Genre	%
News	42
Novels	13
Stories	11
Articles	8
Op-ed columns	8
Essays	7
Research reports-surveys	5
Others (e.g., memoirs, course books)	3
Travel essays	2
Interviews	1
Total	100

publishers at the time and limited resources in terms of budget and workforce), the developers collected samples of electronic texts mainly from daily newspapers in the form of news and opinion columns. They were, however, very careful to maintain balance by selecting texts with no bias toward a particular genre or a writer. The corpus consists of texts dated between 1990 and 2002.

MTC is tagged by XCES style annotation using special software developed by the members of the project group as well as its corpus query workbench. A graphic-based browser, aimed at ordinary users of the corpus with its user-friendly features, was developed to be multi-platform compatible (see Özge and Say (2004) for a detailed description of the corpus workbench). The MTC remains today the only linguistically sound, freely distributed written corpus of modern Turkish.

From today's perspective and taking into account recent advances in corpus linguistics, the MTC is a less adequate source in meeting the demands of linguistic research. As of today, any general reference corpus is expected to be no less than 50 million words in size (Teubert and Cermakova 2004, p. 67). The defining aspects of balance and representativeness, as they have been discussed in recent years, became more and more important in evaluating a reference corpus as a reliable data source in the analyses of patterns emerging in language use in different genres, in varied contexts, and by users of different ages and genders, among many others. Even though the internal balance of the MTC is maintained to a certain degree, almost half of the corpus consists of texts from newspapers (single medium) and represents mainly news and columns (limited genres); therefore, its overall balance and representativeness fall short in meeting the standards set for current linguistic corpora. As emphasized by Lew (2009), the text types most commonly overrepresented in reference corpora are newspaper archives and fictions. In the MTC, as indicated above, newspapers as a text type are overrepresented.

It is evident that despite technological advances in capturing data via sophisticated scanning tools and software, an increase in digitization capacities, the ease of finding texts in corpus construction, and common data management in building processes, corpus development is still a very laborious undertaking. The developers of the MTC should be considered as forerunners who have successfully achieved their goals in the face of huge limitations in the resources allocated.

The number of linguistic analyses taking the MTC (also other corpora derived from it) as the major resource grew rapidly in the years following its construction. It has proven its usefulness and still continues to do so for researchers, as a wealth of studies (numerous graduate dissertations and academic articles) in NLP and linguistics make use of the MTC in their analyses (to name a few, see for example, Kawaguchi (2005) for the analysis of participle and infinitive nominalizations; Karaoğlan et al. (2013) for testing metrics in corpus normalization). Given that the MTC is a written corpus with no spoken component and its limitations in extracting quantitative outputs, linguistic studies conducted over the data clustered mostly in the fields of semantics, pragmatics (Ruhi 2009), and language acquisition (Sofu and Altan 2009). Most of these studies simply use the MTC as a naturally-occurring database of Turkish to obtain either sample extracts or frequency counts of linguistic items to validate their hypotheses. There is hardly a linguistic study (e.g., Işık-Güler and Ruhi 2010; Zeyrek 2012) that follows quantitative methods of corpus linguistics and exploits the MTC to describe any issue in Turkish linguistics thoroughly on the basis of a corpus-driven or corpus-based approach.

14.5.2   Turkish National Corpus (TNC)

In the years following the construction of the MTC, the need for a large-scale general reference corpus of Turkish has become more obvious. To meet the challenge, a group of linguists at Mersin University decided to build a reference corpus of Turkish.[7] The project team followed right from the start the so-called best practices at all stages of corpus development. The end product is the Turkish National Corpus (TNC),[8] a well-balanced, representative, and large-scale (50 million words) free resource of a general-purpose corpus of contemporary Turkish.

The design decisions in the construction of the TNC benefited entirely from previous practices. Major design principles were adopted from the experiences of the British National Corpus (BNC) with minor modifications. The simple idea was to follow the BNC model in constructing a linguistic corpus that would represent the language in a well-balanced manner. Considering the labor-intensive nature of the corpus construction task and limitations on reaching and finding relevant data sources, the size of the corpus was decided to be reduced to half of the BNC size where the distribution of the corpus content is proportionally preserved. The number of words in the corpus is distributed proportionally for each medium, time span, and text domain. Since the BNC is commonly accepted as a balanced corpus, many

---

[7]This was supported by the The Scientific and Technological Research Council of Turkey (TÜBİTAK) (Grant no: 108K242).

[8]www.tnc.org.tr (Accessed Sept. 14, 2017).

Table 14.2 Composition of the written component of the TNC (Aksan et al. 2012) (reprinted with permission)

Domain	%	Medium	%
Fiction	19	Books	58
Social sciences	16	Periodicals	32
Art	7	Misc. published	5
Commerce-Finance	8	Misc. unpublished	3
Op-ed pieces	4	Spoken texts[a]	2
World affairs	20		
Applied sciences	8		
Natural sciences	4		
Leisure writing	14		

[a] Material that is written to be spoken, such as political speech, news broadcasts, etc.

other currently available large-sized reference corpora (e.g., the *American National Corpus*, the *Korean National Corpus*, and the *Polish National Corpus*) also adopt the BNC model to achieve balance and representativeness (McEnery et al. 2006, p. 17).

The selection of texts is based on three criteria: text domain, time, and medium. Put simply, the imaginative domain includes mainly works of fiction (novels, short stories, poems, drama) and the informative domain includes texts representing the sciences, the arts, commerce-finance, belief-thought, world affairs, and leisure. Imaginative texts constitute 19% and informative texts 81% of the TNC, following the distribution adopted in the BNC.

The time span of the texts in the TNC covers a 20-year period between 1990 and 2010. The distribution of sample texts from each medium and domain with respect to years in the period is also carefully calculated (Table 14.2). As for matters of size, the time period covered was also decided on the basis of the volume of publications produced in Turkish and consumed by language users in different genres and text types (see Aksan et al. (2012) for more details of text type choices according to the domains and mediums).

The spoken component of the TNC is composed of orthographic transcriptions of spoken language compiled from formal and informal communicative settings. These include spontaneous, everyday conversations on a variety of topics by users of different ages and genders, and samples of spoken communicative events collected from meetings, lectures, and speeches. A total of one million words in the spoken component represent 2% of the TNC.

Morphological analysis and part-of-speech annotation of the TNC has been done by developing an NLP dictionary based on the NooJ_TR module (Aksan and Mersinli 2011). The unique semiautomatic process of developing the NLP dictionary includes the following steps: (1) automatically annotating the type list with the NooJ_TR module, which follows a root-driven, non-stochastic rule-based approach to annotating the morphemes of the given types by using a graph-based finite-state transducer; (2) manually checking and revising the output and eliminating artificial ambiguities and non-occurring, theoretically possible multi-tags.

The TNC lexicon files containing linguistically motivated tag sets were constructed from scratch. Optimization of the NLP dictionary was conducted manually. Unlike previous studies, the remaining ambiguities do not contain artificial ambiguities and thus serve as a good basis for their documentation (Aksan et al. 2012). Unlike the available taggers, the resulting TNC tagger does not include artificial or theoretically possible but non-occurring ambiguities. Additionally, the number of affixes and the assigned tags for them are all valid in terms of current linguistics literature.

The TNC has a platform-independent, user-friendly Web-based user interface for making queries. It provides for multitude of features for the analysis of corpus texts including concordance display, sorting concordance data, creating descriptive statistics for query results over the language-external restriction categories of texts via distribution, and compiling lists of collocates for node words on the basis of several statistical methods. With 48 million words, the TNC-Demo Version represents 4438 different data sources over 9 domains and 34 different genres, and was published as a free resource for noncommercial use in October 2012. The morphologically annotated, complete version of the TNC v3.0 is planned for release in 2018, offering new query options for linguistic analyses.

The number of users and the number of studies using the TNC as the major electronic resource is increasing. While some of the studies use the TNC for compiling naturally-occurring language evidence and for hypothesis-testing (e.g., Sebzecioğlu 2013; Akşehirli 2014), there are still others following a corpus-driven approach that attempts to build hypotheses and describe Turkish on the basis of the TNC (see, e.g., (Erköse and Uçar 2014) for the cognitive semantic analysis of posture verbs in TNC). Since the TNC is a linguistic corpus, and because it is well-balanced and representative, the conclusions based on TNC data provide valid linguistic descriptions of Turkish, both qualitatively and quantitatively. For example, for the first time in Turkish linguistics, we are able to account for patterns of language use that would give hints for formulaicity in Turkish (see Uçar and Kurtoğlu (2012) for semantic patterning of polysemous verbs; Aksan and Aksan (2013) for genre specification through multiword units). It is now possible to derive frequency information of Turkish lexical items and affixes (Aksan and Yaldır 2012; Aksan and Aksan 2014) as well as multiword units (Aksan and Aksan 2012).

14.5.3 Spoken Turkish Corpus (STC)

The Spoken Turkish Corpus (STC) is the only corpus of its kind that is available for linguistic analyses. Given that the challenges faced by builders of spoken corpora are demanding and that they require a different set of measures for the creation of the resource, maintenance, and dissemination (see e.g., Ruhi et al. (2014) for recent debates on best practices for spoken corpora in linguistic research), the STC is a pioneering work undertaken to create and sustain a multimodal spoken corpus that overcomes most of these challenges in order to be published in its demo version. It is also the product of a team of linguists at METU, constructed with contributions from international collaborators.

The STC is the first general-purpose, large-scale corpus of present-day spoken Turkish. The ultimate aim is to reach the size of ten million words, so the corpus is designed accordingly. Ruhi (2011) states that the raw database of the STC currently contains three million words of audio and video recordings from a variety of geographical and social settings and domains. About 440,000 words of these recordings are under transcription control, with partial morphological and speech act annotation processing in the corpus management system. The STC Demo Version consists of 23 communications and represents 2.4 h of interaction, with 18,357 tokens having been published. It is freely available for nonprofit research purposes.[9] Since the STC is a multimodal corpus, the transcriptions are presented in a time-aligned manner with audio and video files. It uses EXMARaLDA (Extensible Markup Language for Discourse Analysis), an open-source system of data models, formats, and tools for the production and analysis of spoken language corpora (Schmidt 2004).[10] Transcriptions are created with EXMARaLDA's Partitur Editor. The project team adapted a revised form of HIAT for the transcriptions (Ruhi et al. 2010b). The partial morphological analysis of the STC data is done with TRmorph (Çöltekin 2010), and the annotation of requestive/directive speech acts is implemented with Sextant (Wörner 2009) (see Ruhi et al. (2011) and Ruhi (2014) for retrieving requestive/directive speech acts). The final aim is to create a spoken resource annotated for morphology, the socio-pragmatic features of Turkish (e.g., address terms, [im]politeness markers, and a selection of speech act realizations), anaphora, and gestures (Ruhi et al. 2010b).

Among its notable features, the STC's pragmatically informed metadata fields make the sociocultural situatedness of communication visible to researchers. While determining the metadata features, the STC has scrutinized and considered the text classification and other metadata parameters proposed in standardization schemes and features implemented in other spoken corpora (e.g., the BNC). At the same time, in order to achieve pragmatically more fine-grained text descriptors, the STC implements a two-layered scheme regarding text type and discourse content.

On the first level, texts are classified according to speaker relations and the major social activity type. The domains for speaker relations are family/relatives, friend, family-friend, educational, service encounter, workplace, media discourse, legal, political, public, research, brief encounter, and unclassified conversations (Ruhi et al. 2010b). These domains are then subclassified according to activities. The class of workplace discourse includes, for instance, meetings, workplace cultural events (e.g., parties), business appointments, business interviews, business dinners, shoptalk, telephone conversations, and chats.

The second layer of metadata annotation is implemented at the corpus assignment stage and involves the annotation of speech acts based on Searle (1975) (e.g.,

---

[9]std.metu.edu.tr (Accessed Sept. 14, 2017).

[10]exmaralda.org (Accessed Sept. 14, 2017).

Table 14.3 Distribution of domains planned for the STC

Domain	%
Conversations among family and/or relatives	25
Workplace conversations	20
Education	15
Broadcasts	15
Conversations among friends and/or acquaintances	12
Service encounter	5
Natural sciences	4
Other	4

offers and requests), on the one hand, and, on the other hand, the annotation of conversational topics (e.g., child care), speech events (e.g., troubles talk,[11]), and ongoing activities (e.g., cooking)—all encoded under the super metadata category, Topic, in the current state of STC. Speech act and Topic annotation are thus two further metadata parameters in STC (Ruhi et al. 2012).

It is possible to overview the content of the corpus in terms of text categories and the distribution of gender and age at the website of the STC and in its demo version. Table 14.3 displays the STC domains and the planned proportion of the samples from them.[12]

With the publication of the STC Demo version, spoken Turkish discourse has been investigated from different perspectives. The *Journal of Linguistics and Literature* published a special issue on corpus-based analysis of interactional markers (e.g., *tamam* 'okay,' *şey* 'thing,' *hayır* 'no') in the demo version and a selection of the publishable version of the STC (Ruhi 2013). The studies in the collection highlight the significance of "corpus-based perspective to analyzing spoken Turkish and to explore the affective dimension of a number of markers especially in regard to relational management in the tradition of (im)politeness theories" (Ruhi 2013, p. 2). Since the STC consists of data collated from a relatively wide range of domains and genres, the articles explore the pragmatic functions of a number of interaction markers in these domains and genres, and thus they display a depth of discourse domains in the analysis of spoken Turkish. Another comprehensive study, Çelebi (2014) aims to develop a methodological framework to analyze impoliteness in a corpus-driven approach. To attain this goal, the study investigates the STC demo and its publishable data thoroughly by emphasizing the empirical and explanatory power of a corpus approach in pragmatics studies. Lastly, the STC demo version is

---

[11] Tannen defines troubles-talk as a conversational event where interlocutors "share their moments of frustration and irritation, but without expecting a solution"—see The Art of Talking and Listening (Philosophy on the Mesa, November 22, 2010): philosophyonthemesa.com/tag/deborah-tannen/ (Accessed Sept. 14, 2017).

[12] See Spoken Turkish Corpus. Main Features of STC Demo Version: std.metu.edu.tr/en/main-features-of-stc-demo-version (Accessed Sept. 14, 2017).

also utilized to annotate explicit discourse connectives of spoken Turkish in line with the Turkish Discourse Bank's style of annotation (Demirşahin and Zeyrek 2014).

It is worth mentioning another attempt to construct a spoken corpus of Turkish. As a product of two research projects conducted at the Institute of Global Studies and Tokyo University of Foreign Studies,[13] a Corpus of Spoken Turkish containing 514,400 tokens compiled from free conversations on a variety of topics is published and distributed freely for academic research purposes.[14]

This second kind of corpora that we have noted above are the small-size specialized corpora or datasets, each designed for the study of a specific problem identified by the researcher/builder. The existence of such corpora can only be discovered when a particular study appears in publication, announcing the results of the analysis based on a special corpora built for that particular problem only. This is a more common practice in discourse analysis (see e.g., Özyıldırım (2010) for genre analysis on a 160,000-word corpus; Oktar and Cem-Değer (1999) for a critical discourse analysis on 15 newspaper articles) or pragmatics studies where the researcher gathers data either for citing natural language use that would provide evidence for a particular type of a text or speech act (see e.g., Ruhi (2006) for politeness in compliment responses on a spoken Turkish dataset) or to document context-specific preferences in confined contexts of use (see e.g., Çubukçu (2005) for constructive back-channels in 30 Turkish conversations recorded during everyday conversations, business, and formal discussions). There are also small-size sub-corpora that are extracted from the datasets of already existing larger corpora. For instance, the spoken sub-corpus of the TNC containing private and public speeches and conversations is used to investigate discourse analytic and corpus-driven features of requests (Aksan and Mersinli 2015) and thanking (Aksan and Demirhan 2015) speech acts in Turkish.

In addition to the major linguistic corpora we have reviewed above, there are also specialized corpora, as we have noted previously. These are constructed to serve as a comprehensive resource for the particularly specified aims of the researchers. Uçar (2014) built a 713,000-word corpus of the popular comedy show *Komedi Dükkanı* (Comedy Shop) to analyze the semantic and pragmatic properties of conversational humor in Turkish (see also Uçar and Yıldız 2015). To examine lexico-grammatical differences and similarities in predicate uses among disciplinary discourses, Yıldız and Aksan (2014) compiled data from the introduction and conclusion sections of 1178 scientific articles published in the humanities, applied sciences, and basic sciences, and built a one- million-word specialized corpus of Turkish scientific text. Similarly, Uzun et al. (2014) conducted their rhetoric structure analysis on a one-million-word corpus of social science academic articles obtained from the Social Science Database of TÜBA ULAKBİM. Here, we should note that these corpora

---

[13]The twenty-first COE Program "Usage-Based Linguistic Informatics" 2002–2006 and the Global COE Program "Corpus-based Linguistics and Language Education" (2007–2011).
[14]Global COE Program, Corpus-based Linguistics and Language Education: cblle.tufs.ac.jp/en/ (Accessed Sept. 14, 2017).

are not available for other users and do not provide any interface for access. They solely provide linguistically significant outcomes for their specialized domains.

The NLP corpora in Turkish easily outnumber the available linguistic corpora.[15] As we have noted above, a corpus linguistic analysis of Turkish in fact was initiated by the work of NLP researchers. Such corpora cannot be defined as "linguistic" corpora and can by no means function as a representative and a well-balanced resource for linguistic analyses. The main reason why this web-harvested collection of texts is not considered linguistically significant corpora is that they lack design principles or a rationale (Wynne 2005) in their creation. The following points specify the results of this shortcoming on the basis of the principles of corpus design:

- They are not representative and balanced in terms of the text samples they contain. A representative and balanced sample of written and/or spoken texts is compiled in a linguistic corpus, and, thus, observations on linguistic behavior of queried items on this corpus constitute both quantitative and qualitative linguistic findings. These findings lead linguists to make generalizations on typical and central properties of that language overall (see Hoffmann et al. 2008). Otherwise, "without representativeness whatever is found to be true of a corpus, is simply true of that corpus—and cannot be extended to anything else" (Leech 2007, p. 135).
- They are not designed and constructed to meet the external criteria (e.g., domain, genre, date of sample texts) of the corpus- creating process. As a result of this, most of them do not carry any metadata information and thus the content of the corpora is not transparent pertaining to documentation. As underscored by Sinclair (2005), the proper stance of corpus compiler is "to be detailed and honest about the contents. From their description of the corpus, the research community can judge how far to trust their results, and future users of the same corpus can estimate its reliability for their purposes" (p. 98).
- Most of them are not available for public use. Even if they are publicly available as datasets (see e.g., Ferraresi et al. (2008) for English ukWaC; Sak et al. (2011) for Turkish BOUNCorpus. Yıldız University provides a variety of Turkish datasets containing Turkish tweets, blogs, poems, etc.[16]), linguists are not able to utilize them as a language resource for their studies since these corpora are not published with user interfaces to process the sample texts they contain and to conduct corpus queries on them.

Obtaining Web content and processing it as an offline, static corpus is described as Web for Corpus (de Schryver 2002). In line with this approach, The

---

[15]In this chapter, we have strictly confined ourselves to corpora constructed following basic design principles that define the products as corpora in the true sense of the term. There are a number of corpora, some of which are even publicly available; however, they neither provide information regarding their design criteria nor follow the general guidelines of legal issues in corpus construction. Such corpora will not be reviewed here.

[16]www.kemik.yildiz.edu.tr/?id=28 (Accessed Sept. 14, 2017).

BOUNCorpus, constructed to exploit Turkish morphology in natural language processing applications, is the largest web-crawled corpus containing 500 million words. It is composed of NewsCor, which contains texts from three major news portals in Turkish, and GenCor, which includes texts from a general sampling of Turkish Web pages. The corpus is encoded by following the XML Corpus Encoding Standard, XCES[17], and is freely available as a language resource (Sak et al. 2011). Compared to the BOUNCorpus, the relatively small size TurCo is a 50-million-word corpus with 90.40% of it compiled from ten different sites with Turkish content. It is widely used to investigate lexical statistical properties of Turkish (Dalkılıç and Çebi 2002) and to test Turkish word n-gram analysis algorithms (Çebi and Dalkılıç 2004).

Along with these web-derived datasets, the 42-million-word TurkishWaC (Ambati et al. 2012), containing texts from Wikipedia entries and built by employing the Corpus Factory Method (Kilgarriff et al. 2010), is accessible through the commercial corpus query tool Sketch Engine.[18] The tool is a web-based program and works on corpora of any language with tokenized, lemmatized, and POS-tagged content. It offers a number of language-analysis functions among which the most significant are concordance outputs and word sketches summarizing the grammatical and collocational behavior of the query items.

It should be noted that the Web is also accessed directly via Internet-based search engines as a dynamic corpus itself and freely available tools like WebCorp[19] (Renouf et al. 2007), providing users options to utilize the Web as a corpus through commercial search engines. WebCorp is developed for studying language on the Web, and in this respect, searches can be performed to find words or phrases, including pattern matching, wildcards, and part-of-speech. Results are given as concordance lines in KWIC format. Post-search analyses are possible, including time series, collocation tables, sorting, and summaries of meta-data from the matched web pages.

14.6 Conclusions

In this short review, we presented the basics of linguistic corpora, and efforts in Turkey in developing different types of linguistic corpora. Still in its infancy, Turkish corpus linguistics is "practical, pragmatic, and opportunistic." The coming years will bring more sophisticated products, tools of analyses, and linguistic research. A thorough evaluation of the current state of research on language technologies on Turkish was previously presented in the final report of a workshop organized

---

[17]Vassar College, Department of Computer Science, NY, USA: www.xces.org (Accessed Sept. 14, 2017).

[18]Lexical Computing CZ s.r.o.: www.sketchengine.co.uk (Accessed Sept. 14, 2017).

[19]Birmingham City University, Research and Development Unit for English Studies: www.webcorp.org.uk/ (Accessed Sept. 14, 2017).

by the Foundation of the National Speech and Language Technologies Platform in October 2011 (Doğan 2011). Among others, in a separate questionnaire, the participants were asked to evaluate "status of tools and resources for Turkish." On a scale of 0–6 points, "reference corpora" received 1.9 for quantity and 2.9 for quality. The other types of corpora also were assigned scores in the same questionnaire, including treebanks, semantic corpora, discourse corpora, parallel corpora, and speech corpora, and they did not fare much better than the reference corpora. The expert participants, some of whom were corpus builders themselves, agreed to score available corpora below average with respect to measuring criteria. It is no surprise that the final report places the insufficiency of data sources and corpora to the very top of the list of negatively evaluated aspects of the field. We believe that, when asked, the evaluation of the present state of corpus studies would score the same by linguists as well.

References

Aksan M, Aksan Y (2012) Multi-word units in informative and imaginative domains. In: Proceedings of the international conference on Turkish Linguistics, Ankara

Aksan M, Aksan Y (2013) Multi-word units and pragmatic functions in genre specification. In: Proceedings of the international pragmatics conference, New Delhi, pp 239–240

Aksan Y, Aksan M (2014) Frequency effects in Turkish: a study on multi-word units. In: Proceedings of the international conference on Turkish Linguistics, Rouen

Aksan Y, Demirhan UU (2015) Expressions of gratitude in the Turkish National Corpus. In: Ruhi Ş, Aksan Y (eds) Exploring (im)politeness in specialized and general corpora: converging methodologies and analytic procedures. Cambridge Scholars, Newcastle upon Tyne, pp 121–172

Aksan M, Mersinli Ü (2011) A corpus-based Nooj module for Turkish. In: Proceedings of the Nooj 2010 international conference and workshop, Komotini, pp 29–39

Aksan M, Mersinli Ü (2015) Retrieving and analyzing requestive forms: evidence from the Turkish National Corpus. In: Ruhi Ş, Aksan Y (eds) Exploring (im)politeness in specialized and general corpora: converging methodologies and analytic procedures. Cambridge Scholars, Newcastle upon Tyne, pp 173–220

Aksan Y, Yaldır Y (2012) A corpus-based frequency list of Turkish: evidence from the Turkish National Corpus. In: Proceedings of the international conference on Turkish linguistics. Gold Press Nyomda Kft, Szeged, pp 47–58

Aksan Y, Aksan M, Koltuksuz A, Sezer T, Mersinli Ü, Demirhan UU, Yılmazer H, Kurtoğlu Ö, Öz S, Yıldız İ (2012) Construction of the Turkish National Corpus (TNC). In: Proceedings of LREC, Istanbul, pp 3223–3227

Akşehirli S (2014) Dereceli karşıt anlamlılarda belirtisizlik ve ölçek yapısı. J. Lang. Linguist. Stud. 10:49–66

Ambati BR, Reddy S, Kilgarriff A (2012) Word sketches for Turkish. In: Proceedings of LREC, Istanbul, pp 2945–2950

Baker P, Hardie A, McEnery T (2006) A glossary of corpus linguistics. Edinburgh University Press, Edinburgh

Barlow M (2011) Corpus linguistics and theoretical linguistics. Int J Corpus Linguist 16:3–44

Biber D (1993) Representativeness in corpus design. Lit Linguist Comput 8:243–257

Carter R, McCarthy M (2004) Talking, creating: interactional language, creativity, and context. Appl Linguist 25(1):62–88

Çebi Y, Dalkılıç G (2004) Turkish word *n*-gram analyzing algorithms for a large-scale Turkish corpus – TurCo. In: Proceedings of international conference on information technology: coding and computing, Las Vegas, NV, pp 236–240

Çelebi H (2014) Impoliteness in corpora: a comparative analysis of British English and spoken Turkish. Equinox, London

Çetinoğlu Ö (2014) Turkish treebank as a gold standard for morphological disambiguation and its influence on parsing. In: Proceedings of LREC, Reykjavík, pp 3360–3365

Çöltekin Ç (2010) A freely available morphological analyzer for Turkish. In: Proceedings of LREC, Valetta, pp 820–827

Çubukçu H (2005) Karşılıklı konuşmada destekleyici geri bildirim. In: Ergenç İ (ed) Dilbilim İncelemeleri. Doğan Yayıncılık, Ankara, pp 289–305

Dalkılıç G, Çebi Y (2002) A 300MB Turkish corpus and word analysis. In: Proceedings of the conference on advances in information systems. LNCS, vol 2547. Springer, Berlin, pp 205–212

Davis M (2008) The 385+ million word corpus of American English (1990–2008+): design, architecture and linguistic insights. Int J Corpus Linguist 14(2):159–190

Demirşahin I, Zeyrek D (2014) Annotating discourse connectives in spoken Turkish. In: Proceedings of the linguistic annotation workshop, Dublin, pp 105–109

de Schryver G (2002) Web for/as corpus: a perspective for the African languages. Nord J Afr Stud 11:266–282

Doğan M (ed) (2011) Multisaund: Ulusal konuşma ve dil teknolojileri platformu kuruluşu ve Türkçede mevcut durum çalıştayı bildirileri, TÜBİTAK-BİLGEM, Gebze

Du Bois J, Chafe W, Meyer C, Thompson S, Englebretson R, Martey N (2005) Santa Barbara corpus of spoken American English, Parts 1–4, Philadelphia, PA

Erköse Y, Uçar A (2014) Türkçedeki dur- konumlama eyleminin derlem temelli bilişsel anlam çözümlemesi. In: Proceedings of the national linguistics conference, Hacettepe University, Kemer, pp 351–358

Ferraresi A, Zanchetta E, Baroni M, Bernardini S (2008) Introducing and evaluating ukWaC, a very large web-derived corpus of English. In: Proceedings of the workshop on web as corpus workshop – Can we beat Google? Marrakech, Morocco

Francis W, Kučera H (1964) A standard corpus of present-day edited American English, for use with digital computers. Brown University, Providence, RI

Granger S (2003) The international corpus of learner English: a new resource for foreign language learning and teaching and second language acquisition research. TESOL Q 37(3):538–546

Greenbaum S (1991) The development of international corpus of English. In: Aijmer K, Altenberg B (eds) English corpus linguistics. Studies in honour of Jan Svartvik. Longman, London, pp 83–91

Greenbaum S, Svartvik J (1990) The London-Lund Corpus of Spoken English. In: Svartvik J (ed) The London-Lund corpus of spoken English: description and research. Lund University Press, Lund, pp 11–45

Hoffmann S, Evert S, Smith N, Lee D, Prytz YB (2008) Corpus linguistics with BNCweb: a practical guide. Peter Lang, Frankfurt

Holmes J, Vine B, Johnson G (1998) Guide to the Wellington corpus of spoken New Zealand English. University of Wellington Press, Wellington

Işık-Güler H, Ruhi Ş Ruhi (2010) Face and impoliteness at the intersection with emotions: a corpus-based study in Turkish. Intercult Pragmat 7:625–660

Karaoğlan B, Dinçer BT, Kışla T, Kumova-Metin S (2013) Derlem normalizasyonu için bir öneri. In: Proceedings of IEEE signal processing and communications applications conference, Magosa

Kawaguchi Y (2005) Two Turkish clause linkages: –DIK- and –mE-: a pilot analysis based on the METU Turkish corpus. In: Takagaki T, Zaima S, Tsuruga Y, Moreno-Fernandez F, Kawaguchi Y (eds) Corpus-based approaches to sentence structures. John Benjamins, Amsterdam, pp 151–177

Kennedy G (1998) An introduction to corpus linguistics. Longman, London

Kilgarriff A, Reddy S, Pomikalek J, Avinesh PVS (2010) A corpus factory for many languages. In: Proceedings of LREC, Valletta, pp 904–910

Kilimci A, Can C (2009) TICLE: Uluslararası Türk Öğrenci İngilizcesi Derlemi. In: Sarıca M, Sarıca N (eds) Proceedings of the national linguistics conference, Yüzüncü Yıl Üniversitesi, Van, pp 1–11

Kırkıcı B (2009) İmparator çizelgesi vs. imparatorlar çizelgesi: on the (non)-use of plural non-head nouns in Turkish nominal compounding. Dilbilim Araştırmaları Dergisi 1:35–53

Köksal A (1976) A first approach to a computerized model for the automatic morphological analysis of Turkish. PhD thesis, Hacettepe University, Ankara

Leech G (1992) Corpora and theories of linguistic performance. In: Svartvik J (ed) Directions in corpus linguistics. Mouton de Gruyter, Berlin, pp 105–122

Leech G (2007) New resources, or just better old ones? The holy grail of representativeness. In: Hundt M, Nesselhauf N, Biewer C (eds) Corpus linguistics and the web. Rodopi, Amsterdam, pp 133–149

Leech G (2011) Principles and applications of corpus linguistics: interview with Geoffrey Leech. In: V V, Zyngier S, Barnbrook G (eds) Perspectives on corpus linguistics. John Benjamins, Amsterdam, pp 155–170

Lew R (2009) The web as corpus versus traditional corpora: their relative utility for linguists and language learners. In: Baker P (ed) Contemporary corpus linguistics. Continuum, London, pp 289–300

Lüdeling A, Kytö M (2008) Introduction. In: Lüdeling A KM (ed) Corpus linguistics: an international handbook. Walter de Gruyter, Berlin, pp v–xii

McEnery T, Hardie A (2012) Corpus linguistics: method, theory and practice. Cambridge University Press, Cambridge

McEnery T, Wilson A (1996) Corpus linguistics. Edinburgh University Press, Edinburgh

McEnery T, Xiao R, Tono Y (2006) Corpus-based language studies. Routledge, London

Meyer C (2004) English corpus linguistics: an introduction. Cambridge University Press, Cambridge

Newmeyer F (1986) Linguistic theory in America. Academic, London

Oflazer K, Say B, Hakkani-Tür DZ, Tür G (2003) Building a Turkish Treebank. In: Treebanks: building and using parsed corpora. Kluwer Academic, Berlin

Oktar L, Cem-Değer A (1999) Gazete söyleminde kiplik ve işlevleri. Dilbilim Araştırmaları Dergisi, pp 45–53

Özge U, Say B (2004) Development of a corpus workbench for the METU Turkish Corpus. In: Proceedings of LREC, Lisbon, pp 223–225

Özyıldırım I (2010) Tür çözümlemesi. Bilgesu Yayınları, Ankara

Renouf A (2007) Corpus development 25 years on: from super-corpus to cyber-corpus. In: Facchinetti R (ed) Corpus linguistics 25 years on. Rodopi, Amsterdam, pp 27–49

Renouf A, Kehoe A, Banerjee J (2007) WebCorp: an integrated system for web text search. In: Hundt M, Biewer C, Nesselhauf N (eds) Corpus linguistics and the web. Rodopi, Amsterdam, pp 47–67

Rissanen M, Kytö M, Kahlas-Tarkka L, Kilpiö M, Nevanlinna S, Taavitsainen I, Nevalainen T, Raumolin-Brunberg H (eds) (1991) The Helsinki corpus of english texts. University of Helsinki, Helsinki

Ruhi Ş (2006) Politeness in compliment responses: a perspective from naturally occurring exchanges in Turkish. Pragmatics 16:43–101

Ruhi Ş (2009) The pragmatics of *yani* as a parenthetical marker in Turkish: evidence from the METU Turkish corpus. In: Working papers in corpus-based linguistics and language education, vol 3, pp 285–298

Ruhi Ş (2011) Creating a sustainable large corpus of spoken Turkish for multiple research purposes. In: Proceedings of Multisaund: Ulusal konuşma ve dil teknolojileri platformu kuruluşu ve Türkçede mevcut durum çalıştayı, TÜBİTAK-BİLGEM, Gebze, pp 70–73

Ruhi Ş (2013) Interactional markers in Turkish: a corpus-based perspective. J Linguist Lit 10:1–7

Ruhi Ş (2014) Sözlü Türkçe Derlemi'nde temel arama ve edimbilimsel açımlama: Yöntem geliştirme. In: Proceedings of the national linguistics conference, Hacettepe University, Kemer, pp 271–279

Ruhi Ş, Eröz-Tuğa B, Hatipoğlu Ç, Işık-Güler H, Acar G, Eryılmaz K, Can H, Karakaş Ö, Karadaş DÇ (2010a) Sustaining a corpus for spoken Turkish discourse: accessibility and corpus management issues. In: Proceedings of the workshop on language resources: from storyboard to sustainability and LR lifecycle management, Valetta, pp 44–48

Ruhi Ş, Işık-Güler H, Hatipoğlu Ç, Eröz-Tuğa B, Karadaş DÇK (2010b) Achieving representativeness through the parameters of spoken language and discursive features: the case of the spoken Turkish corpus. In: Moskowich-Spiegel F, Isabel CG, Begona I, Lareo M, Sandino PL (eds) Language windowing through corpora. Visualización del Lenguaje a Través de Corpus. Universidade da Coruña, Coruña, pp 789–799

Ruhi Ş, Schmidt T, Wörner K, Eryılmaz K (2011) Annotating for precision and recall in speech act variation: the case of directives in the Spoken Turkish Corpus. In: Proceedings of the conference of the german society for computational linguistics and language technology – working papers in multilingualism, Hamburg, pp 203–206

Ruhi Ş, Eryılmaz K, Acar G (2012) A platform for creating multimodal and multilingual spoken corpora for Turkic languages: insights from the Spoken Turkish Corpus. In: Proceedings of the first workshop on language resources and technologies for Turkic languages, Istanbul, pp 57–63

Ruhi Ş, Haugh M, Schmidt T, Wörner K (eds) (2014) Best practices for spoken corpora in linguistic research. Cambridge Scholar, Newcastle upon Tyne

Sak H, Güngör T, Saraçlar M (2011) Resources for Turkish morphological processing. Lang Resour Eval 45(2):249–261

Sampson G (2013) The empirical trend. Int J Corpus Linguist 18:281–289

Say B (2006) Türkçe için bir derlem geliştirme çalışması. In: Bilgisayar Destekli Dilbilim Çalışmaları Bildirileri, TDK, Ankara, pp 81–88

Say B, Zeyrek D, Oflazer K, Özge U (2004) Development of a corpus and a treebank for present-day written Turkish. In: Proceedings of the international conference on Turkish linguistics, Magosa, pp 183–192

Schmidt T (2004) Transcribing and annotating spoken language with EXMARaLDA. In: Proceedings of the workshop on XML-based richly annotated corpora, Lisbon, pp 69–74

Searle JR (1975) A taxonomy of illocutionary acts. In: Mind and knowledge. Minnesota studies in the philosophy of science. University of Minnesota Press, Minneapolis, pp 344–369

Sebzecioğlu T (2013) Anlık oluşum ve Türkçe anlık sözcüklerin oluşum süreçleri üzerine bir betimleme. J Lang Linguist Stud 10:17–47

Sinclair JM (2005) Appendix to chapter one: how to make a corpus. In: Wynne, M (ed) Developing linguistic corpora: a guide to good practice. ota.ox.ac.uk/documents/creating/dlc. Accessed 3 July 2017

Sofu H, Altan A (2009) Partial reduplication: revisited. In: Proceedings of the international conference on Turkish linguistics, Wiesbaden, pp 63–72

Svartvik J (2007) Corpus linguistics 25+ years on. In: Facchinetti R (ed) Corpus linguistics 25 years on. Rodopi, Amsterdam, pp 11–25

Teubert W (2005) My version of corpus linguistics. Int J Corpus Linguist 10:1–13

Teubert W, Cermakova A (2004) Corpus linguistics: a short introduction. Continuum, London

Uçar A (2014) Özel amaçlı derlemi çeviriyazmak: Bir çeviriyazı modeli. Dilbilim Araştırmaları Dergisi 1:1–30

Uçar A, Kurtoğlu Ö (2012) A corpus-based account of polysemy in Turkish: a case of ver-'give'. In: Kincses-Nagy E, Biacsi M (eds) Proceedings of the international conference on Turkish linguistics. Gold Press Nymoda Kft, Szeged, pp 539–552

Uçar A, Yıldız İ (2015) Humor and impoliteness in Turkish: a corpus-based analysis of the television show Komedi Dükkânı 'comedy shop.'. In: Ruhi Ş, Aksan Y (eds) Exploring (im)politeness in specialized and general corpora: converging methodologies and analytic procedures. Cambridge Scholars, Newcastle upon Tyne, pp 40–81

Uzun L, Erk-Emeksiz Z, Turan ÜD, Keçik İ (2014) Sosyal bilimler alanında Türkçe yazılan özgün araştırma yazılarında uslamlama türlerine göre sav şemaları. In: Proceedings of the national linguistics conference, Kemer, pp 305–321

Wörner K (2009) Werkzeuge zur flachen Annotation von Transkriptionen gesprochener Sprache. PhD thesis, Bielefeld University, Bielefeld

Wynne M (2005) Developing linguistic corpora: a guide to good practice. icar.univ-lyon2.fr/ecole_thematique/contaci/documents/Baude/wynne.pdf. Accessed 14 Sept 2017

Yıldız İ, Aksan M (2014) Türkçe bilimsel metinlerde eylemler: Derlem temelli bir inceleme. In: Proceedings of the national linguistics conference. Hacettepe University, Kemer, pp 247–253

Zeyrek D (2012) Thanking in Turkish: a corpus-based study. In: Ruiz de Zarobe L, Ruiz de Zarobe Y (eds) Speech acts and politeness across languages and cultures. Peter Lang, Bern, pp 53–88

Zeyrek D, Turan ÜD, Bozşahin C, Çakıcı R, Sevdik-Çallı A, Demirşahin I, Aktaş B, Yalçınkaya İ, Ögel H (2009) Annotating subordinators in the Turkish Discourse Bank. In: Proceedings of the Linguistic annotation workshop, Singapore, pp 44–47

Zeyrek D, Demirşahin I, Sevdik-Çallı A, Çakıcı R (2013) Turkish Discourse Bank: porting a discourse annotation style to a morphologically rich language. Dialogue Discourse 4(2):174–184

Chapter 15
Turkish Wordnet

Özlem Çetinoğlu, Orhan Bilgin, and Kemal Oflazer

Abstract Turkish Wordnet is a lexical database for Turkish, built at Sabancı University in Istanbul, Turkey, between 2001 and 2004 as part of the Balkanet project. It currently contains 20,345 lexical items organized into 14,795 synonym sets (synsets hereafter), which are linked to each other via semantic relations such as hypernymy, antonymy, and meronymy. Turkish Wordnet uses the same concept pool as Princeton Wordnet, the eight wordnets of the Euro Wordnet project, and the five other wordnets of the Balkanet project. Synsets were added in several phases, starting with the most basic concepts at the top of the concept hierarchy. Monolingual resources were used to automatically extract semantic relations. Some semantic relations were extracted using the regular morphology of Turkish. Turkish Wordnet is available to researchers in the form of an XML file.

15.1 Introduction

This chapter provides an overview of Turkish Wordnet, a lexical database for Turkish, built at Sabancı University between 2001 and 2004 as part of the Balkanet project (Stamou et al. 2002), a 3-year, EU-funded project for the development of medium-sized wordnets for six languages: Bulgarian, Czech, Greek, Romanian, Serbian, and Turkish.

Ö. Çetinoğlu
University of Stuttgart, Stuttgart, Germany
e-mail: ozlem@ims.uni-stuttgart.de

O. Bilgin
Zargan Ltd., Istanbul, Turkey
e-mail: orhan@zargan.com

K. Oflazer (✉)
Carnegie Mellon University Qatar, Doha-Education City, Qatar
e-mail: ko@cs.cmu.edu

© Springer International Publishing AG, part of Springer Nature 2018
K. Oflazer, M. Saraçlar (eds.), *Turkish Natural Language Processing*,
Theory and Applications of Natural Language Processing,
https://doi.org/10.1007/978-3-319-90165-7_15

317

A wordnet is an electronic lexical database where lexical items (words and phrases) are organized into synonym sets ("synsets"), each representing one underlying concept. Synsets are linked to other synsets by various semantic relations including hypernymy, meronymy, and antonymy. The original wordnet for the English language was built at Princeton University starting in 1990, and currently contains 155,287 unique lexical items grouped into 117,659 synsets (Fellbaum 1998). In response to the success of Princeton Wordnet, wordnets have been developed for more than 50 languages including Catalan, Chinese, Dutch, French, Greek, Hebrew, Hindi, Italian, Japanese, Kurdish, Persian, Russian, Spanish, and Turkish (Global Wordnet Association 2014).

During the 36-month Balkanet project, the Turkish team at the Human Language and Speech Technologies Laboratory of Sabancı University designed and developed a basic wordnet consisting of 20,345 lexical items organized into 14,795 synsets. The basic structure of Turkish Wordnet is largely based on Princeton Wordnet, and design decisions were jointly made by the Balkanet Consortium, of which the Turkish team was a member.

The following sections describe the design and development of Turkish Wordnet. We first provide an overview of the basic structure of Turkish Wordnet and then summarize the design decisions made by the Balkanet Consortium and the Turkish team. We provide basic statistics about the status of the wordnet as of the end of the project, and describe a series of validation tasks that were performed after the end of the development process to ensure consistency and quality. We then list work done by others that have utilized this resource and end with concluding remarks and some directions for future work.

15.2 Basic Structure of Turkish Wordnet

Like in all other wordnets built along the lines of Princeton Wordnet, the basic building block of Turkish Wordnet is a "synset," an abstract entity that acts as a container of lexical items (single words or multi-word phrases) which can be used to refer to the same concept in a given context. All lexical items that belong to the same synset have the same part of speech. Each synset has a unique identifier used to distinguish it from other synsets, a part-of-speech tag which is inherited by all synset members, and an optional definition (gloss) used to describe the concept the synset refers to.

15.2.1 Semantic Relations

So far, the structure described above is not much different from a traditional thesaurus or synonym dictionary. What distinguishes a wordnet from these traditional language resources is that each synset can be linked to one or more other synsets to represent the semantic relations between the relevant concepts.

Table 15.1 Semantic relations used in Turkish Wordnet

Relation	Example
HYPERNYM	*kedi - hayvan* (cat - animal)
HOLO_MEMBER	*filo - deniz kuvvetleri* (fleet - navy)
HOLO_PART	*yarımküre - Dünya* (hemisphere - Earth)
HOLO_PORTION	*kar tanesi - kar* (snow flake - snow)
CAUSES	*koyulaştırmak - koyulaşmak* (to thicken (trans.) - to thicken (intrans.))
BE_IN_STATE	*konforlu - konfor* (comfortable - comfort)
STATE_OF	*konfor - konforlu* (comfort - comfortable)
NEAR_ANTONYM	*iyi - kötü* (good - bad)
SUBEVENT	*horlamak - uyumak* (to snore - to sleep)
ALSO_SEE	*enerjik - aktif* (energetic - active)
VERB_GROUP	*hayal etmek - anlamak* (to imagine - to understand)
CATEGORY_DOMAIN	*mahkeme - hukuk* (court house - law)
SIMILAR_TO	*antidemokratik - otoriter* (undemocratic - authoritarian)
USAGE_DOMAIN	*Aspirin - marka* (Aspirin - brand)

The HYPERNYM (or IS-A) relation is the basic semantic relation used to organize concepts into a hierarchical structure. For example, since a cat is a type of animal, the synset that the word *kedi* "cat" belongs to is linked to the synset that the word *hayvan* "animal" belongs to, via the HYPERNYM relation. Other major semantic relations used in Turkish Wordnet include NEAR_ANTONYM[1] (which links, for instance, *iyi* "good" to *kötü* "bad" to encode the antonymy relation), HOLO_PART (which links, for instance, *yumurta sarısı* "egg yolk" to *yumurta* "egg" to encode the part-whole relation), and CATEGORY_DOMAIN (which links, for instance, *mahkeme* "court house" to *hukuk* "law" to encode the fact that the concept of a court house belongs to the domain of law). Table 15.1 below lists all semantic relations used in Turkish Wordnet, along with examples.

15.2.2 Linking Wordnets to Each Other

Although an isolated wordnet in a single language can be a valuable resource in itself, it cannot be used in multilingual tasks such as cross-language search or

---

[1]Instead of the more straightforward relation name ANTONYM, Turkish Wordnet uses the name NEAR_ANTONYM to link two synsets with opposing meanings to each other. This is because antonymy is, strictly speaking, a relation that holds between individual *lexical items*, not between *concepts*. Consider the synset {*ascend, go up*}, where *ascend* and *go up* are synonyms in their relevant senses. But the two words have different antonyms: *descend* in the case of *ascend*, and *go down* in the case of *go up*. It would not be appropriate to link entire synsets to each other using the antonymy relation in its strict sense. Thus, along with the Euro Wordnet project (see Vossen (1998, p. 32)), Turkish Wordnet used the broader NEAR_ANTONYM relation to link synsets to each other.

machine translation unless two to more wordnets are mapped to each other. A simple way of mapping wordnets to each other is to use the same set of concepts, by using the same unique identifiers for the synsets.

This idea was first implemented by the Euro Wordnet project, which developed wordnets for Czech, Dutch, English, Estonian, French, German, Italian, and Spanish. All wordnets that were part of this project used the same set of synsets adopted from Princeton Wordnet 1.5. The so-called Inter-Lingual Index (ILI) assigns each synset in Princeton Wordnet a unique identifier based on the file offset of the relevant synset in the original Princeton Wordnet data files. This ensures that all eight wordnets of the Euro Wordnet project are connected to each other (see Vossen (1998, p. 39)).

The same method was adopted by the Balkanet project. Hence, Turkish Wordnet is perfectly mapped to Princeton Wordnet, the eight wordnets of the Euro Wordnet project, the five other wordnets of the Balkanet project, and any other wordnet that explicitly uses the concept pool of Princeton Wordnet. Figure 15.1 below depicts the basic structure of Turkish Wordnet as described above.

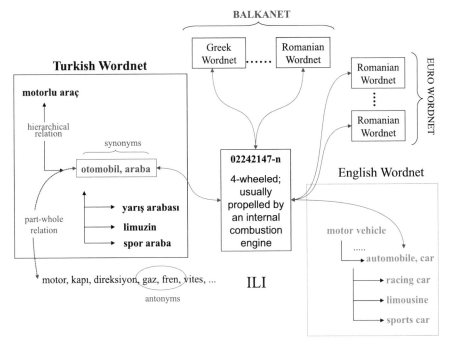

Fig. 15.1 Basic structure of multiple wordnets. Turkish Wordnet is linked to other wordnets through the Inter-Lingual Index (ILI)

15.3 Design Decisions

In the development of the wordnets in the Balkanet Project, many decisions were made by the Balkanet Consortium and these were adopted by all six wordnet teams that were part of the project. Some were made locally by the Turkish team, taking into account the nature of the Turkish language and the tools and resources available. In both cases, however, the decisions were mainly based on the additional experience gained of the Princeton Wordnet and Euro Wordnet projects.

15.3.1 Merge vs. Expand

Projects that aim to construct several interconnected wordnets usually prefer one of the two methodologies known as the "expand model" and the "merge model" in the literature (Vossen 1999). In the expand model, which is considerably simpler to implement, a fixed set of concepts (synsets) is taken from an existing wordnet, and each team translates the lexical items within these synsets into its local language. In the merge model, different synset collections built independently by each partner are combined into a single structure. The cost of the expand model is that the resulting wordnets are biased by the original wordnet, but the benefit is that all wordnets are linked to each other without extra effort. The benefit of the merge model is that the individual wordnets better reflect the structures of the individual languages, but the cost is the difficulty of combining the independent concept pools into a single, coherent structure. Since the Balkanet project aimed at maximum overlap with Princeton Wordnet and Euro Wordnets, it was decided at the outset to follow the expand model: Each team translated a fixed set of synsets from Princeton Wordnet 1.5.

15.3.2 Parts-of-Speech, Definitions, and Sense Numbers

The Turkish Wordnet contains nouns, verbs, and adjectives. Considering that the project aimed at creating medium-sized wordnets covering the relatively more important synsets with the highest possible number of relations to other synsets, adverbs, which do not have a hierarchical structure and have relatively fewer semantic relations to other concepts, have not been included.

Another decision to be made is whether or not to provide brief definitions (glosses) for each synset. The Balkanet Consortium decided that this is a useful feature, and adopted it to the extent resources were available.

Since a given word or phrase can have several meanings, and can thus be a member of more than one synset, each word or phrase in a wordnet must have a unique sense number. The decision to be made at this point is whether sense numbers should be taken from an existing monolingual dictionary, or alternatively, be assigned automatically, and thus randomly. Since it is an extremely labor-intensive and error-prone task to map the senses of two separate lexicons to each other, we decided to assign sense numbers automatically. Consequently, unlike in a traditional monolingual dictionary where senses are ordered according to importance and frequency, sense numbers in Turkish Wordnet do not reflect an order.

15.3.3   Lexical Gaps

Each language organizes its lexicon in a different way than other languages. For example, English uses the single word "uncle" to denote one's father's or mother's brother, whereas Turkish uses the two different words *amca* and *dayı*, respectively, to cover the same conceptual space. This phenomenon, known as a "lexical gap," must be taken into account when designing a wordnet that shares its concept pool with one or more other wordnets. Since Turkish Wordnet is based on Princeton Wordnet, concepts that exist in Princeton Wordnet but are not lexicalized in Turkish create a problem. One option is to create "empty synsets" that have an ID, a position in the hierarchy, and even a part-of-speech tag, but no lexical content. Another possibility is to avoid empty synsets and provide EQ_HAS_HYPONYM and EQ_HAS_HYPERNYM links to the hyponyms and hypernyms of the lexical gap (see Euro Wordnet General Document (Vossen 1999, p. 38)). Turkish Wordnet has adopted the former approach. Currently, there are 1269 lexical gaps in Turkish Wordnet.

15.3.4   No Dangling Nodes or Relations

Another important design decision adopted in the Balkanet project is that, when a new synset taken from Princeton Wordnet is added to a local wordnet, all its hypernyms, up to and including a top node, have to be included in that local wordnet too. In other words, there should be no "dangling nodes"; it should be possible to reach a topmost node from any given node in the concept hierarchy.

As for semantic relations, since in the expand model adopted by Turkish Wordnet, all semantic relations were imported from Princeton Wordnet, adding a new synset to the wordnet resulted in the automatic addition of new semantic relations. In some cases, the synset(s) to which the new synset is linked through a semantic relation was already a part of the local wordnet. However, in some cases, there were "dangling relations" where the synset at one end of the semantic relation was missing. To avoid this, whenever an existing synset involved semantic relations that point to certain other synsets that were not yet part of the wordnet, such other synsets were also included in Turkish Wordnet.

15.3.5 Validating Semantic Relations

When one imports semantic relations from another wordnet, it is possible that some of those relations do not hold between the lexical items of the importing language. Theoretically speaking, synsets correspond to concepts. So if everything goes well, there should be no cases where a semantic relation between those concepts ceases to be meaningful when expressed in another language. But there are certain reasons that give rise to such mismatches. First of all, "concepts" do not have an existence independent of a particular language. They have to be lexicalized. And every lexicalization inevitably involves some kind of attitude, culture, history, and pragmatic restriction. Secondly, concepts do not have well-defined boundaries. In many cases, a wordnet lexicographer translating a word or phrase from another language has to make do with a partial overlap of meaning. Therefore, each semantic relation imported from another wordnet has to be manually validated. The relations hold most of the time (around 95% according to our experience), but the few cases where they don't hold were eliminated.

15.4 The Development Process

This section summarizes, in chronological order, the 36-month development process of Turkish Wordnet, based on the design decisions above adopted by the Balkanet Consortium and the Turkish team, both in view of the past experience of the Princeton Wordnet and Euro Wordnet projects, and the individual characteristics of the languages involved in the Balkanet project.

15.4.1 First Set of Concepts (Subset I)

Having made the basic design decision of using the existing concept pool of Princeton Wordnet, the next logical step is to select an initial set of "important" concepts that will constitute the core of the new wordnet. One option would be to use local resources like a basic dictionary or word frequency list of Turkish. Although this would make sense from a monolingual point of view, the cost would be reduced overlap with other wordnets and increased difficulty of combining the six new wordnets that were being developed as part of the Balkanet project. In order to avoid these costs and maximize overlap with existing wordnets, the Balkanet Consortium decided that each team initially translate the 1310 "Base Concepts" (1010 nouns and 300 verbs) of the Euro Wordnet project (see Vossen (1999, p. 53)). Base concepts are concepts that rank high in the concept hierarchy and have the highest possible number of hyponyms.

15.4.2 Extracting Semantic Relations from Monolingual Resources

After the translation of the first set of 1310 synsets, we made an effort to increase average synset size by adding synonyms, and link these synsets to other synsets via the two basic semantic relations of hyponymy and antonymy. A machine-readable monolingual dictionary of Turkish (Türk Dil Kurumu 1983) was used to semiautomatically extract such relations.

15.4.2.1 Synonyms

The monolingual Turkish dictionary we used for this purpose contained entries in the form $hw : w_1, w_2, \ldots, w_n$, where hw is a headword and w_i is a single word. In these cases, the dictionary definition merely consisted of a list of synonyms. This allowed us to extract 11,126 sets of potential synonyms, using a script to parse dictionary entries. The first row of Table 15.2 exemplifies a single-word definition that produces a two-member synset.

There were also entry patterns in the form $hw : w_1 w_2 \ldots w_n(, w)+$. In these cases, a multi-word definition is followed by one or more synonyms, separated by a comma. These patterns gave us synsets in the form $hw(, w)+$. A total of 10,846 such forms were extracted using a script. These automatically extracted synonyms were then filtered to cover the Base Concepts Subset I only and synonyms that were not already present in the existing synsets are selected. 196 such synset members were added to existing synsets, increasing Turkish Wordnet's average synset size from 1.20 to 1.35. The second row of Table 15.2 gives a definition where the last two words are the synonyms of the word *benzer* "similar" and produce a three-member synset.

Table 15.2 Sample synsets automatically extracted from a Turkish monolingual dictionary

Pattern	Example	Synset
$hw : w_1, w_2, \ldots, w_n$	*Fonksiyon: işlev*	*{işlev, fonksiyon}*
	Function: role	{role, function}
$hw : w_1 w_2 \ldots w_n(, w)+$	*Benzer: Nitelik, görünüş ve yapı bakımından bir başkasına benzeyen veya ona eş olan, müşabih, mümasil*	*{benzer, müşabih, mümasil}*
	Similar: That which resembles another in terms of appearance or structure, alike, homologous	{similar, alike, homologous}

15.4.2.2 Hypernyms

The existence of the phrases *bir tür* or *bir çeşit* "a kind of" in a dictionary definition potentially indicates a hypernymy relation between the headword and the lexical item that follows these phrases. 625 hyponym-hypernym pairs were extracted in this way. The first two rows of Table 15.3 show two such extractions.

In cases where the definition contains the phrase *genel adı* "general term for," more than one hyponym-hypernym pair can be extracted from a single definition. For example, four different hypernymy relations can be extracted from the definition in the third row of Table 15.3. A total of 81 such sets were extracted from the Turkish monolingual dictionary.

Finally, the Turkish suffix *-giller* "member of" is usually used to construct taxonomic terms. Definitions of animals and plants usually contain this suffix, which allowed us to extract 889 hyponym-hypernym pairs, as exemplified in the last row of Table 15.3.

Table 15.3 Sample hypernym relations automatically extracted from a Turkish monolingual dictionary

Pattern	Example	Hypernyms
bir çeşit	*barbut: Zarla oynanan **bir çeşit** kumar*	*barbut – kumar*
	Craps: **A kind of** gambling played with dices	Craps—gambling
bir tür	*vermut: Birçok bitkilerle özel koku verilmiş **bir tür** şarap*	*vermut – şarap*
	Vermouth: **A kind of** wine flavored with various herbs	Vermouth—wine
genel adı	*erdem: Ahlakın övdüğü iyilikçilik, alçakgönüllülük, yiğitlik, doğruluk gibi niteliklerin **genel adı**, fazilet*	*iyilikçilik – erdem* *alçakgönüllülük – erdem* *yiğitlik – erdem* *doğruluk – erdem*
	Virtue: **General term for** ethically praisable characteristics such as righteousness, integrity, purity, decency	Righteousness—virtue Integrity—virtue Purity—virtue Decency—virtue
-giller	*mercimek: Bakla**giller**den, beyaz çiçekli bir tarım bitkisi (Lens culinaris)*	*mercimek – baklagil*
	Lentil: An agriculturally important **member of** legumes, having white flowers (Lens culinaris)	Lentil—legumes

Table 15.4 Sample near-antonym relations automatically extracted from a Turkish monolingual dictionary

Pattern	Example	Near-antonyms
karşıtı	*çirkin: Göze veya kulağa hoş gelmeyen, güzel* **karşıtı**	*çirkin – güzel*
	Ugly: That which does not appeal to the eye or the ear, **opposite of** beautiful	Ugly—beautiful
olmayan	*temiz: Kirli, lekeli, pis, bulaşık olmayan*	*temiz – kirli*
		temiz – lekeli
		temiz – pis
		temiz – bulaşık
	Clean: **That which is not** dirty, soiled, polluted, contaminated	Clean – dirty
		Clean—soiled
		Clean—polluted
		Clean—contaminated

15.4.2.3 Near-Antonyms

Existence of the word *karşıtı* "opposite of" or *olmayan* "that which is not" in a dictionary definition indicates a potential antonymy relation between the headword and the lexical item preceding the words *karşıtı* or *olmayan*. In both cases, one or more near-antonyms can be derived from the definition. Table 15.4 shows a pair extracted from a definition including *karşıtı* and four pairs extracted from a definition including *olmayan*. A total of 235 antonym pairs were extracted in this way.

15.4.3 Second Set of Concepts (Subset II)

Having completed the translation of the first set of 1310 synsets and having enriched these synsets using monolingual resources, the Balkanet Consortium then decided to expand the wordnets to 5000 synsets during a second phase. Each team proposed a set of synsets, using various criteria (corpus frequencies, defining vocabularies, monolingual dictionaries, polysemy, etc.) to determine this new subset.

While choosing the candidates for the second set, the Turkish team followed two different approaches. One of them was to find the so-called "missing hypernyms," and the other was to construct a set of candidates which would be usable by all languages of the Balkanet project. The resulting set of synsets has been formed by combining the results of these two approaches.

- **240 Missing Hypernyms:** These are the 240 hypernyms of Subset I synsets which are not members of Subset I themselves. The idea here is to fill all gaps between members of Subset I up to the relevant topmost nodes in Princeton

Wordnet, so that the expanded set becomes a set of several "chains," where it is always possible to reach a topmost node of Princeton Wordnet by moving up in the hierarchy.

- **1228 Additional Synsets:** While constructing this set of synsets, our aim was to choose concepts that are frequent, rank high on the concept hierarchy, are richly linked to other concepts, and would ensure maximum overlap between all languages represented in the project. As a starting idea, we proposed that the concept of a "defining vocabulary" was well suited to the task of determining such concepts. We used the defining vocabulary of the Longman Dictionary of Contemporary English (Quirk 1987). As a second source, we used the list of most frequent words in the English language, based on the British National Corpus (BNC Consortium 2001). We identified those entries in the Longman Defining Vocabulary which do not already exist among our extended set of synsets (1310 from Subset I and the additional 240 "missing hypernyms"), and we then found their intersection with the most frequent words of the English language. This intersection also allowed us to rank the new entries in terms of their frequencies, so entries higher on the list could be considered more important than those lower. The result was a list of 712 lexical items. We then extracted all Princeton Wordnet synsets that contain these lexical items, obtaining 3114 synsets. Then, we reduced this set by taking only those synsets whose hypernyms are Subset I synsets. The final product is a collection of 1228 synsets. In this way, we eliminated all "dangling nodes" from our hierarchy. The resulting hierarchy contains 247 separate trees of varying length.

This methodology is completely independent of the Turkish language. The motivation is that, at this relatively high level of the hierarchy, the most frequent words of English would be important for all languages. In addition, the task we are faced with is the selection of synsets in the English language, since the Balkanet wordnets were based on Princeton Wordnet. So, the idea was that basing the selection on English would not be misleading. The assumption is that language-specific information gets more important as one moves down the hierarchy.

15.4.4 Shifting to Princeton Wordnet 1.7.1

Before starting the translation of Subset II, the Consortium decided to shift from Princeton Wordnet 1.5 to Princeton Wordnet 1.7.1 as the basic resource. The aim was to avoid certain problems involved in Princeton Wordnet 1.5, such as incorrect links, low-quality and missing glosses, and artificially divided synsets.

15.4.5 Third Set of Concepts (Subset III)

After all partners finished the translation of Subset I and Subset II, the Balkanet Consortium decided that all wordnets should reach 8000 synsets at the end of a third phase. It was decided that this phase should cover an additional 3000 synsets that exist in at least five Euro Wordnets. The criteria of "avoiding missing hypernyms" was again applied.

15.4.6 Shifting to Princeton Wordnet 2.0

During the translation of Subset III, Princeton University released Wordnet 2.0, which contained thousands of additional synsets, verb groups, domain information for synsets, and links between morphologically related items. Having observed that shifting from Version 1.7.1 to Version 2.0 would require minimal effort, the consortium decided to shift to Princeton Wordnet 2.0. Due to the structural changes introduced in Princeton Wordnet 2.0, some synsets in Balkanet wordnets had to be merged, divided, or deleted, mostly automatically but sometimes also manually. Due to the shift to Princeton Wordnet 2.0, the number of Base Concepts in the Balkanet project is not equal to the number of Base Concepts in the Euro Wordnet project.

15.4.7 Adding Balkanet-Specific Concepts

Since the Balkanet project involved six languages from the Balkans and Eastern Europe, the expectation was that there existed a large number of regional/culture-specific concepts that the developers of Princeton Wordnet would not be expected to include in a wordnet of the English language.

Consequently, once the development of the core wordnets was finished using the existing concept pool of Princeton Wordnet, the consortium decided to shift to the "merge model." Initially, each team worked separately to develop its own set of language-specific concepts. The Turkish team developed 299 synsets, comprising 286 nouns, 10 verbs, and 3 adjectives. All Turkish synsets were equipped with brief definitions in English, and 141 synsets also had a picture. 285 of the Turkish synsets were linked to a Princeton Wordnet 2.0 synset via a hypernymy relation.

In the second step, all six teams came together to combine their individual contributions into a single repository called the "Balkanet Inter-Lingual Index" (BILI). The local synsets developed by each partner were checked by all the other partners; identical concepts were determined and assigned a single BILI number. The resulting set consisted of 332 Balkan specific synsets. As would be expected, most BILI concepts belong to culture- and region-specific domains such as the administrative system, religion, wedding traditions, architecture, food, animals, plants, traditional clothes, occupations, traditional arts, music, and tools. Some examples of the Turkish team's contribution are shown in Table 15.5.

Table 15.5 Some language-specific concepts contributed by the Turkish team

Lexical item	English definition
incir reçeli	Jam made of unripe wild figs.
neyzen	Person who plays the musical instrument *ney*.
dayı	Brother of one's mother.
nazar boncuğu	Charm made of blue, white, and yellow glass to protect you from the evil eye.
mescit	Small mosque where Friday prayers and special prayers on holy days are not held.

15.4.8 Final Expansion

The purpose of this final expansion phase was to further increase coverage by adding those concepts that are frequently used in Turkish but do not yet exist in Turkish Wordnet. In order to determine these important and missing synsets, we took the 50,000 most frequent words of a 13-million-word in-house corpus compiled from six different domains of newspaper text. We then manually selected 2575 words that were decided to be important for Turkish and did not exist in Turkish Wordnet at that point.

This process was especially important for adjectives and certain closed classes such as cardinals, ordinals, and names of months, which were not represented in Turkish Wordnet.

15.5 Current Status of Turkish Wordnet

Table 15.6 provides basic statistics on Turkish Wordnet from the October 2014 release. The first three rows show the number of synsets, synset members, and average synset size. Note that average size is calculated simply as the ratio of synset members to synsets. It also includes those synsets that have zero members due to lexical gaps, which occur while trying to add an English synset to Turkish Wordnet via translation, as explained in Sect. 15.3.3 above. The current version of Turkish Wordnet contains 1269 such zero-member synsets. When these are ignored, average synset size rises to 1.50. 8792 of the synsets have only one member, while 3318 have two members, and 971 have three members. The two largest synsets of Turkish Wordnet have 10 members.

6717 of the synsets have a definition. 332 synsets have an SNOTE field that contains an English definition (for the Balkan-specific concepts). 141 of these English definitions are additionally associated with photos of concepts, with a SEE PICTURE identifier in the SNOTE field.

Table 15.7 shows the breakdown of Turkish Wordnet's synsets into the three Base Concept subsets, and into parts of speech. Note that the numbers of the original Base

Table 15.6 Basic statistics on Turkish Wordnet

Basic statistics	Number
Synsets	14,795
Synset members	20,345
Average synset size	1.38
Lexical gaps	1269
Definitions	6717

Table 15.7 Distribution of base concept subsets and parts of speech

Synset type	Count	Part-of-speech	Count
Subset I	1219	Nouns	11,227
Subset II	3470	Verbs	2736
Subset III	3782	Adjectives	792

Table 15.8 Semantic relations

Relation type	Number	Relation type	Number
HYPERNYM	12,908	CATEGORY_DOMAIN	403
SIMILAR_TO	2497	BE_IN_STATE	327
HOLO_PART	1816	STATE_OF	290
NEAR_ANTONYM	1613	HOLO_PORTION	234
HOLO_MEMBER	1245	CAUSES	100
ALSO_SEE	1021	SUBEVENT	131
VERB_GROUP	923	USAGE_DOMAIN	32
		Total	23,540

Concepts described in Sect. 15.4 do not match the numbers in the final version. This is due to the restructuring that occurred when we shifted from Princeton Wordnet 1.5 to 1.7.1, and then to 2.0. All three Base Concept subsets are 100% covered. As for the distribution of parts of speech, nouns dominate Turkish Wordnet with 75.9%, followed by verbs, which account for 18.6%, and adjectives, which constitute only 5.5%.

Table 15.8 lists the number of occurrences of each relation. Naturally, HYPERNYM is by far the most frequent relation. It is followed by SIMILAR_TO, HOLO_PART, and NEAR_ANTONYM. 7646 synsets have only one relation. 4077 of them have two, followed by 769 synsets with three relations. At the most highly connected end of the spectrum, there is one synset each with 29, 30, 32, 40, and 46 relations.

15.6 Quality Validation and Coverage Tests

Following the completion of the development phase, we performed a series of quality validation tasks. For the syntactic quality of the XML file, we used internally-developed scripts and the VisDic tool developed by the Czech team

(Horak and Smrz 2004b,a). VisDic, which is developed to visualize wordnets, also provides a set of tests for checking the consistency of wordnet XML files, such as duplicate IDs, duplicate lexical items, and duplicate links. The latter prevents a lexicographer from linking two synsets via more than one relation. For instance, a synset cannot be both the hypernym and antonym of another synset. A final VisDic test checks if the same lexical item with the same sense number occurs in more than one synset.

As for structural quality, we identified dangling nodes and dangling relations and added the respective missing synsets. We ensured all members of Base Concepts were present in the wordnet. In terms of content quality, we first passed the linguistic content of Turkish Wordnet (synset members, glosses, and usage examples, if any) through a spelling corrector. Then we manually, semiautomatically, or automatically validated all semantic relations imported from Princeton Wordnet. In 95% of the cases, the semantic relations imported from Princeton Wordnet were valid in Turkish as well.

As part of another major validation task, we measured the lexical coverage of Turkish Wordnet by checking the occurrence of high-frequency words of Turkish among synset members. The frequency word lists came from two different resources: The first one is a Turkish translation of George Orwell's novel *Nineteen Eighty-Four* and the second one is an in-house corpus.

While building the frequency lists, we morphologically analyzed and disambiguated all words using a morphological analyzer (Oflazer 1994) and applied the same procedure to synset members. While creating the list, we attached part-of-speech tags to the words, to avoid counting unmatching pairs as covered. As a result 76.6% of the *Nineteen Eighty-Four* lexical items were among synset members when we calculated the ratio of weighted sum of the successfully found lexical items to the total weighted sum. As expected, function words ranked high in the word list, and given that they were not included in the wordnet, they caused a reduction in the overall percentage. When we omitted function words, the percentage rose to 87.40%.

Similarly, we took the 50,000 most frequent words from the 13-million-word corpus mentioned above, excluding function words, and performed the same test. Coverage was 85.94%. When we considered the 20,000 most frequent words, it reached 86.45%. We then limited our list to the 1000 most frequent words of the corpus, and coverage rose to 87.32%.

15.7   Applications of Turkish Wordnet

This section provides an overview of projects and publications that are related to Turkish Wordnet and appeared either during or after the initial development phase.

15.7.1 Capturing Semantic Relations Through Morphology

The basic idea of this application is to effectively utilize morphological processes in a language to enrich individual wordnets with semantic relations. In a scenario where synsets of Wordnet A and Wordnet B are mapped to each other, simple morphological derivation processes in Language A can be used (1) to extract explicit semantic relations in Language A, and use these to enrich Wordnet A; (2) to verify existing semantic relations and detect mistakes in Wordnets A and B; and most importantly (3) to discover implicit semantic relations in Language B, and use these to enrich Wordnet B.

In this study, we focused on Turkish to extract morphological relations in the monolingual context, and propose relation extraction and verification both on Turkish and English in the multilingual context.

In the monolingual context, using morphologically-related word pairs to discover semantic relations is by far faster and more reliable than building them from scratch, especially in a morphologically-rich language with regular morphotactics. Productive affixes facilitate the derivation of lists of pairs using simple rules and improve the internal connectivity of a wordnet. In Bilgin et al. (2004), we identified 12 productive Turkish suffixes as candidates and proposed possible semantic relations for nine of them: WITH, WITHOUT, ACT_OF, ACQUIRE, MANNER, BECOME, BE_IN_STATE, CAUSES, PERTAINS_TO. Only the last three of these relations are defined in Princeton Wordnet and Euro Wordnet.

In the multilingual context, there are two cases: In the first case, semantically-related lexical items in both the exporting and the importing languages are morphologically related to each other, as can be seen in Fig. 15.2. Here, the importing language (Turkish) could have discovered the semantic relation between *deli* "mad" and *delilik* "madness," for instance, by using its own morphology. So, importing the relation from English does not bring an extra benefit. Yet, it can serve as a useful quality-control tool for the importing wordnet, and this has indeed been the case for Turkish.

Using the "expand model" in building Turkish Wordnet resulted in importing a set of relations together with the translated Princeton Wordnet synsets they belong to. Since Turkish employs a morphological process to encode, for example, BE_IN_STATE relations, the list of Turkish translation equivalents contains sev-

Fig. 15.2 Both languages involve morphology

Fig. 15.3 Importing language does not involve morphology

eral morphologically-related pairs like *deli-delilik* "mad-mad**ness**," *garip-gariplik* "weird-weird**ness**," etc. Pairs that violate this pattern potentially involve incorrect translations or some other problem, and the translation method provides a way to detect such mistakes.

In the more interesting case, semantically-related lexical items in the importing language are not morphologically related to each other. For example, the causation relation between the lexical items *yıkmak* and *yıkılmak* is obvious to any native speaker (and morphological analyzer) of Turkish, while the corresponding causation relation between "tear down" and "collapse" is relatively more opaque and harder to discover for a native speaker of English, and impossible for a morphological analyzer of English (Fig. 15.3). Our method thus provides a way of enriching a wordnet with semantic information imported from another wordnet.

We conducted a pilot study on two semantic relations to observe if this relation discovery procedure helps enrich Princeton Wordnet 2.0. We looked into the CAUSES (e.g., kill–die) and BECOME (e.g., stone–petrify) relations. CAUSES is a semantic relation that is present in Princeton Wordnet 2.0; BECOME on the other hand is not directly present, and is only represented by the underspecified relation ENG DERIVATIVE. 80 synset pairs in Turkish Wordnet have synset members related by a causative suffix that corresponds to the CAUSES relation. Only 18 of those pairs have a CAUSES relation in Princeton Wordnet 2.0. Similarly, 83 Turkish synsets were linked via the BECOME relation, by looking at the morphology of the lexical items. Only 11 of them were already linked in Princeton Wordnet 2.0.

Some of the new links proposed involve morphologically unrelated lexical items which cannot be possibly linked to each other automatically or semiautomatically. Interesting examples in the case of the BECOME relation include pairs such as *soap-saponify, good-improve, young-rejuvenate, weak-languish, lime-calcify, globular-conglobate, cheese-caseate, silent-hush, sparse-thin out, stone-petrify*. Interesting examples in the case of the CAUSES relation include pairs such as *dress-wear, dissuade-give up, abrade-wear away, encourage-take heart, vitrify-glaze*.

15.7.2 Turkish Wordnet in Use

Following its distribution, Turkish Wordnet has been used by several researchers as a basic lexico-semantic resource, either alone or in conjunction with another wordnet (usually Princeton Wordnet) or other NLP tools and resources.

Durgar-El Kahlout and Oflazer (2004) propose a meaning-to-word system for Turkish that finds a set of words matching the definition entered by the user. The system uses Turkish Wordnet to expand queries for the purpose of improving the coverage of matches. The use of the synonymy information in Turkish Wordnet increases the system's success rate from 60% to 68%. In another study, Durgar-El Kahlout and Oflazer (2005) take advantage of the links between Turkish Wordnet and Princeton Wordnet to construct a bilingual "root word alignment dictionary," which is used in the word-level alignment module of a statistical machine translation system. They report that the use of the wordnets significantly reduces noisy alignments. Oflazer et al (2006) report on the development of LingBrowser, a set of intelligent, active and interactive tools for helping linguistics students inquire and learn about lexical and syntactic properties of words and phrases in Turkish text. The system also incorporates information from Turkish Wordnet.

In his master's thesis, Boynueğri (2010) uses the definitions, synonyms, and semantic relations in Turkish Wordnet in a word-sense disambiguation task. Pembe and Say (2004) use synonymy information obtained from Turkish Wordnet to build the "lexico-semantic expansion" module of their linguistically motivated information retrieval system. Yücesoy and Öğüdücü (2007) use the hypernym hierarchy of Turkish Wordnet to propose an improved semantic similarity measure, which in turn is used in a document clustering task. Ambati et al. (2012) use the synsets of Turkish Wordnet to generate "coherent topics," which are then used to evaluate the performance of a "word sketches" system.

Özsert and Özgür (2013) use Turkish Wordnet in conjunction with Princeton Wordnet to improve the performance of their graph-based word polarity detection algorithm. The use of wordnets improves their accuracy from 84.5% to 95.5% in the case of Turkish, and from 91.1% to 92.8% in the case of English. In a related study, Demir (2014) uses semantic relations in Turkish Wordnet and Princeton Wordnet in a "valence shifting" task, which aims to "rewrite a text towards more/less positively or negatively slanted versions."

15.8 Conclusion and Directions for Future Work

In this chapter, we have described the design and development of Turkish Wordnet, a semantic network containing 20,345 lexical items organized into 14,795 synsets. Compared to Princeton Wordnet, Turkish Wordnet is a small-scale wordnet developed as part of an international project whose principal purpose was to produce six interconnected core wordnets that would also be linked to Princeton Wordnet and the eight wordnets of the Euro Wordnet project. Since an existing concept/relation pool in another language (English) was being used, the development process was largely a translation process, which had to be performed manually. During the 10 years that have elapsed since the conclusion of the Balkanet project, automatic and semiautomatic methods have been proposed and used for wordnet creation, which can inform future efforts to expand and enrich Turkish Wordnet.

Yıldırım and Yıldız (2012), for example, report the results of an experiment to automatically extract hypernym-hyponym pairs from a Turkish corpus, using lexico-syntactic patterns. Şerbetçi et al. (2011) extract a wider range of semantic relations from Turkish dictionary definitions, once again using lexico-syntactic patterns. They report having extracted more than 58,000 relations at 86.85% accuracy. These performance metrics suggest automatic or semiautomatic methods would facilitate inserting new synsets and relations to Turkish Wordnet.

The current version of Turkish Wordnet exclusively contains synset-to-synset relations, following decisions made on EuroWordNet. However, Princeton Wordnet has defined morphosemantic relations starting from version 2.0 (Miller and Fellbaum 2003) which establishes links between words that are connected to each other through derivational morphology. As explained in Sect. 15.7.1 above, the rich morphology of Turkish allows the automatic creation of a substantial number of word-to-word relations. Adopting Princeton Wordnet morphosemantic relations and using the proposed techniques in Bilgin et al. (2004) create an opportunity for the rapid automatic enrichment of Turkish Wordnet with semantic relations.

To summarize, we think that future efforts to expand and enrich Turkish Wordnet might benefit from automatic and semiautomatic methods that rely more on language resources in Turkish and specific features and mechanisms that are peculiar to the Turkish language.

The XML distribution of Turkish Wordnet is available for research purposes at bitbucket.org/ozlemc/twn/ (Accessed Sept. 14, 2017), together with the VisDic configuration files to visualize and edit the wordnet.

References

Ambati BR, Reddy S, Kilgarriff A (2012) Word sketches for Turkish. In: Proceedings of LREC, Istanbul, pp 2945–2950

Bilgin O, Çetinoğlu Ö, Oflazer K (2004) Morphosemantic relations in and across wordnets: a preliminary study based on Turkish. In: Proceedings of the second global WordNet conference, Brno, pp 60–66

BNC Consortium (2001) British national corpus. www.natcorp.ox.ac.uk/. Accessed 3 July 2017

Boynueğri A (2010) Cross-lingual information retrieval on Turkish and English texts. Master's thesis, Middle East Technical University, Ankara

Demir Ş (2014) Generating valence shifted Turkish sentences. In: Proceedings of the eighth international natural language generation conference, Philadelphia, PA, pp 128–132

Durgar-El Kahlout İ, Oflazer K (2004) Use of wordnet for retrieving words from their meanings. In: Proceedings of the second global WordNet conference, Brno, pp 118–123

Durgar-El Kahlout İ, Oflazer K (2005) Aligning Turkish and English parallel texts for statistical machine translation. In: Proceedings of ISCIS, Istanbul, pp 616–625

Fellbaum C (1998) WordNet: an electronic lexical database. MIT Press, Cambridge, MA

Global Wordnet Association (2014) Wordnets in the world. www.globalwordnet.org/wordnets-in-the-world. Accessed 3 July 2017

Horak A, Smrz P (2004a) New features of wordnet editor VisDic. Rom J Inf Sci Technol 7(1–2):1–13

Horak A, Smrz P (2004b) VisDic- wordnet browsing and editing tool. In: Proceedings of the global WordNet conference, Brno, pp 136–141

Miller GA, Fellbaum C (2003) Morphosemantic links in wordnet. Traitement Automatique de Langue 44(2):69–80

Oflazer K (1994) Two-level description of Turkish morphology. Lit Linguist Comput 9(2):137–148

Oflazer K, Erbaş MD, Erdoğmuş M (2006) Using finite state technology in a tool for linguistic exploration. In: Proceedings of FSMNLP, Helsinki, pp 191–202

Özsert CM, Özgür A (2013) Word polarity detection using a multilingual approach. In: Proceedings of CICLING, Samos, pp 75–82

Pembe FC, Say ACC (2004) A linguistically motivated information retrieval system for Turkish. In: Proceedings of ISCIS, Kemer, pp 741–750

Quirk R (1987) The Longman American defining vocabulary. www.longmandictionariesusa.com/res/shared/vocab_definitions.pdf. Accessed 3 July 2017

Şerbetçi A, Orhan Z, Pehlivan İ (2011) Extraction of semantic word relations in Turkish from dictionary definitions. In: Proceedings of the workshop on relational models of semantics, Portland, OR, pp 11–18

Stamou S, Oflazer K, Pala K, Christoudoulakis D, Cristea D, Tufis D, Koeva S, Totkov G, Dutoit D, Grigoriadou M (2002) Balkanet: a multilingual semantic network for Balkan languages. In: Proceedings of the first global WordNet conference, Mysore

Türk Dil Kurumu (1983) Türkçe Sözlük. Türk Dil Kurumu, Ankara

Vossen P (ed) (1998) Euro WordNet: a multilingual database with lexical semantic networks. Kluwer Academic Publishers, Dordrecht

Vossen P (1999) EuroWordNet general document. www.vossen.info/docs/2002/EWNGeneral.pdf. Accessed 3 July 2017

Yıldırım S, Yıldız T (2012) Automatic extraction of Turkish hypernym-hyponym pairs from large corpus. In: Proceedings of COLING, Mumbai, pp 493–500

Yücesoy B, Öğüdücü ŞG (2007) Comparison of semantic and single term similarity measures for clustering Turkish documents. In: Proceedings of the international conference on machine learning and applications, Cincinnati, OH

Chapter 16
Turkish Discourse Bank: Connectives and Their Configurations

Deniz Zeyrek, Işın Demirşahin, and Cem Bozşahin

Abstract The Turkish Discourse Bank (TDB) is a resource of approximately 400,000 words in its current release in which explicit discourse connectives and phrasal expressions are annotated along with the textual spans they relate. The corpus has been annotated by annotators using a semiautomatic annotation tool. We expect that it will enable researchers to study aspects of language beyond the sentence level. The TDB follows the Penn Discourse Tree Bank (PDTB) in adopting a connective-based annotation for discourse. The connectives are considered heads of annotated discourse relations. We have so far found only applicative structures in Turkish discourse, which, unlike syntactic heads, seem to have no need for composition. Interleaving in-text spans of arguments appears to be only apparently-crossing, and related to information structure.

16.1 Introduction

Discourse is not a haphazard collection of sentences. It is a unit of language above the sentence level, coherently organized around a topic. A sequence of linguistic material is regarded as coherent to the extent that the entities mentioned in the text (in the Hallidayan sense of the term) are connected by presupposition, information structure, and anaphora, as well as lexical links (repetition, metonymy, meronymy, antonymy, etc.). Parts of text such as clauses and sentences are related to each other by such elements, where anaphora can be seen to operate at two levels, the phrasal level and the word level. Pieces of discourse are also connected by what are

D. Zeyrek · I. Demirşahin
Middle East Technical University, Ankara, Turkey
e-mail: dezeyrek@metu.edu.tr; e128500@metu.edu.tr

C. Bozşahin (✉)
Department of Cognitive Science, Institute of Middle East Technical University, Ankara, Turkey
e-mail: bozsahin@metu.edu.tr

© Springer International Publishing AG, part of Springer Nature 2018   337
K. Oflazer, M. Saraçlar (eds.), *Turkish Natural Language Processing*,
Theory and Applications of Natural Language Processing,
https://doi.org/10.1007/978-3-319-90165-7_16

commonly referred to as *discourse relations*, also known as rhetorical or coherence relations, such as additive, causal, contrastive, concessive relations.

The literature on Turkish NLP is quite rich in terms of investigating certain aspects of discourse anaphora (e.g., Tın and Akman (1994); Yüksel and Bozşahin (2002); Yıldırım et al. (2004); Tüfekçi and Kılıçaslan (2005); Tüfekçi et al. (2007)). However, there are no publicly available corpora which would allow investigation of other aspects of discourse. The TDB has been created primarily to address this gap. We start with the connectives and then we hope to add more aspects in the future.

In a series of papers since 2009, the TDB team has reported on the annotation procedure and the annotation tool, on consistency of annotations among the annotators, and on how subordinators are annotated (Zeyrek et al. 2009, 2010; Aktaş et al. 2010; Demirşahin et al. 2012; Zeyrek et al. 2013). The aim of this chapter is to summarize the important aspects of these papers. We first describe the annotation methods and tools we developed in creating the corpus, and summarize the difficulties we encountered during annotations. After a brief comparison with other discourse approaches, we present our investigations on the TDB structures in the current release.[1]

Discourse relations are made explicit in our annotation by discourse connectives, e.g. *ve* "and", *çünkü* "because", *ne var ki*.

"however/even so", *sonra* "after", *ondan sonra* "after that/then". Coherence is not single-handedly caused by explicit connectives. It may also be inferred from the adjacency of textual spans. The PDTB group argues (Prasad et al. 2014) that in many cases discourse relations are implicit, and in such cases adjacency of the clauses provides a hint about the category of the discourse relation. Given the fact that discourse coherence is a multifaceted phenomenon, we have chosen to restrict ourselves in the annotations in certain ways. The TDB annotates explicit discourse connectives and phrasal expressions, that is, connecting devices that contain a deictic anaphor in them along with the text spans they relate, leaving the annotation of implicit discourse connectives to future work (see Example (7) below).

A real example from the corpus, (6), is an excerpt from an interview with an archaeologist excavating the ancient site of Boğazköy. *Çünkü* "because" (underlined) relates the clause on its left (in italics) to the clause on its right (in bold) with the *cause* discourse relation. The clause which is syntactically related to the connective is always called the second argument or Arg2. The other clause is referred to as the first argument or Arg1. Examples (6) and (7) are from Zeyrek and Webber (2008).

(6) Yapılarını kerpiçten yapıyorlar, ama sonra taşı kullanmayı öğreniyorlar. *Mimarlık açısından çok önemli* <u>**çünkü**</u> **bu yapı malzemesini başka bir malzemeyle beraber kullanmayı, ilk defa burada görüyoruz.**

They constructed their buildings first from mudbricks but then they learned to use the stone. *Architecturally, this is very important* <u>**because**</u> **we see the use of this construction material with another one at this site for the first time.**

---

[1]The tools and the data are available to researchers free of charge by applying to the TDB research team through medid.ii.metu.edu.tr (Accessed Sept. 14, 2017).

The TDB follows the principles of the PDTB (Prasad et al. 2014). All discourse connectives are marked as heads. There are no high-level discourse structures such as rhetorical structure trees or dependency graphs. All connectives are assumed to be binary, manifesting always an Arg1 and Arg2. For easy exposition, the connective is underlined, Arg1 is italicized, and Arg2 is rendered in bold.

Example (7) presents implicit discourse relations, which we show with the inferred discourse connectives in parenthesis.

(7) Yürüyor, (sonra) oturuyor, resim yapmaya çalışıyor ama yapamıyor, tabela yaz-
     maya çalışıyor ama yazamıyor, (en sonunda) sıkılıp sokağa çıkıyor, bisikletine
     atladığı gibi pedallara basıyor.
     "He walks around (then), sits down, tries to draw, but he can't, he tries to
     inscribe words on the wooden plaque, but again he can't, (consequently) he
     gets bored, goes out, hops on his bike and pedals."

16.2   The TDB Annotation Cycle

The TDB is approximately a 400,000-word sub-corpus of METU Turkish Bank (Say et al. 2004), balanced with respect to its genre distribution. The first step in the annotation process was the preparation of annotation guidelines and the training of the annotators in terms of discourse and annotation issues. We had three annotators who remained in the research team until the release of the first version of the corpus in 2011. The annotators first studied the annotation guidelines thoroughly, and then they were trained in a semester-long seminar series. The annotation guidelines were not strict rules but included some general principles that were constantly updated as the need arose. The annotators were always asked to reflect their native speaker intuitions on the annotations. The only principle which they were asked to obey was the *minimality principle* of the PDTB Group, which requires them to select a text span as an argument to a connective if it can be interpreted as minimally necessary and sufficient to establish the discourse relation (Prasad et al. 2014).

The annotation cycle includes three steps. First, the annotators independently annotated a given set of connectives then, inter-coder agreement was measured, and disagreements were noted and resolved by the team. Fleiss' Kappa (Fleiss 1971) was run to determine if there was agreement among the coders on argument span boundaries. The average K values are good for discourse-level phenomena ($K = 0.76$ for Arg1, $K = 0.82$ for Arg2) (Zeyrek and Kurfalı 2017). The second step involved revisions in the annotation guidelines. In the final step, the annotations were checked fully by one annotator and by an expert to ensure they were compatible with the annotation guidelines (Zeyrek et al. 2010).

We asked the annotators to identify discourse relations by providing them with a preliminary list of possible discourse connectives. However, the annotators were free to expand the initial list if they had a strong intuition that an additional lexical

or phrasal device that acts as a discourse connective. At least two independent annotators went through the whole data, manually distinguishing between the discursive
and nondiscursive functions of the connectives and annotating them according to
the annotation manual. As the annotators were blind to the annotations of the other
annotator(s), the resulting annotations were free from a jointly conceived abstract
structure of discourse. In subsequent steps, we took the completed annotations as
the basis on which to investigate structures (Demirşahin et al. 2013).

As in the PDTB, we took a connective as having a discursive function when it
relates text spans that convey a proposition, fact, event, situation, etc. Usually, such
abstract objects (Asher 1993) are expressed in clauses, though in certain cases, they
may also be conveyed by nominalizations (Zeyrek et al. 2013).

In order to provide a preliminary set of possible connectives for the annotators,
discourse connectives were compiled from three major syntactic classes: coordinating conjunctions (*ve* "and", *ama* "but, yet", *fakat* "but"), complex subordinators
(postpositions co-occurring with a converb, for example, *için* "for, so as to", *karşılık*
"although" (Zeyrek et al. 2013), and discourse adverbials (*öte yandan* "on the other
hand", *ayrıca* "in addition, separately"). Phrasal expressions, which are phrases with
a postposition and a deictic anaphor, such as *buna rağmen* "despite this", *bundan
ötürü* "due to this", were added to this list when complex subordinators (*rağmen*
"despite", *ötürü* "due to") were being annotated. Phrasal expressions fall outside the
zone of connective proper due to the deictic anaphor in their composition. Yet, the
decision to annotate phrasal expressions together with discourse connectives was
taken because of the frequency of such phrases in the language, and their limited
syntactic compositionality. More importantly, the annotators tended to take them at
an equal footing with discourse connectives proper, and we did not want to tamper
with this intuition.

The complete annotation scheme used in the TDB is given in Table 16.1 (Zeyrek
et al. 2013). There could be richer tagsets but in the TDB 1.0, we chose to keep
the scheme as simple as possible, to avoid potential difficulties that might arise if
the annotators were to pay attention to a wider set of discourse issues during the
annotations.

Table 16.1 The annotation scheme of the TDB

Label	Denotation
Conn	The connective's head
Arg1	First argument of the connective
Arg2	Second argument of the connective
Supp1	Supplement to the first argument
Supp2	Supplement to the second argument
Shared	The subject, object, or adverbial phrase shared by a relation
Shared supp	Supplement for the shared material
Mod	Modifier of the connective or the modifier of the relation

16.2.1 Major Sources of Disagreements Among Annotators

As the inter-coder reliability among three annotators stabilized, we shifted to a new procedure called *pair annotation* (Demirşahin et al. 2012; Demirşahin and Zeyrek 2017). Two annotators created the annotations as a pair, sitting next to each other before the screen and working collaboratively. In time, one of the annotators began to act as the leader and the other as the navigator, as in pair programming (Williams et al. 2000), though our pair annotation emerged independently of the latter. In the TDB 1.0, the pair annotation procedure has been used as a supplementary means to independent annotations (also see Zeyrek et al. (2013)). Our observation is that it seems to have improved the inter-annotator agreement and speed of annotation. The lessons we learned can benefit others, so we report them here in some detail.

Zeyrek et al. (2010) reported major sources of disagreements in the TDB by examining 60 connectives and 6,873 annotations, which comprise 75.94% of the total connective tokens and 81.01% of the total number of annotations in the TDB 1.0. Eight of these connectives gave particularly low κ values (< 0.80) for Arg1 and Arg2.[2]

The reported inconsistencies were largely due to partially overlapping annotations for Arg1 (63.98%). The remaining inconsistencies arose because the annotations for Arg1 (9.74%), or Arg2 (10.17%), had no overlap. There were other inconsistencies due to simple human errors, for example, errors in selecting spaces, leaving characters out, etc. (9.75%). On the other hand, the lack of adequate definitions in the guidelines (3.39%) and annotators' errors in following the linguistic definitions in the guidelines (2.97%) were observed at negligible percentages. We concluded that the coverage of the guidelines was comprehensive, and the annotators were well-trained in linguistic issues.

Zeyrek et al. (2010) attribute the inconsistencies to various sources. Here we will deal with two of the reasons discussed in the original article: (i) different interpretations of the minimality principle, and (ii) taking nominalizations as arguments to a connective. The first case seems to arise from the delayed suffixation property of Turkish, where several clauses are linked, allowing the predicative morpheme to appear only at the last clause ("delayed" in earlier ones), as in (8), with the delayed predicative morpheme *-tır* in clause (c) underlined, and the discourse connective in clause (b) shown in bold.

(8) (a) Onlara sunulan kurbanlar, başlangıçta insanlardı
   "At the beginning, it was humans that were sacrified for them."
   (b) **Fakat** bu âdet sonraları hafifletilerek, insan yerine hayvanlar kurban edilmeğe başlanmış,

---

[2]The connectives that gave low κ values are *amaçla* "for this purpose", *ayrıca* "in addition/separately", *dolayısıyla* "in consequence of", *fakat* "but", *oysa* "however", *rağmen* "despite/despite this", *tersine* "in contrast", and *yandan* "on the one hand/on the other hand".

"**But** later on, loosening this tradition, (they) started to sacrifice animals instead of humans,"

(c) sonunda da bu hayvanları temsil eden bazı şeylerin (...) kâğıt hayvan figürlerinin (...) yahut da bir taşın suya atılmasının yeterli olacağına inanılmıştır.

"finally, it <u>was</u> believed that it would be sufficient to throw a stone or paper animal figures to the water, as well as other objects that represent these animals."

One annotator takes clauses (b) and (c) in the scope of the predicative morpheme, and annotates both clauses as Arg2 to *fakat* "but". Others annotate only (b) as Arg2.

Regarding point (ii) above, we can look at the discourse connective *ve* "and", the most frequently occurring connective in the TDB 1.0.[3] In Turkish, nominalization is encoded by various suffixes forming nonfinite clauses. While some of the resulting nominalizations are unambiguously interpreted as an abstract object, some are harder to do so. For example, in the clauses with the infinitive suffix *-mAk* as potential arguments to a connective in (9), there was no disagreement.

(9) *18. yüzyılın yaptığı, 17. Yüzyılın yarattıklarını çoğaltmak* **ve yaymaktır.**
     "*What the 18th century did was* to *increase* **and** extend *what the 17th century created*."[4]

Other suffixes (e.g., *-mA, -yHş*) also cause inconsistencies. This is attributed to the fact that these suffixes productively derive common nouns, making it difficult for annotators to take them as nominalized clauses with an abstract object interpretation. For example, in (10), *geliş-me* (improve-mA) "improvement" and *yapılaş-ma* (construct-mA) "construction" have caused disagreements. In (11), there was less disagreement. The final decision was to annotate nominalizations when they can be interpreted as derived from a clause with a subject, as in the PDTB. With this principle, only (11) would be annotated as a discourse relation.

(10) Deprem bölgesinde yeniden *gelişme* ve *yapılaşmanın* planlanması gibi ciddi bir sorun bulunmaktadır.
     "There is the important issue of planning the *improvement* and *re-construction* of the areas affected by the earthquake."

(11) Artık onu *beklemenin* **ve aramanın** boşuna olduğunu anlamıştır.
     "He has already figured out that it was futile *to wait for her* **and to search for her**."

In summary, certain disagreements seem to arise from morphological aspects of Turkish. Others we could keep under control by having a comprehensive guideline. This aspect speaks of a need to employ morphological components not only for syntax but for discourse as well.

---

[3]Of the 7,486 *ve* "and" tokens in the TDB, 2,111 are annotated as discourse connectives.
[4]The disagreed text spans are rendered in both italics and boldface.

Table 16.2 Allomorph inflections of the suffix *-dik* and additional inflections

-dığım	-diğim	-duğum	-düğüm	-tığım	-tiğim	-tuğum	-tüğüm
-dığın	-diğin	-duğun	-düğün	-tığın	-tiğin	-tuğun	-tüğün
-dığı	-diği	-duğu	-düğü	-tığı	-tiği	-tuğu	-tüğü
-dığımız	-diğimiz	-duğumuz	-düğümüz	-tığımız	-tiğimiz	-tuğumuz	-tüğümüz
-dığınız	-diğiniz	-duğunuz	-düğünüz	-tığınız	-tiğiniz	-tuğunuz	-tüğünüz
-dıkları	-dikleri	-dukları	-dükleri	-tıkları	-tikleri	-tukları	-tükleri

16.2.2 The Discourse Annotation Tool for Turkish

The annotations were created by an annotation tool particularly designed for the TDB project (Aktaş et al. 2010). The Discourse Annotation Tool for Turkish (DATT) allows the annotators to search for connective candidates in the source text and to create stand-off XML annotations.

The search feature of the DATT supports regular expressions for searching through Turkish morphological patterns taking into account allomorph variations. The annotators can search for the inflections of a connective candidate. For example, one can search for *-dığı için* "for—result driven" and *-mak için* "for—goal driven", and their inflections separately. For example, the query string -DH(ğH|k)(m|n|lArH)(|Hz) stands for all possible inflections and allomorphs of *-dik* given in Table 16.2.

After a search, DATT retrieves the list of files that include the search item. When a file is selected, the annotators can optionally highlight the search item on the source text.

For every discourse relation, the connective, the first argument and the second argument are obligatorily marked. The remaining parts of a discourse relation captured in TDB (the modifier, the shared material, the material that supplements the arguments or the shared material) are marked if they are present. Annotators can select discontinuous text spans. In addition, there is a *notes* box where annotators can type free text. They use this field for their comments on the annotations.

In the course of annotations, a sense annotation module was added to DATT, offering the sense tags in the PDTB sense hierarchy (Prasad et al. 2014) in a drop-down menu.

The annotations are kept as valid XML files. For the annotated spans, the begin and end character offsets as well as the text content are kept in span elements, whereas the sense is saved as an attribute of the relation.

16.3 Connectives and Discourse Structure

Most work on structure in discourse assumes or imposes a hierarchical tree structure (Hobbs 1985; Mann and Thompson 1988; Polanyi 1988; Asher 1993; Webber 2004). However, many also report deviations from a strict tree structure. Hobbs

(1985) proposes trees that connect or intertwine at the boundaries. Egg and Redeker (2008) report on genre-restricted multi-parenting in Rhetorical Structure Theory proposed by Mann and Thompson (1988). Wolf and Gibson (2004, 2005) claim that crossing dependencies are abundant in discourse, and offer a much more permitting structure. A grammar-based approach, that of Nakatsu and White (2010), which makes use of a theory that is well-equipped to deal with crossing dependency computation of the syntactic kind, is not designed with this extension in mind. They use cue threading within a single grammar to handle discourse links, similar to the D-LTAG-style (Webber 2004) connective approach. We are agnostic on the issue of having grammars *for* discourse. But, not employing dependency computation of the crossing kind when we can in fact do so is suggestive of a mechanism in which there is no need for semantics above function application. This is also our finding in the TDB.

The PDTB was annotated locally for each discourse connective. It does not impose a global structure on discourse in the process of annotation, which is followed also in the TDB. Lee et al. (2006, 2008) analyzed the dependencies observed in the PDTB, and came up with a list of discourse relation dependencies. Specifically they observed *independent relations, full embedding, shared argument, properly contained argument, partially overlapping arguments*, and *pure crossing*.

Independent relations and *full embedding* conform to a tree structure, whereas *shared argument, properly contained argument, partially overlapping arguments*, and *pure crossing* deviate from it. They claim that only *shared arguments* and *properly contained arguments* (see Sect. 16.4) should be considered as contributing to the complexity of discourse structure; the reason being that the instances of partially overlapping arguments and pure crossing can be explained away by anaphora and attribution, both of which are non-discourse-structural phenomena.

Along the same lines, Aktaş et al. (2010) observe similar dependencies in the TDB. In addition to the dependencies in Lee et al. (2006), Aktaş et al. (2010) identify *nested relations* and *properly contained relations*. Demirşahin et al. (2013) offer a quantitative analysis of the dependency configurations observed by Aktaş et al. (2010) in order to give an overview of Turkish discourse structure and to find out the extent to which Turkish discourse structure deviates from tree structures. We compile these findings next.

16.4 Discourse Relation Configurations in the TDB

We provide attested examples from the TDB, along with their classification and some assesment. We refer to the first public release, the TDB 1.0.

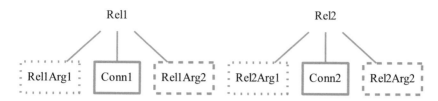

Fig. 16.1 Independent relations

16.4.1 Independent Relations

When two relations are found sequentially in the text, with no discourse-level dependencies between them, these are called *independent relations*. Of the 8483 explicit relations in the TDB, 5010 (59.05%) occur in isolation, making them the most common configuration. Figure 16.1 schematizes the independent relations configuration. An example from the corpus is given below:

(12) (a) *Sen de haberdar değildin* <u>ve</u> **ben hayatımda ilk kez yıkmaya değil aşmaya çalışıyordum**. İzin vermiyor, engeller koyuyordun. Dikenli tellerle çeviriyordun bu duvarı. Yaralanıyordum tırmanırken, kanıyordum. Kırılıyordum, acıyordum, ama bırakmıyordum.
   "You were not aware either, **and for the first time in my life I was trying to construct things instead of de-constructing them**. You were not letting me but preventing me. You were fencing this wall with barbwire. I was being injured and bleeding while climbing. I was breaking down, aching, yet not letting go."

(b) Sen de haberdar değildin ve ben hayatımda ilk kez yıkmaya değil aşmaya çalışıyordum. İzin vermiyor, engeller koyuyordun. Dikenli tellerle çeviriyordun bu duvarı. *Yaralanıyordum tırmanırken, kanıyordum. Kırılıyordum, acıyordum,* **ama bırakmıyordum**.
   "You were not aware either, and for the first time in my life I was trying to construct things instead of de-constructing them. You were not letting me but preventing me. You were fencing this wall with barbwire. *I was being injured and bleeding while climbing. I was breaking down, aching,* <u>yet</u> not letting go."

16.4.2 Full Embedding

When a discourse relation as a whole is the argument of another discourse relation, the resulting configuration is *full embedding*. There are 2548 nonindependent dependency configurations in the TDB, and 695 (27.28 %) are full embedding.

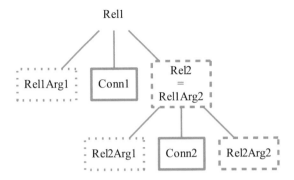

Fig. 16.2 Full embedding

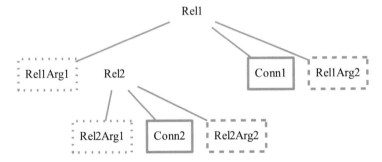

Fig. 16.3 Nested relations

Figure 16.2 graphically represents their configuration. For example below, relation (b) is fully embedded under relation (a).

(13) (a) *Gün ağarana dek uğraşıyor* **ve kadın terasa çıkmadan önce kaçıyordu**.
     "He struggled until sunrise **and left before she got out to the terrace**."
  (b) Gün ağarana dek uğraşıyor ve **kadın terasa çıkmadan önce** *kaçıyordu*.
     "He struggled until sunrise and *left* **before she got out to the terrace**."

16.4.3 Nested Relations

Nested relations are one of the structures identified by Aktaş et al. (2010). This type of structure is not reported in Lee et al. (2006) for English. In the nested relations configuration, one relation occurs linearly between the arguments of the other relation. Only 138 (5.42 %) of the nonindependent configurations are nested relations (see Fig. 16.3), for example in (14), relation (b) is nested in relation (a).

(14)(a) Bir süre kapısında bir köpek gibi süründüm. Benden sonra âşık olduğu adamı gece gündüz izledim. ıçim kıskançlık, acı, kin ve nefretle doluydu. Anlatması güç duygular bunlar. Adam onu dövüyordu. *Bazı geceler kulağımı kapısına dayar, dayak yerken attığı çığlıkları dinlerdim.* **Sonra barışırlardı.** Ne tuhaf bir şeydi bu! Sonra da bu parka düştüm işte.

"For a while I crept like a dog in front of her door. Day and night I followed the guy she fell in love after me. I was filled with jealousy, pain, revenge, and hatred inside. These feelings are hard to talk about. The guy was beating her. *Some nights I would put my ear to her door and listen to her screams.* **Then they would make up.** What a weird thing it was! Then I ended up in this park, you see."

(b) *Bir süre kapısında bir köpek gibi süründüm. Benden sonra âşık olduğu adamı gece gündüz izledim.* ıçim kıskançlık, acı, kin ve nefretle doluydu. Anlatması güç duygular bunlar. Adam onu dövüyordu. Bazı geceler kulağımı kapısına dayar, dayak yerken attığı çığlıkları dinlerdim. Sonra barışırlardı. Ne tuhaf bir şeydi bu! **Sonra** da **bu parka düştüm işte.**

"*For a while I crept like a dog in front of her door. Day and night I followed the guy she fell in love after me.* I was filled with jealousy, pain, revenge, and hatred inside. These feelings are hard to talk about. The guy was beating her. Some nights I would put my ear to her door and listen to her screams. Then they would make up. What a weird thing it was! **Then I ended up in this park, you see.**"

16.4.4 Shared Argument

Shared arguments occur when the same text span is taken as an argument by two distinct discourse connectives. This is the most frequent tree-violating configuration in the PDTB, and the second most frequent (489 , or 19.19%) in the TDB (see Fig. 16.4 for its schema). For example in (15), the Arg2 span of relation (a) is precisely the Arg1 span of relation (b).

(15)(a) *Vazgeçmek kolaydı, ertelemek de.* **Ama tırmanmaya başlandı mı bitirilmeli!** Çünkü her seferinde acımasız bir geriye dönüş vardı.

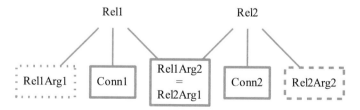

Fig. 16.4 Shared argument

"*It was easy to give up, so was to postpone.* **But once you start climbing you have to go all the way!** Because there was a cruel comeback every time."

(b) Vazgeçmek kolaydı, ertelemek de. Ama *tırmanmaya başlandı mı bitirilmeli!* **Çünkü her seferinde acımasız bir geriye dönüş vardı.**
   "It was easy to give up, so was to postpone. But *once you start climbing you have to go all the way!* **Because there was a cruel comeback every time.**"

16.4.5 Properly Contained Argument

When one argument of a relation is contained in an argument of another relation without fully overlapping with it, the resulting configuration is a *properly contained argument*. There are 194 properly contained arguments in the TDB, comprising 7.61% of nonindependent relations. As in the PDTB, attribution may be the cause of properly contained arguments. Relative clauses and complements of verbs—when they can be interpreted as abstract objects—may also be the cause of properly contained arguments (Demirşahin et al. 2013). In the example below the first argument of (a) is the complement of *biliyor* "he knows", and is properly contained in the first argument of (b) (see Fig. 16.5).

(16) (a) Biliyor, *bir zaman sonra batı yönüne sapacak.* **East Village ve Lower East Side'ın kalabalık mahallelerinden geçmek, B ve A caddelerinde, Houston, Essex, Hester sokaklarında yürümek için.** [...] Beş aydır hep bu yolu izliyor çünkü.
   "He knows that *some time later he will turn towards west* **in order to pass through the crowded neighborhoods of the East Village and Lower East Side and to walk along the B and A streets, and Houston, Essex, and Hester roads.** [...] Because he has been following the same route for five months."

(b) *Biliyor, bir zaman sonra batı yönüne sapacak.* East Village ve Lower East Side'ın kalabalık mahallelerinden geçmek, B ve A caddelerinde, Houston,

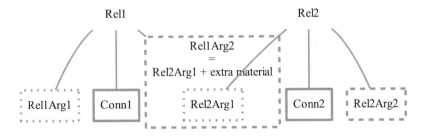

Fig. 16.5 Properly contained argument

Essex, Hester sokaklarında yürümek için. [...] **Beş aydır hep bu yolu izliyor <u>çünkü</u>**.

"He knows that some time later he will turn towards west in order to pass through the crowded neighborhoods of the East Village and Lower East Side and to walk along the B and A streets, and Houston, Essex, and Hester roads. [...] **<u>Because</u> he has been following the same route for five months**."

16.4.6 Properly Contained Relation

When a discourse relation as a whole is contained in an argument of another relation without fully overlapping with it, the resulting configuration is a *properly contained relation*. Properly contained relations were not attested in Lee et al. (2006). With 1,018 instances, they are the most frequent tree-violating configuration in the TDB, making up 39.95% of all nonindependent configurations. Whether this difference is due to the differences between the two languages or due to the multi-genre nature of the TDB, which contains narratives that are likely to contain this structure, is open to further research. Similar to the properly contained arguments, attribution, relative clauses, and verbal complements can result in properly contained relations. For example, below, relation (a), which has a relative clause as its Arg1, combines with some other text and forms the Arg1 of relation (b). Thus, relation (a) is properly contained in relation (b) (see Fig. 16.6).

(17) (a) Sabah çok erken saatte **bir önceki akşam gün batmadan <u>hemen önce</u>** *astığı* çamaşırları toplamaya çıkıyordu ve doğal olarak da gün batmadan o günkü çamaşırları asmak için geliyordu.

"She used to go out to gather the clean laundry *she had hung to dry* **<u>right before</u> the sun went down the previous evening**, and naturally she came before sunset to hang the laundry of the day."

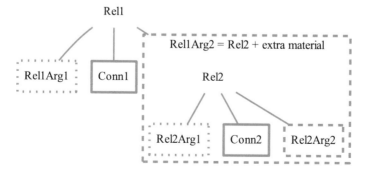

Fig. 16.6 Properly contained relation

(b) *Sabah çok erken saatte bir önceki akşam gün batmadan hemen önce astığı çamaşırları toplamaya çıkıyordu* **ve doğal olarak da gün batmadan o günkü çamaşırları asmak için geliyordu.**

"*She used to go out to gather the clean laundry she had hung to dry the previous evening right before the sun went down,* **and naturally she came before sunset to hang the laundry of the day.**"

16.4.7 Partially Overlapping Arguments

If an argument of one relation and an argument of another relation partially overlap, we have *partially overlapping arguments*. They are rare in the TDB, occurring only 12 times.

In (18) the second argument of *but* in (a) contains only one of the two conjoined clauses, whereas the first argument of *after* in (b) contains both of them. The most probable cause for this difference in annotations is the combination of the blind annotation with the minimality principle (see 16.2.1). This principle guides the participants to annotate the minimum text span required to interpret the relation. Since the annotators cannot see the previous annotations, they have to assess the minimum span of an argument again when they annotate the second relation. Sometimes the minimal span for one relation is annotated differently than the minimal span required for the other, resulting in partial overlaps (Fig. 16.7).

(18) (a) *Yine istediği kişiyi bir türlü görememişti,* **ama aylarca sabrettikten sonra gözetlediği bir kadın soluğunu daralttı,** tüyleri diken diken oldu.
   "*Once again he couldn't see the person he wanted to see,* **but after waiting patiently for months, a woman he peeped at took his breath away,** gave him goosebumps."

   (b) Yine istediği kişiyi bir türlü görememişti, ama **aylarca sabrettikten sonra** *gözetlediği bir kadın soluğunu daralttı, tüyleri diken diken oldu.*
   "Once again he couldn't see the person he wanted to see, but **after waiting patiently for months,** *a woman he peeped at took his breath away, gave him goosebumps.*"

Fig. 16.7 Partially overlapping arguments

16.4.8   Pure Crossing

Pure crossing occurs when one argument of a relation falls between the arguments of another relation, resulting in a crossing dependency. There are only two such instances in the TDB. One of them is an anaphoric relation, anchored by the phrasal expression *o zaman* "then" (see example (6) in Demirşahin et al. (2013)). The other one involves two structural connectives *için* "because" and *ve* "and" and will be discussed below in example (19).[5]

The crossing relations in (19) calls for a detailed analysis.

(19) (a) *Ceza*, **Telekom'un iki farklı internet alt yapısı pazarında tekel konumunu kötüye kullandığı için** ve uydu istasyonu işletmeciliği pazarında artık tekel hakkı kalmadığı halde rakiplerinin faaliyetlerini zorlaştırdığı için *verildi.*

"*The penalty was given* **because Telekom abused its monopoly status in the two different internet infrastructure markets** and because it caused difficulties with its rivals' activities although it did not have a monopoly status in the satellite management market anymore."

(b) Ceza, Telekom'un *iki farklı internet alt yapısı pazarında tekel konumunu kötüye kullandığı* için **ve uydu istasyonu işletmeciliği pazarında artık tekel hakkı kalmadığı halde rakiplerinin faaliyetlerini zorlaştırdığı** için verildi.

"The penalty was given because *Telekom abused its monopoly status in the two different internet infrastructure markets* **and** because **it caused difficulties with its rivals' activities although it did not have a monopoly status in the satellite management market anymore.**"

(c) *Ceza*, Telekom'un iki farklı internet alt yapısı pazarında tekel konumunu kötüye kullandığı için ve **uydu istasyonu işletmeciliği pazarında artık tekel hakkı kalmadığı halde rakiplerinin faaliyetlerini zorlaştırdığı için** *verildi.*

"*The penalty was given* because Telekom abused its monopoly status in the two different internet infrastructure markets and **because it caused difficulties with its rivals' activities although it did not have a monopoly status in the satellite management market anymore.**"

---

[5]Here we follow Forbes-Riley et al. (2006) who argue that discourse adverbials and other connectives such as coordinating and subordinating conjunctions differ in how they take their arguments. Discourse adverbials only take their second argument structurally, their first argument being anaphoric. Other kinds of discourse connectives take both of their arguments structurally. Thus, we use the term "structural discourse connective" for coordinating and subordinating conjunctions, "anaphoric discourse connective" for discourse adverbials as well as expressions that contain a deictic anaphor (i.e., phrasal expressions).

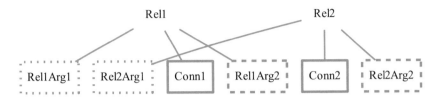

Fig. 16.8 Pure crossing

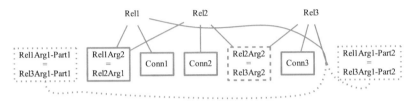

Fig. 16.9 The structure of (19)

Figure 16.8 represents the hypothetical pure crossing configuration, where Arg1 of the second relation (Rel2) is interspersed between the arguments of the first relation (Rel1). However, the crossings in (19) are of a different kind. The structure of (19) is shown in Fig. 16.9.

The solid crossing lines in Fig. 16.9 are due to the repetition of the causal connective *için* "because". An alternative analysis would be to assume that the relation headed by *ve* "and" takes the relations headed by *için* "because" under its scope. This kind of annotation was not allowed by the TDB guidelines; we leave it to further machine learning research to identify such cases. The actual dependencies in this example are shared arguments between (a) and (b), (a) and (c), and (b) and (c). Had the author opted for using a single *için*, which is perfectly grammatical, the configuration would be full embedding.

The dotted lines in Fig. 16.9 show another kind of apparent crossing, which should be considered separate from the other discourse dependencies. It is the result of a noncontinuous argument span, and is realized within the relation rather than across two relations. In other words, crossing occurs within a discourse relation and is not the result of two interacting discourse relations. The noncontinuous spans of the first arguments in (a) and (c) in Fig. 16.9 "wrap" around their second arguments. When an adverbial subordinate clause occurs in the middle of the matrix clause, this is called wrapping, which is quite common in Turkish. There are 479 instances of wrapping in the TDB. Figure 16.10 shows wrapping in isolation.

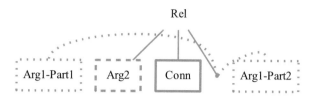

Fig. 16.10 Wrapping

As discussed in Demirşahin et al. (2013), in Turkish discourse, wrapping appears to be motivated by information structure. In the unmarked cases, a subordinate clause precedes the matrix clause and introduces the theme. In (14) (a) and (c), wrapping takes *ceza* "penalty" away from the matrix clause (rheme) and makes it part of the theme.

There is no function composition in this example. Despite the apparent crossing, all connectives in (19) have applicative semantics, that is, the meaning of the relations depends only on two arguments, and there is no need for function composition to compose over multiple connectives and their text spans to derive the meaning of the whole. This is unlike syntax, where composition is commonplace: *This is the man who John thinks Mary claims I like*, where the heads *think, claim* force composition to reach down to the argument structure of *like*. Genuine crossings, of the kind we see in Swiss German, are forced to compose as well: *Jan säit das mer₁ d'chind₂ em Hans₃ es Huus₄ lönd₁,₂ hälfe₂,₃ aastriiche₃,₄* (we the children-ACC Hans-DAT the house-ACC let help paint) "Jan says that we let the children help Hans paint the house" (Shieber 1985).[6] There is no composition in discourse in this sense.

16.5   Results and Conclusions

Table 16.3 summarizes the distribution of all discourse configurations attested in the TDB.

It should be noted that independent relations are by and large the most frequent configuration for the explicit connectives in the TDB. 59.05% of the explicit connective relations are independent. Among the nonindependent relations, 67.31% seem to be tree-violating configurations. However, about half of them are due to anaphoric connectives (Table 16.4). If one of the arguments in a configuration is the argument of an anaphoric connective, we treat the configuration as anaphoric, that is, as being realized as a result of the anaphoric characteristic of that connective.

In addition to the shared arguments and properly contained arguments that were accepted in discourse structure by Lee et al. (2006), we identified properly contained

---

[6]Indices show argument-taking.

Table 16.3 Distribution of nonindependent configurations in the TDB 1.0

Configuration	*Occur.*	%
Full embedding	695	27.28
Nested relations	138	5.42
Total non-violating configurations	**833**	**32.69**
Shared argument	489	19.19
Properly contained argument	194	7.61
Properly contained relation	1018	39.95
Pure crossing	2	0.08
Partial overlap	12	0.47
Total tree-violating configurations	**1715**	**67.31**
Total	**2548**	**100.00**

Table 16.4 Distribution of anaphoric relations among tree-violating configurations

Configuration	Structural (%)	Anaphoric (%)	*Total*
Shared argument	158(32.31%)	331(67.69%)	489
Properly contained argument	65(33.51%)	129(66.49%)	194
Properly contained relation	547(53.73%)	471(46.27%)	1018
Partial overlap	9(75.00%)	3(25.00%)	12
Pure crossing	1(50.00%)	1(50.00%)	2
Total	780(45.48%)	935(54.52%)	1715

relations. Some of these structures have a common characteristic: They occur due to non-discourse-level phenomena such as attribution, relative clauses, and verbal complements. The properly contained relation or argument is realized within an argument which is not related to it on a discourse level. The semantics of these elements do not seem to be dependent on each other, and as a result they have applicative semantics.

The few partial overlaps we have encountered could mostly be explained away by wrapping and by different interpretations of annotation guidelines by the annotators, especially the minimality principle. Recall that wrap has applicative semantics. Of the two pure crossing examples we have found, one was anaphoric, and the other one could be explained in terms of information structurally motivated relation-level interleaving without composition, rather than genuine crossing dependency. In other words, if we leave the processing of information structure to other processes, the need for more elaborate annotation disappears.

In Joshi's (2011) terminology, immediate discourse in the TDB appears to be an applicative structure, which, unlike syntax, is in no need of currying or function composition. We think that pure crossing (i.e., crossing of the arguments of structural connectives) is not genuinely attested in the TDB. The annotation scheme need not be enriched to allow more complex algorithms to deal with unlimited use of crossing. There was a reason in every contested case to go back to the annotation, and revise it in ways to keep the applicative semantics, without losing much of the connective's meaning.

If applicative semantics is all we need, we will move on to further processing issues, such as word-level relations, information structure, and anaphora. We have seen in the last decade of twentieth century some narrow theorizing on these aspects, for example, Centering Theory for certain kinds of anaphora and information structure, Wordnet for word-level anaphora, and CCG for grammar-information structure interface. We anticipate that their combined effort, along with connectives, might help explain discourse structure.

References

Aktaş B, Bozşahin C, Zeyrek D (2010) Discourse relation configurations in Turkish and an annotation environment. In: Proceedings of the linguistic annotation workshop, Uppsala, pp 202–206

Asher N (1993) Reference to abstract objects in discourse. Kluwer Academic Publishers, Dordrecht

Demirşahin I, Zeyrek D (2017) Pair annotation as a novel annotation procedure: the case of Turkish Discourse Bank. In: Pustejovsky J, Ide N (eds) Handbook of linguistic annotation. Springer, Dordrecht

Demirşahin I, Yalçınkaya İ, Zeyrek D (2012) Pair annotation: adaption of pair programming to corpus annotation. In: Proceedings of the linguistic annotation workshop, Jeju, pp 31–39

Demirşahin I, Öztürel A, Bozşahin C, Zeyrek D (2013) Applicative structures and immediate discourse in the Turkish Discourse Bank. In: Proceedings of the linguistic annotation workshop, Sofia, pp 122–130

Egg M, Redeker G (2008) Underspecified discourse representation. In: Benz A, Kuhnlein P (eds) Constraints in discourse. John Benjamins, Amsterdam, pp 117–138

Fleiss JL (1971) Measuring nominal scale agreement among many raters. Psychol Bull 76(5):378

Forbes-Riley K, Webber B, Joshi A (2006) Computing discourse semantics: the predicate-argument semantics of discourse connectives in D-LTAG. J Semant 23(1):55–106

Hobbs JR (1985) On the coherence and structure of discourse. Tech. Rep. CSLI-85-37, CSLI, Stanford, CA

Joshi A (2011) Some aspects of transition from sentence to discourse. Keynote address, Informatics Science Festival, Middle East Technical University

Lee A, Prasad R, Joshi A, Dinesh N, Webber B (2006) Complexity of dependencies in discourse: are dependencies in discourse more complex than in syntax. In: Proceedings of the international workshop on treebanks and linguistic theories, Prague

Lee A, Prasad R, Joshi A, Webber B (2008) Departures from tree structures in discourse: shared arguments in the Penn Discourse Treebank. In: Proceedings of the third workshop on constraints in discourse, Potsdam

Mann WC, Thompson SA (1988) Rhetorical structure theory: toward a functional theory of text organization. Text 8(3):243–281

Nakatsu C, White M (2010) Generating with discourse combinatory categorial grammar. Linguist Issues Lang Technol 4(1):1–62

Polanyi L (1988) A formal model of the structure of discourse. J Pragmat 12(5):601–638

Prasad R, Webber BL, Joshi A (2014) Reflections on the Penn Discourse TreeBank, comparable corpora, and complementary annotation. Comput Linguist 40(4):921–950

Say B, Zeyrek D, Oflazer K, Özge U (2004) Development of a corpus and a treebank for present-day written Turkish. In: Proceedings of the international conference on Turkish Linguistics, Magosa, TRNC, pp 183–192

Shieber S (1985) Evidence against the context-freeness of natural language. Linguist Philos 8:333–343

Tın E, Akman V (1994) Situated processing of pronominal anaphora. In: Proceedings of the Konferenz, Verarbeitung natürlicher Sprache, Vienna, pp 369–378

Tüfekçi P, Kılıçaslan Y (2005) A computational model for resolving pronominal anaphora in Turkish using Hobbs' naïve algorithm. Int J Comput Intell 2(1):71–75

Tüfekçi P, Küçük D, Yöndem MT, Kılıçaslan Y (2007) Comparison of a syntax-based and a knowledge-poor pronoun resolution systems for Turkish. Poster presented at international symposium on computer and information sciences (ISCIS)

Webber B (2004) D-LTAG: Extending lexicalized TAG to discourse. Cognit Sci 28(5):751–779

Williams L, Kessler RR, Cunningham W, Jeffries R (2000) Strengthening the case for pair programming. IEEE Softw 17(4):19–25

Wolf F, Gibson E (2004) Representing discourse coherence: a corpus-based analysis. In: Proceedings of COLING, Geneva, pp 134–140

Wolf F, Gibson E (2005) Representing discourse coherence: a corpus-based study. Comput Linguist 31(2):249–287

Yıldırım S, Kılıçaslan Y, Aykaç RE (2004) A computational model for anaphora resolution in Turkish via centering theory: an initial approach. In: Proceedings of the international conference on computational intelligence, Istanbul, pp 124–128

Yüksel Ö, Bozşahin C (2002) Contextually appropriate reference generation. Nat Lang Eng 8(1):69–89

Zeyrek D, Webber BL (2008) A discourse resource for Turkish: annotating discourse connectives in the METU corpus. In: Proceedings of the workshop on Asian language resources, Hyderabad, pp 65–72

Zeyrek D, Turan ÜD, Bozşahin C, Çakıcı R, Sevdik-Çallı A, Demirşahin I, Aktaş B, Yalçınkaya İ, Ögel H (2009) Annotating subordinators in the Turkish Discourse Bank. In: Proceedings of the 3rd linguistic annotation workshop, Singapore, pp 44–47

Zeyrek D, Demirşahin I, Sevdik-Çallı A, Balaban Hö, Yalçınkaya İ, Turan ÜD (2010) The annotation scheme of the Turkish Discourse Bank and an evaluation of inconsistent annotations. In: Proceedings of the 4th linguistic annotation workshop, Uppsala, pp 282–289

Zeyrek D, Demirşahin I, Sevdik-Çallı A, Çakıcı R (2013) Turkish Discourse Bank: porting a discourse annotation style to a morphologically rich language. Dialogue Discourse 4(2):174–184

Zeyrek, D, Kurfalı, M (2017) TDB 1.1: Extensions on Turkish Discourse Bank. In: Proceedings of the 11th linguistic annotation workshop, Valencia, pp 76–81

Index

© Springer International Publishing AG, part of Springer Nature 2018
K. Oflazer, M. Saraçlar (eds.), *Turkish Natural Language Processing*,
Theory and Applications of Natural Language Processing,
https://doi.org/10.1007/978-3-319-90165-7